Roman Jackiw
80th Birthday Festschrift

Roman Jackiw
80th Birthday Festschrift

Editors

Antti Niemi
University of Stockholm, Sweden

Terry Tomboulis
University of California, Los Angeles, USA

Kok Khoo Phua
Nanyang Technological University, Singapore

We **World Scientific**

NEW JERSEY · LONDON · SINGAPORE · BEIJING · SHANGHAI · HONG KONG · TAIPEI · CHENNAI · TOKYO

Published by

World Scientific Publishing Co. Pte. Ltd.

5 Toh Tuck Link, Singapore 596224

USA office: 27 Warren Street, Suite 401-402, Hackensack, NJ 07601

UK office: 57 Shelton Street, Covent Garden, London WC2H 9HE

Library of Congress Control Number: 2020937448

British Library Cataloguing-in-Publication Data
A catalogue record for this book is available from the British Library.

ISBN 978-981-121-066-2 (hardcover)
ISBN 978-981-121-067-9 (ebook for institutions)
ISBN 978-981-121-228-4 (ebook for individuals)

For any available supplementary material, please visit
https://www.worldscientific.com/worldscibooks/10.1142/11562#t=suppl

Desk Editor: Ng Kah Fee

Preface

Roman Jackiw is a towering scientific figure whose impact extends over a wide area of contemporary theoretical and mathematical physics. He has made enormous contributions to our understanding of quantum field theory, both as the foundational language of physics and as a powerful tool applied to physical problems ranging from high energy to condensed matter physics. He is particularly renowned for his many fundamental contributions to the development of functional integral techniques and his discoveries of topological and geometrical phenomena in the field theory context. His scientific results will continue to influence and inspire generations of theoretical and mathematical physicists.

Roman Jackiw received his scientific education at Cornell University, where Hans Bethe and Kenneth Wilson were jointly his advisors. After receiving his PhD Roman joined Harvard University as a Junior Fellow. Subsequently, he moved to the Center for Theoretical Physics at MIT, where he stayed most of his career as Jerrold Zacharias Professor of Physics; he currently has the status of Emeritus Professor. He also spent long periods at CERN, UCLA and Rockefeller University.

During a stay at CERN, Roman together with John Bell discovered the celebrated triangle anomaly, now recognized as one of the most profound examples of the relevance of quantum field theory to the real world, and a corner stone for our understanding of symmetries and their breaking in the quantum context. The concept of anomalies originating in the work of Roman and John Bell, and, independently, Steven Adler, besides showing the way to the construction of the Standard Model, has been central to all attempts at field theories going beyond it as well as the construction of consistent string theories. It also plays a prominent role within the broad area of topological phases in condensed matter physics, which, especially after the discovery of topological insulators, has blossomed in recent years leading even to the design of new materials.

Among his numerous other highly influential original contributions is the derivation of diagrammatic rules for finite-temperature quantum field theory, the development of effective actions for composite operators, the discovery of fractional charge and spin in field theories, and the introduction of the topological (Chern–Simons) mass term in three-dimensional gravity and gauge theories as a way of breaking parity and providing a mass. Among his other, more recent contributions is the introduction of classical field theory techniques to fluid mechanics problems.

Roman Jackiw's scientific contributions have been recognized by many awards including the Dannie Heineman Prize for Mathematical Physics and the Dirac Medal of the ICTP. He is a member of the US National Academy of Sciences and has received honorary doctorates from Kiev, Montreal, Tours, Turin and Uppsala universities.

Roman is our teacher, mentor, friend, and colleague. We dedicate this volume to him as a celebration of his scientific life on his 80th birthday. It includes both personal recollections and original scientific contributions by his many friends and colleagues, covering many aspects of his life and prolific scientific achievements.

Contents

Part 1
Personal Recollections

Chapter 1

Recent Path Crossings with Roman and Anomalies

Stephen L. Adler*
*Institute for Advanced Study, Einstein Drive,
Princeton, NJ 08540, USA.*

I begin with an anecdote, and then discuss my recent work on anomalies in spin-$\frac{3}{2}$ theories.

1. Introduction and an anecdote about Roman

The discovery just over 50 years ago of chiral anomalies by me [1] and by Bell and Jackiw [2] is now featured in a section or chapter of the standard quantum field theory textbooks, and my account of work on anomalies during the years from 1968 to 1976 has been written elsewhere [3,4]. In this essay I turn to recent interactions with Roman, and with anomalies. I begin with an anecdote. In 2015 I was invited to attend a workshop held in August at Galiano Island, off Vancouver, on "Probing the Mystery: Theory & Experiment in Quantum Gravity". I was scheduled to speak in a session devoted to topics related to wave function collapse models, as was my friend and occasional collaborator Angelo Bassi. In discussing with Angelo whether to attend, one of the things that appealed to us, apart from the prospect of good physics and interesting participants, was the fact that Galiano Island is conveniently accessed by seaplane, which intrigued us both. So we decided to go. When we showed up to embark on the seaplane, Roman was also among the group waiting to board. When we got onto the plane, Roman turned to me and said "I know the plane won't crash; Adler and Jackiw can't die at the same time." We did get to the island safely, but the flight was in fact interesting in that there is a high tension line strung across the entrance to the island harbor at which we were landing. To clear this, the seaplane descended to perhaps 10 to 20 feet above the water, and skimmed along this way for quite a distance, passing under the wires, and then settling down into the water. The procedure was reversed on departure, again the plane skimming along just over the water, before climbing to altitude once safely under the wires. A memorable flight, as was Roman's remark before taking off.

---

*Electronic address: adler@ias.edu

2. Spin-$\frac{3}{2}$ anomalies and my second recent path crossing with Roman

My recent encounters with anomalies have all revolved around the question of whether a spin-$\frac{3}{2}$ field can be gauged outside of a supergravity context, and if it can be, what is its chiral anomaly? I was led to this by a conjectured grand unified model [5] that I have been studying for several years based on the gauge group $SU(8)$. The model is suggested in part by rearranging the multiplet content of $N{=}8$ supergravity, in a way that preserves boson–fermion balance (that is equal numbers of boson and fermion degrees of freedom) while not insisting on full supersymmetry. The fermion content of the model is one left chiral spin-$\frac{1}{2}$ 56 of $SU(8)$, two left chiral spin-$\frac{1}{2}$ $\overline{28}$ of $SU(8)$, and one left chiral spin-$\frac{3}{2}$ 8 of $SU(8)$. If one adopts the standard rule [6] that the chiral anomaly of a spin-$\frac{3}{2}$ field in a representation $\mathcal{R}$ is 3 times the chiral anomaly for a spin-$\frac{1}{2}$ field in the same representation $\mathcal{R}$, then this representation content cancels $SU(8)$ gauge anomalies [7].

However, in taking this counting seriously, a puzzle arises. From old work of Johnson and Sudarshan [8] , and of Velo and Zwanziger [9], it was known that the Dirac bracket for a gauged spin-$\frac{3}{2}$ theory is singular at small gauge field $\vec{B}$. Hence perturbation theory in the gauge coupling breaks down, so how then can one talk about a one-loop chiral anomaly? My original thought was that this problem is just a reflection of the fact that when a Rarita–Schwinger field is gauged the free field fermionic gauge invariance is broken. I found a nice way to restore exact fermionic gauge invariance by adding an auxiliary field [10] and hoped this would improve things. But a careful analysis of the extended gauged model with Henneaux and Pais [11] shows that the singularity is just moved to the Dirac bracket for the auxiliary field, and the model remains non-perturbative.

I then noted that in my original unification model [5] the group representations allow a direct coupling of the Rarita–Schwinger field to a spin-$\frac{1}{2}$ field, with a coefficient with dimensions of a mass. A simplified Abelian version of this coupling has an interaction term

$$S_{\text{interaction}} = m \int d^4x (\bar{\lambda}\gamma^\nu \psi_\nu - \bar{\psi}_\nu \gamma^\nu \lambda) \quad , \tag{1}$$

with ψ_ν the spin-$\frac{3}{2}$ field and λ the spin-$\frac{1}{2}$ field. For this model, one finds that the singular denominator in the Dirac brackets $1/[g\vec{\sigma} \cdot \vec{B}(\vec{x})]$ is replaced by $1/[m^2 + g\vec{\sigma} \cdot \vec{B}(\vec{x})]$, with g the gauge coupling. So a perturbation expansion is possible, making it possible to compute the chiral anomaly, and I carried this computation out in [12].

The anomaly computation for the coupled model is where my second recent path crossing with Roman occurred. Some of the new textbooks compute chiral anomalies by dimensional regularization, in which the triangle diagram is kept in its original form with three vertices and three propagators, continued away from dimension 4. The problem with using this method in the coupled model is that the propagators and vertices are complicated matrix structures, and the algebra becomes complicated. The original anomaly papers [1] and [2] noted that the chiral anomaly arises because when applying the zeroth order Ward identity to the divergence of a current in the triangle, the formal shift of integration variable needed to get a divergence of zero involves a linearly divergent integral, leading to a nonzero shift when evaluated carefully. There is a very nice treatment of the shift method in Roman's lectures in [13], which is readily adapted to the coupled model calculation. This method has the advantage that after using the zeroth order Ward identity, one is left with calculating the shift of an expression with only two vertices and two propagators, a considerable simplification. Carrying out this computation [12] shows that the anomaly contribution from the fermion triangle is 5 times the standard spin-$\frac{1}{2}$ anomaly. Because the coupling term in Eq. (1) leads to a secondary constraint with a nonvanishing (and hence Dirac second class) constraint bracket with determinant $\det[(m^2 + g\vec{\sigma} \cdot \vec{B}(\vec{x}))\delta^3(\vec{x} - \vec{y})]$, exponentiation of this constraint with ghost fields gives a non-propagating ghost with an anomaly contribution of 0. Hence the total chiral anomaly in the coupled model remains 5.

Since second class constraints are less familiar than first class ones, I repeated this calculation with Pais [14] in the extended coupled model, in which an auxiliary field is added to give an exact fermionic gauge invariance. In this case the constraints generate the fermionic gauge transformation, have vanishing brackets, and so are Dirac first class. Now the standard Faddeev–Popov method applies; one adds a gauge fixing action following Nielsen [15], which gives a ghost contribution to the chiral anomaly of -1. But when one computes the triangle graph contribution, again following the shift method in Roman's lectures, one finds an anomaly of 6 times the standard spin-$\frac{1}{2}$ anomaly. There are other diagrams in the extended model arising from the auxiliary field construction, and they all give an anomaly contribution of 0. So the total anomaly is 5, agreeing with the initial calculation I did in [12].

What is interesting about this result is that it is *not* what one gets by applying the standard lore [6], which gives an anomaly of 3 for the spin-$\frac{3}{2}$ field ψ_ν, plus 1 for the spin-$\frac{1}{2}$ field λ, for a total anomaly of 4. Hence in the only version of Rarita–Schwinger that I can find which admits a gauging outside of the supergravity context, the standard lore about spin-$\frac{3}{2}$ anomalies gives the wrong answer!

Because the fermion content of the model of [5] was based on the standard anomaly lore, the model must be modified to cancel $SU(8)$ chiral anomalies. The natural way to do this is to add a left chiral spin-$\frac{1}{2}$ $\overline{8}$ field to the model. I am now starting to think about the implications of doing this [16].

3. Happy 80th

As just described, after a long hiatus of not working on anomalies, my recent studies of spin-$\frac{3}{2}$ brought me back to Roman's excellent lectures on the subject. To conclude, very best wishes to Roman on the occasion of his 80th birthday!

References

1. S. L. Adler, Phys. Rev. **177**, 2426 (1969).
2. J. S. Bell and R. Jackiw, Nuovo Cimento A **60**, 47 (1969).
3. S. L. Adler, "Commentaries, Sec. 3", in *Adventures in Theoretical Physics*, V37 in the World Scientific Series in 20th Century Physics, World Scientific Publishing Company, Singapore (2006).
4. S. L. Adler, "Anomalies to All Orders", in G. 't Hooft, ed., *50 Years of Yang-Mills Theory*, World Scientific Publishing Company, Singapore (2005).
5. S. L. Adler, Int. J. Mod. Phys. A **29**, 1450130 (2014).
6. M. J. Duff, "Ultraviolet Divergences in Extended Supergravity", in S. Ferrara and J. G. Taylor, eds., *Supergravity '81*, Cambridge University Press, London (1982), p. 197 (also S. Ferrara and J. G. Taylor, CERN TH. 3232, p. 41); L. Alvarez-Gaumé and E. Witten, Nucl. Phys. **B234**, 269 (1984); N. K. Nielsen and H. Römer, Phys. Lett. B **154**, 141 (1985); S. M. Christensen and M. J. Duff, Phys. Lett. B **76**, 571 (1978); S. M. Christensen and M. J. Duff, Nucl. Phys. **B154**, 301 (1979); N. K. Nielsen, M. T. Grisaru, H. Römer, and P. Van Nieuwenhuizen, Nucl. Phys. **B140**, 477 (1978).
7. N. Marcus, Phys. Lett. B **157**, 383 (1985).
8. K. Johnson and E. C. G. Sudarshan, Ann. Phys. **13**, 126 (1961).
9. G. Velo and D. Zwanziger, Phys. Rev. **186**, 1337 (1969).
10. S. L. Adler, Phys. Rev. D **92**, 085022 (2015).
11. S. L. Adler, M. Henneaux, and P. Pais, Phys. Rev. D **96**, 085005 (2017).
12. S. L. Adler, Phys. Rev. D **97**, 045014 (2018).
13. R. Jackiw, "Field Theoretic Investigations in Current Algebra", in S. B. Treiman, R. Jackiw, and D. J. Gross, eds., *Lectures on Current Algebra and Its Applications*, Princeton University Press, Princeton (1972).
14. S. L. Adler and P. Pais, Phys. Rev. D **99**, 095037.
15. N. K. Nielsen, Nucl. Phys. **B140**, 499 (1978).
16. S. L. Adler, Int. J. Mod. Phys. A **34**, 1950230 (2019).

Chapter 2

Roman Jackiw and the Family Structure

Luis Álvarez-Gaumé*
Simons Center for Geometry and Physics
SUNY at Stony Brook,
Stony Brook, NY 11794-3636, USA

Contribution to the volume *Roman Jackiw - 80th Birthday Festschrift*

I recall a paper by Roman Jackiw and David Gross where they studied the consistency of gauge theories with chiral matter. By exhibiting the delicate cancellation of the gauge anomalies between quarks and leptons they provided fundamental theoretical support to the family structure of the Standard Model (SM).

1. Some recollections

I first met Roman Jackiw in 1980, when my PhD advisor Dan Freedman moved to MIT from Stony Brook, and he was kind enough to take me with him. I had arrived from Spain as a graduate student in August 1978 to SUNYSB. For me it was the discovery of a new world in many ways. Spain has just started to wake up from the forty years of the Franco regime, and the constitution that brought democracy was being elaborated.

Coming to Stony Brook was to a large extent due to serendipity. I read the article in Scientific American by Dan Freedman and Peter van Nieuwenhuizen [2], and was fascinated by it. My intention since finishing high school was to go to study in the United States after the university and the army (mandatory at the time). It was a lucky coincidence that Stony Brook was among the few US universities with contacts with the univesity where I studied in Madrid. Following the advice of some professors, I applied, and fortunately was admitted as a student, and joined the physics department as a teaching assistant at the end of August 1978. I did not have a very orthodox education in physics. During the death knells of the "ancient regime" there was a lot of social unrest, especially at the university. When the students were not on strike, it was the professors, nearly in an alternating pattern. Hence those in my class and in those adjacent to mine, ended up with a rather fragmentary knowledge of the standard curriculum. Mostly obtained by

---

*Emeritus Member, Theory Department, CERN, CH-1211 Geneva 23.

studying in the library and at home, and often in discussions with other students or professors who were willing to provide some guidance. Being able to follow classes from beginning to end was a great experience.

Soon after passing the qualifying exams, I went to see Professor Dan Freedman, who gave me a problem in the study of supersymmetric non-linear sigma models, concerning their geometric and ultraviolet properties. He was not working in supergravity at the time, and since Peter van Nieuwenhuizen was stranded in Europe with some visa or passport problem, I had to abandon my expectations to work on supergravity. I am priviledged that both became lifelong friends. Curiously enough, I am now writing this contribution in Stony Brook, where I am back after so many years. It really feels like coming home.

MIT was a bit intimidating, but it was exciting to join the physics community in the Boston area. I remained in Boston till 1988. Meeting Roman was a great experience. He has an eclectic knowledge of physics, he is happy to share it with others. It is unfortunately that over those years we never managed to work on any project together. We had plenty of discussions on anomalies, [3–5], and their Hamiltonian formulation in terms of cocycles and abelian extensions of the current algebra presented in the work of Faddeev and Shtashvili [6]. We had discussions on the topology of gauge fields, large-N methods, low-dimensional gravity ... but no collaboration. Pity. I learned a great deal from him, not only in physics. Roman loves conversation, and has a great sense of humor. We all enjoyed his wonderful irreverence. After the joint seminar on Wednesdays, that alternated between Harvard and MIT, we used to go for supper with the speaker. The dinner was hosted by Roman at MIT, and by Sydney Coleman at Harvard. I have very fond memories of those evenings.

Since that time, we have only met very few times. Once in Boston, for Shelly Glashow's 60th birthday, and later on in a few European conferences. The last one at the ICTP in Trieste. Making a contribution to this volume dedicated to Roman gives me the oportunity to express my appreciation for the interactions and discussions we had in the past.

Among the many jewels in Roman's production, my aim is to highlight a paper written with David Gross [1], and related to the title of this essay. The paper presented a lucid analysis of the quantum field theoretic consequences of having anomalies in gauged symmetries, and the need to cancel them. Let's briefly recall some landmarks in the making of the SM [7].

The formulation of electroweak unification started in [8], where the $SU(2) \times U(1)$ theory is formulated. The model for leptons was presented in [9] and [10] where spontaneous symmetry breaking and the Higgs mechanism are introduced to generate the vector boson masses and to accommodate the correct masses for the leptons compatible with gauge invariance. At the end of 1963 Gell-Mann introduced the quark fields [11]. He noted the parallel between the leptonic weak current and the suggested one for the (u, d) isospin-doublet. Soon after Y. Hara [12] introduced a fourth quark, and also drew the analogies with the weak interactions. Glashow and Bjorken noted Hara's work, and the fourth quark, named "charmed quark" by them [13] played a crucial role in Glashow's later work with Iliopoulos and Maiani (GIM) [14]. They showed that the introduction of the fourth quark provides a natural explanation for the suppression of $\Delta S = 2$ and flavor-changing neutral current processes.

If one studies the SM as known at the time, it had non-trivial gauge anomalies, and this was fatal for the consistency of the theory. Gross and Jackiw, and independly the authors of [15] realized that a minimal way to achieve consistency was to cancel the lepton anomalies with those of the quarks. This required the introduction of the fourth quark with the correct quantum numbers, providing fundamental theoretical basis for the family structure of quarks and leptons. In the words of John Iliopoulos [18],

> ... However, the mechanism [GIM] does not tie together leptons and quarks, in spite of the fact that the title of the original GIM paper was: *Weak interactions with lepton–hadron symmetry.* Such a symmetry is not implied by the requirement for the suppression of processes with FCNC. It is remarkable that such a symmetry is imposed by the mere mathematical consistency of the theory and it was discovered immediately after the renormalizability of general Yang–Mills theories, with or without spontaneous symmetry breaking, was proven.

Probably, the first time the terminology "Standard Model" was used we in [21]. The charm quark was discovered soon after the works, as well as the τ-lepton. The family structure then implied that another doublet of quarks should be found. This was brilliantly confirmed with the discovery of the bottom and top quarks (1977, 1995 respectively). Details can be found in [7] and [17].

2. The anomaly cancellation within a family

Rather than following the original computations in [1, 15], we present the current textbook version of the anomaly cancellation in the SM. There are three gauge groups, and we will take the liberty of extending the color gauge group from $SU(3)$ to $SU(N)$. Hence the gauge symmetry of this theory is: $SU(N) \times SU(2) \times U(1)$. The family structure contains a rather intricate chiral structure. The left- and right-handed fermions of the matter sectors have the following qauntum number assignments:

$$(N, 2)^L_{q_L} \oplus (1, 2)^L_{l_L} \tag{1}$$
$$(N, 1)^R_{u_R} \oplus (N, 1)^R_{d_R} \oplus (1, 1)^R_{e_R}. \tag{2}$$

The superscripts correspond to left- and right- handed fermions. The anomaly cancellation conditions coming from the triangle diagrams are listed below. The first three originate from the three gauge currents associated to the groups listed on the left. The fourth one corresponds to a triangle where in one vertex we have a $U(1)$ current and energy–momentum tensors in the other two [20][a]:

1) $U(1)\,SU(2)^2$, $2Nq_L + 2l_L = 0$

2) $U(1)\,SU(N)^2$, $2q_L - (u_R + d_R) = 0$

3) $U(1)^3$, $2Nq_L^3 + 2l_L^3 - Nu_R^3 - Nd_R^3 - e_R^3 = 0$

4) $U(1), T, T$ $2Nq_L + 2l_L - N(u_R + d_R) - e_R = 0$ (3)

To solve these equations, two cases need be considered. The first corresponds to $e_R \neq 0$. Since the equations are homogeneous, we can fix the scale by simply taking $e_R = -1$ as in the Standard Model. Once this choice is made, the solution is unique up to a permutation of the u_R, d_R values. We will not explore the second solution with $e_R = 0$.

Some simple algebra shows that with the above condition, we obtain the following hypercharge assignments, reproducing the SM for $N = 3$:

$$e_R = -1 \quad l_L = -\frac{1}{2} \quad q_L = \frac{1}{2N} \quad u_R = \frac{N+1}{2N} \quad d_R = -\frac{N-1}{2N}. \quad (4)$$

The other non-trivial solution corresponds to swapping the u and d values. It is interesting to notice that without the fourth condition, even if we set $e_R = -1$, there is a one-parameter family of solutions depending on l_L, but most of them are not rational. Fixing the electric charge of the left- handed fermions determines the solution completely independently of the fourth condition.

In a recent paper [22] the authors make a number of interesting remarks regarding the anomaly cancellation in the SM and the structure of Fermat-like diophantine equations. In the early eighties, when anomalies beyond four dimensions were explored, I remember Edward Witten making the interesting remark that Fermat's last theorem would be equivalent to showing that a $U(1)$ gauge theory with three chiral fermions with non-vanishing integer charges, is always anomalous in any number of even dimensions. Since then, the theorem was proven by Andrew Wiles, but the methods hardly resemble anything in quantum field theory...

3. Parting comments

It is a pleasure to pay a small tribute to Roman's brilliant contributions to fundamental physics. I made this choice because of its beauty, its apparent simplicity and the depth of its consequences. We are still far from understanding the origin

---

[a]The same triangle diagram was evaluated first in [19] as the gravitational correction to PCAC, making a negligible contribution to π^0-decay. In [20] the possible non-vanishing value of this diagram in the context of local currents provides a signal that energy–momentum conservation could be violated, if we insist on the covariant conservation of the $U(1)$-current. Thus a new anomaly condition follows. The trace of any generator of the gauge group should vanish. While this is true for any simple or semi-simple Lie group automatically, the vanishing of the trace does not necessarily call for unification.

of the family structure and the low-energy properties of the SM. Perhaps the new paradigm(s) will be discovered in the decade now beginning, or perhaps they are already in front of our eyes, but we have not yet been able to see them. Let us hope so, and stay healthy to enjoy the new insights.

Happy birthday Roman.

4. Acknowledgements

It is a pleasure to thank Álvaro de Rújula for discussions on the making of the SM. He was both a witness and an actor in such a remarkably fertile period in the exploration of fundamental physics.

References

1. D. J. Gross and R. Jackiw, "Effect of anomalies on quasirenormalizable theories," Phys. Rev. D **6**, 477 (1972).
2. D. Z. Freedman and P. van Nieuwenhuizen, "Supergravity and the unification of the laws of physics", Sci. Am. **238** 2, 126–143 (1978).
3. S. L. Adler, "Axial vector vertex in spinor electrodynamics," Phys. Rev. **177**, 2426 (1969).
4. J. S. Bell and R. Jackiw, "A PCAC puzzle: $\pi^0 \to \gamma\gamma$ in the σ model," Nuovo Cim. A **60**, 47 (1969).
5. S. L. Adler and W. A. Bardeen, "Absence of higher order corrections in the anomalous axial vector divergence equation," Phys. Rev. **182**, 1517 (1969).
6. L. D. Faddeev and S. L. Shatashvili, "Algebraic and Hamiltonian methods in the theory of nonabelian anomalies," Theor. Math. Phys. **60**, 770 (1985) [Teor. Mat. Fiz. **60**, 206 (1984)]. doi:10.1007/BF01018976.
7. An excellent account of the Standard Model and its history, can be found in: M. J. G. Veltman, *Facts and Mysteries in Elementary Particle Physics* (World Scientific, 2003).
8. S. L. Glashow, "Partial symmetries of weak interactions," Nucl. Phys. **22**, 579 (1961).
9. S. Weinberg, "A model of leptons," Phys. Rev. Lett. **19**, 1264 (1967).
10. A. Salam, "Weak and electromagnetic interactions," Conf. Proc. C **680519**, 367 (1968).
11. M. Gell-Mann, "A schematic model of baryons and mesons," Phys. Lett. **8**, 214 (1964).
12. Y. Hara, "Unitary triplets and the eightfold way," Phys. Rev. **134**, B701 (1964).
13. J. D. Bjorken and S. L. Glashow, "Elementary particles and SU(4)," Phys. Lett. **11**, 255 (1964).
14. S. L. Glashow, J. Iliopoulos and L. Maiani, "Weak interactions with lepton–hadron symmetry," Phys. Rev. D **2**, 1285 (1970).
15. C. Bouchiat, J. Iliopoulos and P. Meyer, "An anomaly free version of Weinberg's model," Phys. Lett. **38B**, 519 (1972).
16. A. Pais and S. B. Treiman, Phys. Rev. Lett. **35**, 1556 (1975).
17. M. Tanabashi *et al.* [Particle Data Group], "Review of Particle Physics," Phys. Rev. D **98**, 030001 (2018). doi:10.1103/PhysRevD.98.030001.
18. Jean Iliopoulos, "Glashow–Iliopoulos–Maiani mechanism", *Scholarpedia* **5**, 7125 (2010).

19. R. Delbourgo and A. Salam, "The gravitational correction to PCAC," Phys. Lett. **40B**, 381 (1972).
20. L. Alvarez-Gaume and E. Witten, "Gravitational anomalies," Nucl. Phys. B **234**, 269 (1984).
21. A. Pais and S. B. Treiman, Phys. Rev. Lett. **35**, 1556 (1975).
22. N. Lohitsiri and D. Tong, "Hypercharge Quantisation and Fermat's last theorem," *SciPost Phys.* **8**, 009 (2020) [arXiv:1907.00514 [hep-th]].

Chapter 3

Romaniana: A Student's Appreciation of Roman Jackiw, on his 80th Birthday

Michael Bos

I was Roman's PhD student at MIT while in graduate school from 1984 to 1988. In those days physics undergraduate education in The Netherlands, where I grew up, was a five-year program consisting solely of physics and supporting math, longer and more focused than a U.S. college physics major. I took advantage of that head start by taking the written parts of MIT's PhD qualifying exam soon after arriving. A bit later, while filing into the old Center for Theoretical Physics common room for a seminar, Roman turned to me and asked, more or less in public, "Are you still coming to work for me?" I had merely made a preliminary round among potential advisors in the CTP and didn't recall a previous suggestion to that effect. But I inferred from his question that I was welcome and eagerly accepted on the spot.

In the CTP Roman stood at what seemed to me the optimal point of seniority. The Center had its old guard of purple names who still had offices but rarely took on students, as well as some relative newcomers whose long-term prospects were as yet unclear. In between, Roman was both established and active; his signal contributions to field theory were relatively recent and well-known.

On the other hand, his reputation among students and postdocs was somewhat forbidding. He was said to be a tough supervisor and his demeanor elicited a certain apprehension. To be frank, to those who did not know him, Roman did not always project a warm and cuddly feeling, at least in those days. It was the attitude of one who did not suffer fools gladly, not uncommon among his generation (and caliber) of physicists. I would soon find out about his distinctive brand of humor, typically in the form of a mildly sarcastic comment, delivered in deliberate diction and accompanied by what I can only call the Roman grin. Anyone who has spent time under his tutelage knows that grin well.

The apprehension quickly vanished once you entered his orbit. His stern engagement reflected at heart an old-fashioned professional paternalism based on reasonable expectations in both directions, and he took his part of the bargain seriously. He was liberal with his time and constructive in his feedback. Even when we had disagreements he never ceased to be encouraging and supportive. He always made sure I got into the summer schools and conferences I wanted, and he was (so I was told) generous in job recommendations.

That Roman was willing to take on new students was a statement in its own right. To those who came of age during the 1980's it felt like particle theory was on the edge of stasis. From the vantage point of the time, the 1960's and early 1970's had been a great era of progress for particle physics and a golden age of field theory. The taming of non-abelian gauge theories, the creation of the Standard Model and the field-theoretic understanding of phase transitions had coined a number of sterling reputations, Roman's among them. Importantly, it seemed that even physicists whose names would not rise all the way to the canon were able to develop careers off the collateral challenges the new frameworks threw off and the variations they invited. But by the time my cohort of students entered graduate school, the reverberations of those original impacts were down to their fourth or fifth echo and the remaining challenges seemed intractable. As we plaintively told ourselves, there was "not much left to do".

Against that, Roman would point out that every generation goes through doubts of the sort, and that in the seemingly golden period of the 1960's many in the profession had in fact believed that progress had stalled and the end was near. Once when I expressed concern about finding good projects he countered with a paraphrase of Thomas Mann: to be a successful author you need to write every day, no matter your confidence or enthusiasm; you might not always like the result, but at least you will be sure that when true inspiration calls, you will be ready to receive it. His willingness to keep giving projects and pointers earned him the respect of students across the CTP, even (perhaps especially) of those he did not supervise.

As a teacher, by instruction or by example, Roman stood out for his emphasis on clarity, precision and good sense. There is perhaps no experience more commonly shared among his students than of being challenged in those regards, and no context in which the Roman grin was more frequently on display. In doing so he implicitly held up the standards of his own written work, which stands out for its pedagogical excellence. Papers or reviews in the Romanesque style proceed in a measured, careful phrasing in which one easily hears Roman's actual voice. The presentation stresses logical continuity over intuitive jumps, whatever the actual genesis of the argument. A building is to some extent demystified if the architect points out the location of all the beams.

A related virtue of Roman's teaching is his care about credit. I myself had to unlearn the young man's tendency to assume that the latest is the greatest, which led me to an occasional case of attributional myopia. Roman would insist that a footnote is a stake in the ground, that if you don't know the history of an idea with confidence you should not be firm in attributing it, that it is best to be inclusive – but also that, as a practical matter, an established body of knowledge is often best referenced through a review.

Along the way I developed an impression of what one might call Roman's taste in physics, such as it was by the mid-1980's. One of my PhD projects involved some tedious diagrammatic computations that ended up filling a hundred or so pages of algebra; I felt that it earned me respect in his eyes but I did not sense that at that stage of his career he still had a personal affinity for such work. Likewise, in spite of his having been Ken Wilson's student at Cornell[a], I never heard him express a special interest in scaling, phase transitions, or lattice field theory. At the time, among his many contributions to field theory, it was mainly the anomaly phenomenon in all its avatars, the topological effects that attended them and the instantiation of those effects in (mostly) lower-dimensional models that seemed to hold the focus of his own attention.

While his work has been the pioneering conduit into physics for some profound mathematical ideas and while he maintained excellent relationships with prominent mathematicians, his mathematics came in careful doses. The boom in string theory, which had its second coming during my graduate study, and its numerous satellite boomlets, chartered a thriving business importing mathematical wares into physics wholesale, which was not really Roman's thing. In fairness, the MIT theory group as a whole was not quick to embrace this.

Roman did not strive to exude, by nature or affectation, the kind of how-did-he-do-that brilliance I sensed in some other physicists of his stature. No Mozart or Beethoven, no effortless virtuoso or brooding genius, the professional persona he presented rather fit the model of a Johann Sebastian Bach, the Bach of the Leipzig years, firmly rooted in the traditions of his craft, strongly committed to passing those on, relentless at feeding the liturgy and producing, in the process, a steady stream of work, always interesting, often great, at times outstanding.

****

As I said above, my work under Roman took place at the onset of an unsettling time to be a young particle physicist. I did not last; after getting my PhD I stuck around for some postdoctoral jobs before leaving for Wall Street. My regular interactions with Roman came thereby to an end.

The contribution he made to my education did not. While I no longer have much daily use for anomalies or renormalization, I continue to benefit from the professional discipline Roman instilled. I like to think that his advice has served me well.

In due course I made the mistake of mentioning to my wife some of the pointed phrases Roman used on me and other students. She did not hesitate to make them her own, and thus the Roman grin persists in our household like the smile of the Cheshire cat. To this day, if I equivocate, I continue to be asked if I "could be a little bit more explicit".

---

[a]I once asked Roman why Cornell, for being an out-of-the-way place, has employed and trained such an exceptional number of first-rate physicists. He replied that not everybody feels the need to live in a big city, and that to have a first-rate physics department, you don't have to convince everyone to live in your location – you just need enough good people for a first-rate department.

Lately I sometimes cross paths with Roman again, either back at MIT, or in a different context, which I shall briefly explain. One evening in 1985 I was walking with my friend and fellow-student Stefano Forte along Massachusetts Avenue when we crossed Roman in a state of uncommon elation: his wife So-Young Pi had just given birth to their son Stefan. We congratulated him wholeheartedly. A few years later – it must have been late 1989 or early 1990 – I came across young Stefan during a party given by Roman and So-Young in the residential tower of Rockefeller University in New York. Roman was spending a sabbatical at Columbia, where I was a postdoc at the time; So-Young was likewise at Rockefeller. Roman mentioned casually that Stefan, whose proximate ambition was to meet the then-novel Teenage Mutant Ninja Turtle movie characters (or, as Roman put it with his grin, "he actually wants to meet the costumes"), had started to learn the violin. Well, why not? Fast forward a few decades and Stefan has become one of the world's leading violinists, and attending a concert by him now holds the promise of finding Roman in the audience. By the way, as artists are invariably more famous than scientists, Stefan has displaced his father in the pecking order of Jackiws as displayed, dispassionately, by a Google search.

****

I would like to end by pointing to an elephant in the room.

As Roman's former student I harbor a rooting interest in the annual communiqués from Stockholm, in which Roman's name has so far gone unmentioned. I have no particular standing in these matters but I believe a compelling case can be made. The chiral anomaly is an essential theoretical insight, elegant and surprising. It is directly vindicated by actual experiment as the conclusive explanation for otherwise puzzling meson decays. It is a key element of the Standard Model (and all its candidate extensions) by providing verifiable constraints on the content of elementary particle families. Its descendants and relatives are finding applications well beyond its original context. It seems, in short, the kind of important discovery for which Nobel Prizes have historically been awarded. Like many discoveries it had its precursors, but for all practical purposes the phenomenon was analyzed, settled and socialized conclusively only in the wave of activity that Roman co-initiated, and if credit is apportioned he deserves a lion's share. Add to that the gravitas of a distinguished career, and what boxes remain to be checked?

Most of those who made fundamental contributions to the Standard Model have by now been recognized, and Roman, like all of us, is not getting any younger. If respectful advocacy is permitted here, Roman's 80th birthday might be a good occasion to consider this once more.

Chapter 4

Electron Fractionalization in Celebration of Roman Jackiw's 80th Birthday

Claudio Chamon

Physics Department, Boston University,
Boston, MA 02215, USA

The concept of fractional fermion number emerged from the pioneering work of Roman Jackiw and Claudio Rebbi in 1976. Their work, in addition to a long list of other contributions by Roman, led to a number of developments in modern condensed matter physics that are tied to zero modes and topology. In this contribution, I briefly trace the influence of Roman's ideas. I also present a puzzle regarding rational vs. irrational charges originating from my works with Roman, and its resolution.

1. Charge fractionalization

The seminal paper by Jackiw and Rebbi [1] showed that zero energy solutions exist when one-dimensional fermions couple to a bosonic field background that interpolates between two symmetry broken states, i.e., a soliton connecting two vaccua. (They also showed the same effect in a three-dimensional model.) They showed that the fermion number associated with this localized solution is $-1/2$ when the state is empty, and $+1/2$ when occupied. For charged fermions, their result is the first example of the phenomenon that now became familiar in modern condensed matter physics, that of charge fractionalization.

Independently, Su, Schrieffer, and Heeger [2, 3] discovered in 1979 a lattice realization of the one-dimensional case in a model of polyacetylene that captures the relevant interactions between phonons and electrons in the polymer chains. With that example, the notion of charge fractionalization started to become familiar in physics, and bridged different of the discipline subfields [4].

In those first years there were a number of fundamental questions regarding whether or not the fractional charge was a sharp quantum number; not only should its expectation value be fractional, but its variance zero. These questions are relevant to the treatments in condensed matter systems, which are finite in size and where zero modes must appear in pairs, as opposed to the infinite system of Ref. 1 where a single soliton can exist. The question found a positive answer, that the variance of the charge vanishes, meaning that the charge is an eigenstate of a properly defined charge operator. This operator consists of the local charge density weighted by a smoothly varying function enclosing a region that is large compared to the size of a single zero mode wavefunction but small compared to the separation between the zero modes. Several groups reached this answer independently, including Roman and collaborators [5–8].

It was also shown that exotic fermionic quantum numbers in one-dimension are not restricted to rational values, but can be tuned continuously [9–12] by a small breaking of an energy-reflection symmetry assumed in Refs. 1 and 2. These (potentially) irrational charges are also well defined in the sense discussed above in that their variance vanishes in the same appropriate limits.

Shortly after, the discovery of the fractional quantum Hall effect brought about new ways in which the electron is fractionalized. Laughlin's fractionally charged quasiparticles [13] find their origin in electron-electron interactions that stabilize the gapped quantum Hall liquid state, as opposed to the case above where electrons only interact with their non-trivial background potential. One of the most fruitful angles to understand the physics of the quantum Hall effect is through its description in terms of Chern–Simons theory [14], which was first brought into the physics vernacular by Roman's work with Deser and Templeton [15]. Among the many ideas that emerged from the Chern–Simons description of the quantum Hall effect is Wen's notion of topological order [16], which underscores a new type of order in quantum liquids that exists in the absence of any symmetry breaking. One of the manifestations of topological order is a ground state degeneracy that depends on the genus of the manifold where the liquid resides, and cannot be lifted by any local perturbation.

The fractional charges in the quantum Hall effect are rational multiples of the electron charge. The charge can be tied to the quantization of the Hall conductance, which is established using Laughlin's gauge argument [17, 18]. It can also be tied to the topological degeneracy of the state on, say, a torus. Oshikawa and Senthil [19] argued that if the charge is fractionalized to $1/q$, there must be a ground state degeneracy on a surface of genus g that scales with q^g in general and q^{2g} if the excitations are bosonic or fermionic. Their arguments are quite general, and are based on the algebra between the operation of flux insertion and the operation of creating and annihilating a quasiparticle and quasihole pair around different radii of a torus. The assumption that the excitations are deconfined is needed for the winding of the pair of excitations to cost bounded energy at the intermediate steps.

While the argument of Ref. 19 was directed at systems in spatial dimensions equal or greater than two, one can follow step by step their construction and apply it to 1D systems with fractionalization rooted at spontaneous symmetry breaking. The argument requires one to move deconfined fractionally charged quasiparticles around a loop, as well as being able to insert flux through the loop. But in the case of systems in Refs. 9–12 the ground state is two-fold degenerate irrespective of whether one breaks or not energy-reflection symmetry, say, by inclusion of a staggered chemical potential μ_s. So there is an apparent contradiction, in that the value of the charge can be varied continuously as function of μ_s, and therefore the connection between charge and degeneracy is lost.

In 2010, Masaki Oshikawa, Christopher Mudry, and I overlapped at the workshop in celebration of Eduardo Fradkin's 60th birthday, and had a chance to discuss this problem. The resolution reached was the following. Suppose that one starts in the vacuum associated with a dimerization to which we assign a negative value of the order parameter Δ. (The mass of the fermions is $m = \sqrt{\Delta^2 + \mu_s^2}$ in the 1D system.) Creating a kink/anti-kink pair corresponds to building a small region with positive Δ. Pulling the pair apart increases the size of the positive region, at the expense of the negative one. Going anti-clockwise, the anti-kink would always be to the right of the kink. Now, once one winds these two defects and annihilate them on the side of the loop that is diametrically opposite to where the pair was created, the vacuum will be that associated with a dimerization to which we assign a positive value of the order parameter Δ. Starting with this new positive value, one can no longer create the anti-kink to the right of the kink, but only in the opposite order. So instead of being able to repeat multiple times the same pair excitation that is used in Ref. 19 (combined with the flux insertions), one can only undo the previous step. So instead of exploring a number of different states with the two non-commutative operations of pair creation and flux insertion, one remains in a game of ping-pong between two states. The phase picked up as the fractional charges circle the quantized flux inserted (ping) is canceled by the process of undoing the same thing (pong). Hence any connection between degeneracy and charge value is not there. Deconfined irrational charges in 1D present no contradiction to the simple but general argument.

In what follows, we shall present an example of fractionalization in 2D that resides at the intersection between these two views: 1) fractional charges bound to zero modes associated with a topologically non-trivial background of a bosonic field; 2) fractional charges associated to topologically ordered systems. We shall discuss how irrational charges can appear in the former, and how to reconcile it with the fact they cannot from general gauge arguments appear in the latter.

2. Zero modes in two-dimensions

In another pioneering contribution, Roman and P. Rossi [20] presented in 1981 another example, now in 2D, of a physical system where the Atiyah–Singer index theorem [21] applies. The example consisted of 2+1D Dirac fermions coupled to a Higgs field. A physical realization of this setting exists at the surface of 3D topological insulator placed in contact with an s-wave superconductor [22]. The 2D Dirac fermions are provided by the surface states of the topological insulator, while the Higgs field comes from the pair potential due to the superconductor. Roman showed in Ref. 20 that n zero modes exist in the spectrum if n vortices are present in the Higgs field background. Because of the particle-hole or charge conjugation symmetry in this system, the zero-mode functions are quantized into self-adjoint fermions, i.e., they are Majorana operators bound to a topological defect [23]. While Roman did not call these by that name, they are presented explicitly in equations in his work.

In 2007, Hou, Mudry, and I [24] realized that one of the possible mass terms that could open a gap in graphene could also serve as a bosonic mode that can host vortices, and thus bind zero modes. The term corresponds to a Kekulé dimerization pattern, which connects in k-space the two 2-dimensional representations of Dirac fermions at the K_+ and K_- points of the Brillouin zone. We were not aware of Ref. 20 by Roman at first; we did find the zero-mode solution, but once we learned of his paper before our manuscript was completed, we found his solution not only to precede ours but also to be more general in that he considered multiple vorticity. Charge is a good quantum number in our graphene example (i.e., the order parameter is not associated to superconductivity), and the zero modes correspond to complex fermions, not Majoranas. Charge $1/2$ is bound to these zero modes. Just as polyacetelene with its dimerizations provides a lattice realization of the field theory for the 1D model of Ref. 1, graphene with a Kekulé background provides a lattice realization of fractionalization via zero modes in 2D.

Because in 2D vortices interact logarithmically, the fractional charges in 2D found in Ref. 24 were marginally confined, and they would only be accessible above some Kosterlitz–Thouless temperature. Roman and So-Young [25] resolved this issue by including an axial gauge field (itself connected to lattice distortions in graphene), which screens the logarithmic interaction between the vortices much in the same way that the magnetic flux screens the interactions between superconducting vortices. They thus succeeded in deconfining the fractional charges in this 2+1D model. Notably, in odd space-time dimensions, they did not have to include the effects of the chiral anomaly – to yet again mention a key contribution by Roman, with Bell [26], that finds its way into this tale of fractionalization.

The introduction of the axial gauge field, in addition to the usual gauge potential, allows us to make a connection to the other kind of fractionalization that is tied to topological order. In a series of papers [27–30], some in collaboration with Roman, we sorted out issues concerning when charge can be irrational and when it cannot.

3. Irrational and rational charges in 2D

Consider the Lagrangian density

$$\mathcal{L} = \bar{\psi}\left[\gamma^\nu\left(i\partial_\nu + A_\nu + \gamma_5 A_{5\nu}\right) - (\varphi_1 - i\gamma_5\varphi_2) - \gamma^3\mu_s\right]\psi, \qquad (1)$$

where $\boldsymbol{A}$ and $\boldsymbol{A}_5$ are vector and axial gauge potentials; φ_1 and φ_2 are the real and imaginary parts of a complex scalar field $\varphi = \varphi_1 + i\varphi_2$ that encode the Kekulé distortion; and μ_s is a field that acts like a staggered chemical potential that breaks charge conjugation symmetry. It is useful to define $\varphi_3 \equiv \mu_s$ and to pack the three components in a vector field $\boldsymbol{\varphi} = (\varphi_1, \varphi_2, \varphi_3)$.

In the absence of the axial vector potential, one can compute the conserved current components induced by a non-trivial background of φ [9,31] :

$$J^\nu = \frac{1}{8\pi}\, \epsilon^{\nu\alpha\beta}\, \epsilon^{abc}\, \varphi_a\, \partial_\alpha \varphi_b\, \partial_\beta \varphi_c \tag{2a}$$

and the induced charge

$$Q = \frac{\Omega}{4\pi}, \tag{2b}$$

where Ω is the spherical angle traced by the vector φ far away from the defect. Because the staggered chemical potential tilts the projection of φ away from the Equator, the charge does not need to equal $1/2$, and it can be varied continuously, just as in the case of 1D polyacatelene, when the staggered chemical potential is present. However, in 1D it does not cost energy to separate a kink and an anti-kink, while in 2D a vortex and an anti-vortex are attracted logarithmic. The axial vortex of Roman and So-Young screens that interaction; so a natural question is how it affects the induced charge in conjunction with the addition of a non-zero staggered chemical potential.

Upon including the axial gauge field, the charge can be written as the flux of a field that combines the texture of the complex field φ, the axial vector potential, and the tilt due to the staggered chemical potential:

$$Q = \int d^2r\, (\partial \wedge b)(r), \tag{3}$$

where

$$b_\nu = \frac{1}{2}\partial_\nu \phi - \frac{1}{2}\, (\partial_\nu \phi + 2A_{5\nu})\cos\theta , \tag{4}$$

with θ the zenith angle (equal to π when $\mu_s = 0$) and ϕ the azimuthal angle corresponding to the phase of the complex field φ. When the axial field is turned off in (4), one recovers the result of (2b) when the azimuthal angle ϕ sweeps 2π.

However, when the axial field is turned on so as to precisely screen the logarithmic interactions, the term in parenthesis in (4) vanishes. In that case, the charge $Q = 1/2$, independent of the angle θ. In other words, the same field that deconfines the induced charge rationalizes it. Although for this type of fractionalization mechanism we require a bosonic vector field φ to provide the mass and the non-trivial background for the fermions, we find that, just like in the fractional quantum Hall effect, the charges must be rational if they are deconfined.

Another bridge between the two mechanisms can be made if one integrates out the massive fermions and compute the effective action for the complex field φ and gauge potentials A and A_5 for fixed μ_s. In the limit when the mass $m = \varphi_1^2 + \varphi_2^2 + \mu_s^2 \to \infty$, the field φ is slaved to the axial vector potential, and the effective Lagrangian reads

$$\mathcal{L}^{\text{eff}} = \frac{1}{2\pi}\, \text{sgn}(\mu_s)\, \epsilon^{\mu\nu\rho}\, A_\mu\, \partial_\nu\, A_{5\rho} . \tag{5}$$

With dynamical gauge fields, this double Chern–Simons action has a 4-fold degeneracy on the torus. The re-rationalization of the charge and the connection with a topologically degenerate state on the torus squares with the elegant argument presented in Ref. 19 for systems in two or higher dimensions.

There appeared at first sight to be two separate worlds of fractionalization: 1) that associated to symmetry breaking and topologically non-trivial backgrounds of otherwise non-interacting fermions, and 2) that associated to strong electronic interactions, such as those existing for dispersionless electrons within a flat Landau level. The example of fractionalization in graphene-like systems provides us with a bridge that connects these two worlds.

4. Learning from Roman beyond his work

The first time that I met Roman in person was when I was a 1st year graduate student at MIT working with Xiao-Gang Wen. I was working on chiral edge states, and I had a question on a paper of Roman's which basically provided a way to properly quantize chiral fields [32]. My question was about a factor of $1/2$ that was there for the chiral fields but not for the non-chiral ones. I made an appointment to see Roman. My fellow graduate students warned me that I should come prepared! (I didn't quite know what they meant by prepared.) So prepared I went, having read the paper over and over again, although I could still not figure out a good physical answer for the $1/2$ factor. I met Roman as scheduled, and left the office with a clear, precise, and satisfying answer. Little would I know that I would have many other opportunities to get clear, precise, and satisfying answers from Roman.

When I started collaborating with Roman I was already a full professor, but one never stops learning, and specially from someone like Roman. In addition to physics, I got exposure to the "Roman-style" of clear and concise expression, specially in writing, that is hard to match.

In commemorating Roman's 80th birthday, I want to comment on something special that I witnessed in his 70th birthday celebration. At dinner, there was a large table reserved for Roman and all his former students in attendance. The importance that Roman gives to his students makes it clear that his legacy extends well beyond the remarkable papers that he wrote; his legacy includes the generations of scientists to whom he, in person, taught a great deal.

References

1. R. Jackiw and C. Rebbi, "Solitons with fermion number 1/2," Phys. Rev. D **13**, 3398 (1976).
2. W. P. Su, J. R. Schrieffer, and A. J. Heeger, "Solitons in polyacetylene," Phys. Rev. Lett. **42**, 1698 (1979).
3. W. P. Su, J. R. Schrieffer, and A. J. Heeger, "Soliton excitations in polyacetylene," Phys. Rev. B **22**, 2099 (1980).

4. R. Jackiw and J. R. Schrieffer, "Solitons with fermion number 1/2 in condensed matter and relativistic field theories," Nucl. Phys. B**190**, 253 (1981).

5. S. Kivelson and J. R. Schrieffer, "Fractional charge, a sharp quantum observable," Phys. Rev. B, **25**, 6447 (1982).

6. R. Rajaraman and J. S. Bell, "On solitons with half integral charge," Phys. Lett. **116B**, 151 (1982).

7. J. S. Bell and R. Rajaraman, "On states, on a lattice, with half-integral charge," Nucl. Phys. **B220**[FS8], 1 (1983)

8. R. Jackiw, A. K., Kerman, I. Klebanov, and G. Semenoff, "Fluctuations of fractional charge in soliton anti-soliton systems," Nucl. Phys. **B225**[FS9], 233 (1983).

9. J. Goldstone and F. Wilczek, "Fractional quantum numbers on solitons," Phys. Rev. Lett. **47**, 986 (1981).

10. M. J. Rice and E. J. Mele, "Elementary excitations of a linearly conjugated diatomic polymer," Phys. Rev. Lett. **49**, 1455 (1982).

11. R. Jackiw and G. Semenoff, "Continuum quantum field theory for a linearly conjugated diatomic polymer with fermion fractionization," Phys. Rev. Lett. **50**, 439 (1983).

12. S. Kivelson, "Solitons with adjustable charge in a commensurate Peierls insulator," Phys. Rev. B **28**, 2653 (1983).

13. R. B. Laughlin, "Anomalous quantum Hall effect: An incompressible quantum fluid with fractionally charged excitations," Phys. Rev. Lett. **50**, 1395 (1983).

14. S. C. Zhang, T. H. Hansson, and S. Kivelson, "Effective-field-theory model for the fractional quantum Hall effect," Phys. Rev. Lett. **62**, 82 (1989).

15. S. Deser, R. Jackiw, and S. Templeton, "Topologically massive gauge theories," Ann. Phys. (N.Y.) **140**, 372 (1982).

16. Xiao-Gang Wen, "Topological orders in rigid states," Int. J. Mod. Phys. B4, 239 (1990).

17. R. B. Laughlin, "Quantized Hall conductivity in two dimensions," Phys. Rev. B **23**, 5632 (1981) .

18. B. I. Halperin, "Quantized Hall conductance, current-carrying edge states, and the existence of extended states in a two-dimensional disordered potential," Phys. Rev. B **25**, 2185 (1982).

19. M. Oshikawa and T. Senthil, "Fractionalization, topological order, and quasiparticle statistics," Phys. Rev. Lett. **96**, 060601 (2006).

20. R. Jackiw and P. Rossi, "Zero modes of the vortex-fermion system," Nucl. Phys. B**190**, 681 (1981).

21. M. F. Atiyah and I. M. Singer, "The index of elliptic operators I, II, IV, V'," Ann. of Math. **87**, 484 (1968); **87** 546 (1968); **93** 119 (1971); **93** 139 (1971).

22. L. Fu and C. L. Kane, "Superconducting proximity effect and majorana fermions at the surface of a topological insulator," Phys. Rev. Lett. **100**, 096407 (2008).

23. N. Read and D. Green, "Paired states of fermions in two dimensions with breaking of parity and time-reversal symmetries and the fractional quantum Hall effect," Phys. Rev. B **61**, 10 267 (2000).

24. C. -Y. Hou, C. Chamon, and C. Mudry, "Electron fractionalization in two-dimensional graphenelike structures", Phys. Rev. Lett. **98**, 186809 (2007).

25. R. Jackiw and S.-Y. Pi, "Chiral gauge theory for graphene," Phys. Rev. Lett. **98**, 266402 (2007).

26. J. S. Bell and R. Jackiw, "A PCAC puzzle: $\pi^0 \to \gamma\gamma$ in the σ-model," Il Nuovo Cimento A. **60**, 47 (1969).

27. C. Chamon, C.-Y. Hou, R. Jackiw, C. Mudry, S.-Y. Pi, and A. P. Schnyder, "Irrational

vs. rational charge and statistics in two-dimensional quantum systems," Phys. Rev. Lett. **100**, 110405 (2008).

28. C. Chamon, C.-Y. Hou, R. Jackiw, C. Mudry, S.-Y. Pi, and G. Semenoff, "Electron fractionalization for two-dimensional Dirac fermions," Phys. Rev. B **77**, 235431 (2008).

29. S. Ryu, C. Mudry, C. -Y. Hou, and C. Chamon, "Masses in graphenelike two-dimensional electronic systems: Topological defects in order parameters and their fractional exchange statistics," Phys. Rev. B **80**, 205319 (2009).

30. C. -Y. Hou, C. Chamon, and C. Mudry, "Deconfined fractional electric charges in graphene at high magnetic fields," Phys. Rev. B **81**, 075427 (2010).

31. A. G. Abanov and P. B. Wiegmann, "Theta-terms in nonlinear sigma-models," Nucl. Phys. B **570**, 685 (2000).

32. R. Floreanini and R. Jackiw, "Self-dual fields as charge-density solitons," Phys. Rev. Lett. **59**, 1873 (1987).

Chapter 5

Roman Jackiw and our Physics in Common

John M. Cornwall

Distinguished Professor Emeritus, Department of Physics and Astronomy, UCLA

1. Early years

Roman and I met in 1970 when he visited me at UCLA, so I have known him as a colleague and friend for nearly 50 years. I have taken sabbaticals at MIT on two occasions in the 1970s, during which I got to know and admire both families with which he was engaged. Of course, over the years I have also gotten to know and admire his wide and deep grasp of physics.

I have written only two papers with Roman as a co-author, but it turns out that we (with our respective colleagues) on several occasions were working on essentially the same problems, without knowing of the others' work. Although many of these physics interests date from long ago, their insight is still of great value in such fields as non-perturbative QCD, in which I have worked for a long time. After a little bit of discussion of our early years' work, I turn to several areas of research where we had similar ideas at about the same time, and although much of this work is old I find that it is still relevant to what I am thinking about.

Other authors to this volume will address Roman's wide-ranging and important contributions to other fields, such as condensed matter physics, so I will stick to the issues where we had strong mutual interests.

Roman's first visit to UCLA was to discuss the MIT and UCLA programs in light-cone physics, at the time a hot topic both theoretically and experimentally. My UCLA collaborator Dick Norton and I had been busy since 1968 (for a representative paper of several that we wrote, see [1]). using a spectral representation for forward scattering amplitudes that incorporated causality.* An example occurs in deep-inelastic $e - p$ scattering, then a hot subject in both experiment and theory:

$$S_{\mu\nu}(p,q) = i \int d^4x \, e^{iq\cdot x} \langle p|T[J_\mu(x/2)J_\nu(-x/2)]|p\rangle \tag{1}$$

where p is an on-shell proton momentum and q an asymptotically-large spacelike

---

*See [2] for references to the spectral representation. This work gives an alternative derivation of the Bjorken-Johnson-Low limiting behavior.

momentum transfer; J_μ is the electromagnetic current. Asymptotically-large means that the matrix elements are approaching the region of the light cone.

Roman and I wrote our first paper together [3] during his visit to UCLA. In writing this paper with Roman, I was pleased by his insistence on sticking to the canonical foundations of quantum field theory in his investigations as opposed to making non-perturbative conjectures not necessarily supported in field theory, or abandoning field theory for the popular S-matrix ideas of the day. We presented the canonical light-cone commutators of various electromagnetic current elements, yielding commutators whose matrix elements (when augmented with Schwinger terms) could in principle be isolated from experiments. The two independent (spin-averaged) scalar functions depend on two variables, and when when these are ex-panded in power of the kinematic variable $q^2/2q \cdot p$ expressed infinitely many sum rules for equal-time commutators of the current with its time derivatives. I will not discuss this work in any more detail, but will merely say that, in my opinion, it added substantially to the work being done in light-cone physics.

2. Dynamical symmetry breaking

There is no real symmetry breaking in a locally gauge-invariant theory, but we follow the general abuse of notation in using the term.

Dynamical symmetry breaking (without Higgs-Kibble particles) was a growing interest for many authors in the early 1070s as a means of generating gauge bosons masses. Roman, collaborating with Ken Johnson, and I, collaborating with Dick Norton, wrote virtually simultaneous papers [4,5] on dynamical symmetry breaking (that is, without Higgs particles) in Abelian gauge theories, each group unaware of the other's efforts.

QCD has no symmetry breaking, but there is a kind of gluon mass generation in QCD (see Sec. 4). It comes neither from elementary scalar fields nor from dynamical symmetry breaking. Nonetheless, to generate the mass term one needs a multiplet of massless scalar fields longitudinally coupled to the gauge bosons; these act in much the same way as Nambu-Goldstone fields and participate in such non-perturbative effects as confinement and chiral symmetry breaking. So it turns out that the ideas of [4,5] have applicability well beyond what was originally envisaged.

3. Higher-point effective actions

Roman and I wrote only one more paper together (along with Roman's then-student Terry Tomboulis) [6]; that was in 1974, when I was on sabbatical at MIT. Even so, we often were thinking along similar lines, as I will note later. Roman and his collaborators had been working on summing up graphs for the effective action of a scalar field in terms of a one-particle irreducible one-point function. He and his collaborators studied [7] spontaneous symmetry breaking at large N in the $O(N)$ non-linear σ model, one of many papers that Roman had contributed to since we had last been together. I cite only this example because it appeared in the same volume of Physical Review as [6] and is one small illustration of Romans ability to work on many projects at once. This paper continued the study of dynamical symmetry breaking (without Higgs fields).

Meanwhile, I had been working with Dick Norton on developing perturbative expansions of effective actions in terms of higher-order proper Green's functions [8]. These expansions could yield a variational approach to Schwinger-Dyson equations that always used these higher-;point proper Green's functions. By extending his work on an effective action in terms of one-point fields, Roman produced an elegant way of looking at the most important case, which is expressing an effective action in terms of a proper two-point function, by beginning with a two-point current coupled to the propagating fields in the action. I have used the ideas of [6] on several occasions for the effective action of a non-Abellian gauge theory, and I am sure that there are many more opportunities still unexplored.

4. Gauge boson mass in d=3 non-Abelian gauge theories

Three-dimensional QCD is the non-perturbative part of high-temperature (T) QCD, where its gauge potentials bearing spatial indices are the magnetic potentials of the d-4 theory. The essential non-perturbative feature was the dynamical generation of a mass for this magnetic gluon. This is perhaps not unexpected, since in d=3 the gauge coupling $g_3^2 \sim g_4^2 T$ has dimensions of mass. However, the mass of the magnetic gluon vanishes to all orders of perturbation theory,[†] leaving behind an infinite string of power-law (as well as logarithmic) infrared divergences, of the form $(g^2 \Lambda_{IR}^2)^N$ where Λ_{IR} is an infrared cutoff, to be set to zero in the full theory.

This sort of thing was puzzling long ago, since calculations which might be revealing, such as of the gauge boson proper self-energy, were intrinsically gauge-dependent in perturbation theory, and one could not be sure even of the sign of the one-loop corrections. With the development in 1981 of the Pinch Technique (as reviewed in [9]) it became possible to resum Feynman graphs so that the proper Green's functions were different from their Feynman forms but which summed to the same gauge-invariant S-matrix. In particular, the **gauge-invariant** one-loop d=3 gauge-boson proper self-energy $\Pi(q)$ calculated in the Pinch Technique, with loops calculated with the lowest-order massless gluon propagators, is [10]

$$\Pi(q) = q^2 - \pi b_3 g_3^2 q \qquad (2)$$

where q is the Euclidean (real and positive) momentum and b_3 is a **positive** real number: $b_3 = 15N/32\pi$ for color group $SU(N)$.[‡] One consequence of this positivity is that the one-loop gluon propagator has a tachyonic pole. This disease, coming

---

[†]This happens because there is no locally non-Abelian gauge-invariant mass term that has a finite number of gluon fields; such a term exists but with infinitely many interactions with infinitely many gauge potentials.

[‡]Infrared divergences appear at higher loop orders.

from the "wrong" sign of b_3, is analogous, and in fact derives from, the so-called "wrong" sign of d-4 QCD. The obvious cure would be a dynamical mass term $+m^2$ with the mass sufficiently large to overcome the tachyonic pole, and I began to work on this problem both ind d=3 and d=4, with its "wrong" sign calling for gluon mass generation. To avoid misunderstanding, we are using the term gluon "mass" loosely; it does not refer to a pole mass. It refers to the fact that in the Pinch Technique theory as well as in Landau-gauge lattice simulations[§] the gluon propagator has exactly this behavior. The zero-momentum propagator of the gluon proper self-energy in the Landau gauge is not the same as that of the gauge-invariant Pinch Technique, but they are related by a calculable quantity. each proper self-energy has a **positive** and **finite** value at zero momentum, and does not vanish or become singular there as it does in perturbation theory. We can simply call this value $1/m^2$ to define the gluon mass. (The mass m also defines the now-physical cutoff Λ_{IR}).This is inaccurate in Minkowski space, but acceptable in Euclidean space where it is problematic to extrapolate to imaginary Euclidean momenta.

We have already mentioned how gauge-invariant gluon mass generation in d=3,4 necessarily involves a multiplet of massless scalar fields that participate in gauge transformations that are singular (that is, carrying topological quantities) at large distances (but at small distances divergences are cancelled by contributions from short-ranged fields). The results (reviewed in [9]) include condensates of solitons such as center vortices and nexuses that are responsible for confinement, chiral symmetry breaking, and various topological properties.

Many workers, including myself [11], have worked on generating the needed mass dynamically, with no extra ingredients in the d=3 QCD action.[¶] But around the time that the Pinch Technique was developed, Jackiw and collaborators [14] observed that adding a Chern-Simons term to the usual d=3 gauge action produced a gluon mass at the classical level. This was an intriguing idea, and received much well-deserved attention. It does not seem to work for QCD, because the Chern-Simons term violates parity; QCD does not. But this Chern-Simons mass term has many applications that we do not have room to pursue here.

5. The future

I am too old to carry forward any serious physics with the ideas mentioned here, but I am not old enough to think that there will not still be crucial applications of dynamical mass generation based partially on the ideas of dynamical gauge-symmetry breaking and the construction of powerful—perhaps variational techniques for the Pinch Techniques—based on higher-point effective actions. I hope Roman and I

[§]See [12] for a careful study of Landau-gauge lattice propagators in d=3,4.
[¶]My collaborators in Ref. [9] and their collaborators have spent many years on the gluon mass problem in d=4, which is considerably more difficult technically. Their work is based on a hybrid of the Pinch Technique and use of lattice Landau-gauge data, and leads to promising results; see [13].

are not yet too old to see others continue to make progress in the subjects we both studied, sometimes together and sometimes, unknowing of the other's work. It has been a great pleasure to work with Roman and to watch him produce one important paper after another.

References

1. J. M. Cornwall and R. E. Norton, Phys. Rev. **177**, 2581 (1969).
2. J. M. Cornwall, Current-Commutator Constraints on 3- and 4-Point Functions, Phys. Rev. Lett. **16**, 1174 (1966).
3. J. M Cornwall and R. Jackiw, Canonical Light-Cone Commutators, Phys. Rev. D**4**, 367 (1971).
4. J. M. Cornwall and R. E. Norton, Spontaneous symmetry breaking without scalar mesons, Phys. Rev. D**8**, 3338 (1973).
5. R. Jackiw and K. Johnson, Dynamical model of spontaneously broken gauge symmetries, Phys. Rev. D**8**, 2386 (1973).
6. J. M. Cornwall, R. Jackiw, and E. Tomboulis, Effective Action for Composite Operators, Phys. Rev. D**10**, 2428 (1974).
7. S. Coleman, R Jackiw, and H. D. Politzer, Spontaneous sytmmetry breaking in the O(N) model for large N.
8. J. M. Cornwall and R. E. Norton, On the Formulation of Relativistic Many-Body Theory, Ann. Phys. **91**, 106 (1975).
9. J. M. Cornwall, J. Papavassiliou and D. Binosi, "The Pinch Technique and its Applications fo Non-Abelian Gauge Theories", Cambridge University Press, Cambridge, 2011.
10. J. M. Cornwall, Dynamical mass generation in continuum quantum chromodynamics, Phys. Rev. D**6**, 1453 (1982).
11. J. M. Cornwall, Exploring dynamical gluon mass generation in three dimensions, Phys. Rev. D **93**, 025201 (2016) [arXiv:1510.03453 [hep-ph], October, 2015.
12. A. Maas, Phys. Rev. D **91**, 034502 (2015) [arXiv:1402.5050 [hep-lat]].
13. A. C. Aguilar, D. Binosi and J. Papavassiliou, Front. Phys. (Beijing) **11**, no. 2, 111203 (2016) [arXiv:1511.08361 [hep-ph]].
14. S. Deser, R. Jackiw, and S. Templeton, Three-dimensional massive gauge theories, Phys. Rev. Lett **48**, 975 (1982).

Chapter 6

Fun in 2+1

S. Deser

Walter Burke Institute for Theoretical Physics,
California Institute of Technology
Pasadena, CA 91125, USA
Physics Department, Brandeis University
Waltham, MA 02454, USA
deser@brandeis.edu

To celebrate Roman Jackiw's eightieth birthday, herewith some comments on gravity and gauge theory models in $D = 3$, the chief focus of our many joint efforts.

1. Introduction

It is a pleasure to dedicate this note, on this round birthday, to my long-time friend and collaborator Roman Jackiw. We were born eight years apart at about the same latitude, if slightly different longitudes, in prewar Poland, but first met when he arrived in Cambridge as a Harvard Junior Fellow, after a Cornell PhD — and a co-authored graduate textbook — with Hans Bethe. That guaranteed his technical skill and mastery of theory, all of which have been amply borne out by his subsequent contributions. Roman has been an ornament to MIT — and the Boston area physics community — ever since. Our own collaboration spanned a quarter century, with fifteen papers garnering some 6500 citations to date. Even so, our joint bestseller is only his #2 work (#1 would be the ABJ chiral anomaly). These bare statistics miss all the fun we had, ranging from lecturing at a Brazilian school in far-off Ouro Prieto to soirées in his and So-Young Pi's Beacon Hill apartment, to our many violent discussions while collaborating and to the altogether different tenor of the interactions during our related efforts with Gerard 't Hooft.

Our work over those decades had an unplanned, but retrospectively manifest, theme: Quantum field theories and gravity models in lower — especially $D = 3$ — dimensions. It proved extremely fruitful, giving rise to numerous novel consequences, including massive yet gauge invariant models [1,2] — the first new classical mass generating mechanism — amongst all the other new Chern–Simons (CS) physical and mathematical effects, and separately, what we called the dynamics of flat

space, finding life in — seemingly empty — GR in $D = 3$ [3]. These topics having
continued to snowball in popularity, this may be a good point to comment on some
aspects of their evolution.

2. Why $D = 3$?

As a child, I recall seeing a popular (prewar) book entitled "Why split atoms?", one
that I assumed pled for less cruelty to them. Here I ask "why work in $D = 3$?" in
a rather more positive vein. The obvious Everest defense is of course "because it's
(sometimes effectively) there" but there are good reasons aside from the fun and
learning we derive in seeming detours from reality. For example, as I write this, a
new posting suggests the use of the strong fields in various gravitational collapses to
look for the, slightly generalized $D = 4$ extension of, CS gravity whose antecedents
can also be traced to ($D = 10$!) string theory. The totally different CS application
to the quantum Hall effect is instead about planar condensed matter physics. Then
there is the enormous and still growing, mathematical physics, pure CS industry,
let alone its supersymmetric versions whose existence are a model's sine qua non
these days, and brings the whole CS notion to fermions as well! [4] Our revival of
$D = 3$ GR had another spinoff: the BTZ industry — sourceless black holes entirely
due to identifying points in pure AdS [5]. Just as $D > 4$ physics has been a great
source of $D = 4$ insights, so clearly is $D = 3$.

3. $D = 3$

In $D = 4$, Einstein gravity (GR) is itself a dynamical system whose excitations
must have positive energy for stability; this is guaranteed (only) with the canonical
gravitational sign choice. As a far from trivial dividend, this same choice leads
to attraction between like masses [6]. But $D = 3$ Einstein gravity [3] is neither
dynamical nor does it mediate forces between masses; the former because Riemann
and Einstein tensors are equivalent here (they are double duals of each other), so
empty ($G_{mn} = 0$) space is flat; the latter because space's flatness between sources
means it cannot transmit any local (but only global) interactions between them.
Nevertheless, there remains a rich global source interaction, including their quantum scattering! [7, 8] The cosmological, (A)dS, extensions [9] of $D = 3$ GR are
correspondingly different from their $D = 4$ counterparts as well.

Separately, our totally different, odd-dimensional, $D = 3$ extensions of gauge
theory and of GR, introduced a physical role for their respective first and third
derivative order Chern–Simons terms, and incidentally forcing the sign of the (now
lower derivative) Einstein component to be opposite to that in $D = 4$. These
models have deep "topological" roots: their gauge and gravity terms are the 3D
"descents" K^4 of the 4D axial invariant densities F^*F and R^*R, each of which is a
total divergence $\partial_\mu K^\mu$.

4. 3 Topics in 3D

So much for generalities; let us now briefly consider some specifics, starting with aspects of 3D GR AdS solutions and their relations to the very different nature of the (subsequent) BTZ solutions. Consider the frame in which the source-free AdS metric is

$$ds^2 = -F dt^2 + F^{-1} dr^2 + r^2 d\theta^2, \quad F = 1 + \left(\frac{r}{l}\right)^2. \tag{1}$$

In Ref. 9, where $\Lambda = l^{-2} = 0$, it was shown that a (positive mass m) particle action produces the value

$$F = 1 - 4Gm + 0 \quad \rightarrow \quad F = 1 - 4Gm + \left(\frac{r}{l}\right)^2. \tag{2}$$

Specifically, (2) first shows the $\Lambda = 0$ point source solution, then how it generalizes to AdS for a positive mass point source and gravitational constant G. The conventional ("$D = 4$") sign is $G > 0$. [We actually generalized to more sources, but that is irrelevant here.] Instead, the BTZ solution is given in the same frame by

$$F^* = -M + \left(\frac{r}{l}\right)^2, \tag{3}$$

where M is a positive constant. It is this negative sign choice that provides the black hole (BH) interpretation, a bit like that of the ("1/2") dS intrinsic horizon, where instead the r^2 term turns negative and its "F" $= 1 - (r/l)^2$. The BTZ solution has no explicit matter source; rather, it is pure AdS but with certain points identified. [In $D = 4$, the BH sign is automatically related to overall (necessarily positive) mass, this coefficient of $1/r$ being the total energy of the system, with, or even without, physical matter sources.] Let us instead consider this metric in terms of our physical point sources as in (2). The combination $4Gm$ must exceed unity for negative $-M$ (note that only the product Gm enters, BUT m is necessarily > 0, so only the conventional positive sign of G has a chance). Recall also that there is no gravitational contribution to the total mass in $D = 3$, as noted earlier. As already shown in Ref. 3, these are forbidden, unphysical values. Indeed, here is the relevant explicit quote: "In the above we have taken $m < 1/4G$. When m exceeds this limit, the metric near the particle becomes singular, e.g., the distance from the particle to any other point diverges as r^{1-4Gm}. This limitation on the mass may also be understood in terms of the formal equivalence between $a \leftrightarrow -a$ and the coordinate inversion $r \leftrightarrow l/r$ implied by (2.8). In other words, a high-mass particle at the origin with $m > 1/4G$ is actually a particle at infinity with acceptable mass $1/2G - m$. The geometrical counterpart of this argument is given in Section V, where it is also shown that composite (many-body) sources can have a higher (up to $1/2G$) total mass." Nothing changes in this analysis for AdS gravity; for example the diverging distance is a purely local effect. The special value $4Gm = 1$ is also analyzed there to be a strange cylinder. We conclude that the BTZ solution, while mathematically correct as a pure AdS space with those identifications, cannot be constructed from physical, positive mass, sources. The authors indeed state that theirs are solutions of the source-free Einstein equations, i.e., without matter sources, and specifically without the cone structure emanating from point matter. We have in fact speculated that these solutions are separated from their normally sourced counterparts by a "superselection" rule [10].

Next, a brief look at closed timeline curves (CTC), a.k.a. time travel in pure GR. This subject, first worried about by Einstein, even before completing GR, as a possible disease of dynamical, as against fixed, spacetime, was of course made concrete by Godel's contribution to Einstein's 70th birthday(!), but was not as scary as local CTC discovered later.

In Refs. 11 and 12, 't Hooft, Roman and I responded to a purported Godelian $D = 3$ solution involving string sources [13]. True to Einstein's original intuition that physically admissible sources would not produce CTC, we showed that the threat was indeed unphysical here as well. Ironically, many years later, the local CTC menace was to arise in the more dangerous $D = 4$ massive gravity context [14, 15], this time showing it to be unphysical as well, a welcome echo of our 3D work!

As a final topic in our sampler, we describe the remarkably simple — in first order form, like its SUGRA predecessor — locally supersymmetric TMG (its gauge field counterpart is of course simpler still) [4]. There are obviously separate fermionic companions to the Einstein and CS parts: the former is, also obviously, the massless Rarita–Schwinger action in a gravitational background with *a priori* independent dreibein and connections to allow for torsion as in $D = 4$, and equally obviously as devoid of excitations as GR. Instead, the friend of the CS term must be of second derivative order (always one less than its bosonic equivalent), and all fermionic terms must depend only on the (dual) field strength $f^\mu = \epsilon^{\mu\nu\lambda} D_\nu \psi_\lambda$ (the covariant derivative in the curl is thus only with respect to the spinorial index, namely a connection term) to retain the original local gauge invariance under $\Delta\psi_\mu = \partial_\mu a(x)$, where $a(x)$ is the fermionic gauge parameter. This, together with the requirement that it must share the fermonic equivalent of the conformal-invariance of the CS term, forces it to be

$$L_{3/2}^{CS} \sim m^{-1} \int d^3x f^\mu \gamma_\nu \gamma_\mu f^\nu; \tag{4}$$

the overall coefficient likewise shares its inverse length dimension. Indeed, even the term's descent from a $D = 4\partial_\mu K^\mu$ is nicely reproduced. All this is gratifying, since there was no *a priori* guarantee that a super version would exist, particularly one so nicely matching the topological bosonic part's properties.

5. Envoi

Many more happy years to you and to lower dimension physics progress!

Acknowledgements

This work was supported by the U.S. Department of Energy, Office of Science, Office of High Energy Physics, under Award Number de-sc0011632.

References

1. S. Deser, R. Jackiw, and S. Templeton, Three-dimensional massive gauge theories, *Phys. Rev. Lett.* **48**, 975, (1982).
2. S. Deser, R. Jackiw, and S. Templeton, Topologically massive gauge theories, *Ann. Phys.* **140**, 372, (1982).
3. S. Deser, R. Jackiw, and G. Hooft, Three-dimensional einstein gravity: Dynamics of flat space, *Ann. Phys.* **152**, 220, (1984).
4. S. Deser and J. H. Kay, Topologically massive supergravity, *Phys. Lett.* **120B**, 97, (1983).
5. M. Banados, C. Teitelboim, and J. Zanelli, The black hole in three-dimensional space-time, *Phys. Rev. Lett.* **69**, 1849, (1992). arXiv:hep-th/9204099.
6. S. Deser, How special relativity determines the signs of the nonrelativistic, Coulomb and Newtonian, forces, *Am. J. Phys.* **73**, 6, (2005). arXiv:gr-qc/0411026.
7. S. Deser and R. Jackiw, Classical and quantum scattering on a cone, *Commun. Math. Phys.* **118**, 495, (1988).
8. G. 't Hooft, Nonperturbative two particle scattering amplitudes in (2+1)-dimensional quantum gravity, *Commun. Math. Phys.* **117**, 685, (1988).
9. S. Deser and R. Jackiw, Three-dimensional cosmological gravity: Dynamics of constant curvature, *Ann. Phys.* **153**, 405, (1984).
10. S. Deser and J. Franklin, Is BTZ a separate superselection sector of CTMG?, *Phys. Lett. B.* **693**, 609, (2010). arXiv:1007.2637.
11. S. Deser, R. Jackiw, and G. 't Hooft, Physical cosmic strings do not generate closed timelike curves, *Phys. Rev. Lett.* **68**, 267, (1992).
12. S. Deser and R. Jackiw, Time travel?, *Comments Nucl. Part. Phys.* **20**, 337–354, (1992). arXiv:hep-th/9206094.
13. J. R. Gott, Closed timelike curves produced by pairs of moving cosmic strings: Exact solutions, *Phys. Rev. Lett.* **66**, 1126, (1991).
14. S. Deser and A. Waldron, Acausality of massive gravity, *Phys. Rev. Lett.* **110**, 111101, (2013). arXiv:1212.5835.
15. S. Deser, K. Izumi, Y. C. Ong, and A. Waldron, Massive gravity acausality redux, *Phys. Lett. B.* **726**, 544, (2013). arXiv:1306.5457.

Chapter 7

Roman Jackiw and Gauge Field Theory Reminiscences of MIT Postdoc Days*

Nicholas Manton

DAMTP, Centre for Mathematical Sciences
University of Cambridge
Wilberforce Road, Cambridge CB3 0WA, United Kingdom

I started a 2-year postdoc at the Center for Theoretical Physics, MIT in October 1979. This was after completing a Ph.D. in Cambridge U.K., mainly on Yang–Mills–Higgs monopoles, followed by a one-year postdoc at the Ecole Normale, Paris, working on gauge fields with symmetry and their classification. It was the first time I'd been to America. At Boston Logan airport, the immigration officer saw that I was going to work at MIT and said "That's a good school!" I was amused by the different words that Americans use, compared to British English. For us, school is for children, but I was 27 at the time. Some words were difficult to pick up at first, like "route" pronounced like "rout" and "momentarily" meaning "in a moment".

It took more than a month before I found an apartment. That was at 20 Concord Avenue, near the attractions of Harvard Square — shops, restaurants, newsstands, and Harvard University itself — and I needed to take the Red Line subway two stops to Kendall Square to get to MIT. I imagined Harvard Square would look like the spacious Trocadéro in Paris, with grand institutional buildings on one side, and a lively city scene of café terraces opposite. In fact, it is more homely and rather suburban, because Harvard, by American standards, is very old, with some buildings dating back to the 18th century, and the more modern developments are of limited scale so as not to dwarf them. And curiously, Harvard Square is rather narrow and not square in shape at all.

MIT looks grander than Harvard, at least at the front, with its colonnaded portico on Massachusetts Avenue, and substantial buildings from around 1916. The CTP is in this older part, but one needs to go along an almost endless corridor and upstairs to get there. The professors, including Roman, were on the third floor, with offices off a rather swish lobby, hung with original prints of Ansel Adams photos (these were later too valuable to display, and were replaced). The postdocs and graduate students had offices upstairs, which were still better fitted out than rooms in neighbouring departments. Notably there was air-conditioning, which made CTP a much more pleasant place to work in summer than many other parts of MIT.

I wasn't directly Roman's postdoc, but rather of the whole theoretical physics group. Nevertheless, Roman seemed particularly interested in the kind of work I had done, so naturally we had many discussions. He had built up quite a large group of students and postdocs. Other postdocs included Paolo Rossi, Manoj Prasad, and Luc Vinet. Roman was well known by then for his various contributions to the theory of anomalies, and more recently to the study of fermions coupled to monopoles and other solitons, and the fractional charges that they could have. Roman was also the coauthor, with Nohl and Rebbi, of a clever ansatz (the JNR ansatz) for constructing multi-instantons in SU(2) gauge theory.

Roman was interested in my work in Paris with Peter Forgács on symmetric gauge fields, and was more familiar than I was with the general idea that if a physical system of particles or fields interacts with a fixed background with continuous symmetries, then the system's dynamics has conserved quantities given by Noether's theorem. For example, in a spherically symmetric background, there is a conserved angular momentum. Flat space is spherically symmetric around a chosen origin, and one may also add a spherically symmetric gauge field. The formula for the conserved angular momentum changes when the gauge field is added. Roman knew a very general way of finding the conserved quantity in any symmetric background, a method I have taught to Part III students in Cambridge U.K. in more recent years.

My formalism with Peter Forgács for symmetric gauge fields was also very general, so Roman and I could combine our insights to find conserved quantities of a very general form, which could be applied to many special cases, for example, to particles interacting with background monopoles. There are two variants of the conserved quantities when there is a background gauge field. One is 'canonical', which is what directly follows from the standard version of Noether's theorem, and the other is 'covariant', meaning that it has a more manifestly gauge invariant appearance. Both versions use quantities that were in my paper with Forgács. In the covariant version, this is the Higgs field that we had discussed at length.

Roman worked quickly on this. We had discussions one day in his office, and by the next morning he had already written a draft section of a paper. I contributed too, making technical and conceptual improvements. One result was to make the covariant version look more like the canonical version, by more carefully defining the field variations required for the application of Noether's theorem. These variations were themselves gauge dependent. I also learned from Roman more about the stress–energy tensor of field theories during this project. This has canonical and covariant versions too.

After about two months we had a finalised paper, which we submitted to the locally edited journal, Annals of Physics [1]. They particularly welcomed papers that had some originality, but also a pedagogical flavour. I imagined working further with Roman after this, and we quite often had further discussions in his office, including with visitors or his students. But nothing was promising, and I ended up working more on my own.

MIT had a cafeteria where I often had a hot lunch, but there was also a popular sandwich counter. This was my first experience of the splendid, loaded American sandwiches. Locals would be asked to "shoot" their complicated orders. There was a choice of bread, of meat or cheese slices or both, of lettuce, tomato and sprouts, of butter, mayo, pickles etc. The server making the sandwiches had an amazing memory for what had been requested. I liked ham and swiss on rye most of the time, but tried other variants too. Sandwiches could be taken back to the CTP where a group of the established professors would get together for lunchtime chat. This group included Francis Low, Herman Feshbach, Vicky Weisskopf, Jeffrey Goldstone, Felix Villars, Bob Jaffe and Arthur Kerman. They were nuclear physicists as much as particle physicists, and several were well connected with government, because they had been close to those involved with the Manhattan project.

Near my office were two larger offices for graduate students. Among the students were Andy Strominger, Joe Lykken, Lawrence Krauss, Antti Niemi and Manu Paranjape, all of whom had successful academic careers later. I got to know them reasonably well and met them again on various occasions over the years. The post-doc in the office next to me, Paolo Rossi from Scuola Normale in Pisa, had an interest in monopoles. So had Manoj Prasad, who had earlier with his advisor Charlie Sommerfield found the exact monopole solution of charge 1 in the case of a massless Higgs field. The Prasad–Sommerfield 1-monopole solution was known to solve a first-order Bogomolny equation, and it was realised that static multi-monopole solutions were most likely to be found by investigating this Bogomolny equation further.

In fact, the first new, precise solution was found by Richard Ward, then at Trinity College, Dublin. This was an axially symmetric solution of charge 2, with the monopoles as close together as they can be. He had found this by using some of the subtle geometry derived from his understanding of instantons and twistors. Manoj and Paolo decided to try to develop this further. (Paolo also worked with Claudio Rebbi on this.) I was involved in the discussions briefly, but found I had nothing really to contribute. They exploited a different reformulation of the monopole equations, something called the Yang equations, and managed to rederive Ward's solution. They found further information about the fields, and also found analogous solutions with higher charges. All these were axially symmetric, and surprisingly, they were unique apart from the possibility of rigid shifts. I had imagined in my earlier studies that axially symmetric monopoles would look like chains of charge 1 monopoles, with arbitrary parameters representing the monopole separations along the axis of symmetry, but the actual solutions are best thought of as rings of magnetic charge density.

There were beautiful autumn colours (fall colors) to admire from mid-October to early November. People recommended going to New Hampshire or Vermont to see them, but I didn't have a car so that was not practical. But just a couple of weeks after settling in at MIT, I took the local train from North Station in Boston to Concord, which is a picture-postcard New England town with many buildings from before the Revolution. Here, and also in historic Salem, the colours were particularly fine.

By mid-December it was much colder, with clear skies and overnight temperatures of 15 degrees (Fahrenheit); the views of the Boston cityscape were beautiful, being much clearer in the dry air than anything one experiences in London. The winters were really cold, until early February. My British clothes including a duffel coat were hardly adequate, and I found it essential to walk really fast to the Red Line stations at the Harvard Square and MIT stops. Fortunately, my apartment was very well heated by a large furnace in the basement, close to where one could dry laundry, and occasionally I needed to open windows in the coldest weather to cool the place.

Summer was different. My apartment had no air conditioning, which was not unusual in New England, and it got stiflingly hot and humid in July and August. I bought a large fan that rested in the window opening, but several evenings I still had to walk the streets to get some relief. Fortunately, there was a fine ice cream parlour to visit that was air-conditioned. I also experimented with going to the baseball at Fenway Park, home of the Boston Red Sox. Tickets were easily available for most games, because there is a game every day when the team is at home and not on a road trip. Tickets also get discounted after the game has started. I enjoyed the game, and went once more. It's family entertainment where everyone sits, without the rowdiness and drunkenness associated with some live sports in England at that time. I was surprised to see that there were few away supporters. I politely applauded when the away team scored, but got frowns from neighbours, so stopped that. It isn't so much fun to watch a game when the home team is losing, because the whole crowd loses interest and starts to drift away before the end. The TV coverage is more unbiassed. Quite fun is the 7th inning stretch, between the top and bottom halves of the 7th inning, when everyone stands up to stretch their arms and legs.

The fielders are very skilled to carry out the various plays, including double plays to get two batters out. The batters, during practice, can hit balls easily, but during real games the pitchers make hitting really hard, and this keeps the scoring low. One thing that surprised me tactically was that pitchers are often changed by the manager's decision, although usually too late, after the pitcher has got tired and given away a run or two. Why not earlier? I think it's because the teams have only a limited number of pitchers, and the relief pitcher who comes on near the end can't be brought on earlier, otherwise he couldn't pitch again, game after game, day after day. Possibly the reason is that pitchers are some kind of heroes, like gladiators, who have to be tested to their limit, otherwise the crowd would be disappointed. Obviously a starting pitcher who holds on to win a game without being relieved is some kind of hero. People in England, and foreigners I got to know in the US, were quite surprised at my captivation with baseball, and how I got to learn some of the jargon of the game.

During my two years at MIT, I had several opportunities to travel to other places in the US, mostly to give seminars, and got used to the idea of taking flights every few months. I routinely worried about being late at the airport because of road traffic delays, and missing my flight, but Bob Jaffe advised that if you didn't miss about one flight per year, then you were wasting too much time at airports.

One trip I made was by train from Boston to New Haven CT to give a seminar at Yale. There I met Charlie Sommerfield who had found the exact monopole solution with Prasad. At Yale there is an interesting museum of British art. A little later, at a conference in Trieste on monopoles, I met Evgeny Bogomolny, who had found the key first-order partial differential equations satisfied by several types of solitons, including monopoles. Consequently, I have met all three of the BPS (Bogomolny–Prasad–Sommerfield) physicists personally. BPS is now mentioned frequently and widely, especially in the context of quantized supersymmetric theories, but few are now familiar with who B, P and S are.

Early in 1980, Roman had been for a period to California, where he had contacts at UCLA and at the new Institute for Theoretical Physics (ITP) at UCSB, the University of California at Santa Barbara. He generously arranged for his students and postdocs, about four of us, to spend a month at ITP. At that time, ITP was not very well known, and they had space to welcome visitors.

I went out there in May 1980, flying first to Los Angeles and seeing for the first time the colourful, rocky landscape of the American south-west, as well as Los Angeles and the Pacific Ocean. At Los Angeles' LAX airport I changed to a small plane of the airline Golden West, with two propellers. I had never flown before on a plane like this. It had only about a dozen passengers, together with two pilots and a cabin crew of one. It took off quietly after a very short acceleration on the giant LAX runway. The flight was about half-an-hour along the coast to Santa Barbara, and arriving there was a memorable experience. There was a pretty, Spanish-style terminal building to which one had to walk, and a powerful, very attractive smell of blossom mixed into the sea breeze. The many nearby Eucalyptus trees probably contributed to the smell. Despite some aircraft noise, the atmosphere there seemed very peaceful.

I moved into a small house quite near the airport and also near the university, shared with the others from MIT. It was on the edge of Isla Vista, the student dormitory area of UCSB. I found the Spanish names that abound in California cute, but soon realised that the culture is dominantly English-language based and American, despite the many Mexicans and other Spanish speakers living there. In the evening, there was a lot of noise from frogs in the marshy areas near the house.

I could walk to ITP, and spent most of my time in the UCSB neighbourhood during that month. ITP was on the sixth (top) floor of Ellison Hall, one of the main academic buildings, and the offices had great views either towards the ocean and the islands beyond, or inland towards the quite high mountains just a few miles behind Santa Barbara, running parallel to the coast. The site is close to a lagoon that is almost connected to the ocean, and beyond it is an almost endless sandy beach. I took some walks around there, but only surfers in their wet suits could enjoy going into the fairly cold water at that time of year. Tar on the beach, that seeps out from the oil reserves in the Santa Barbara Channel, is also a hazard.

I certainly liked being in Santa Barbara. One could reach downtown quite easily by bus 24 from the campus. Santa Barbara was largely destroyed by an earthquake in 1925. The city decided on a rigorously controlled rebuilding in the Spanish colonial (Andalusian) style, including Moorish decorations. The courthouse is the masterpiece in this style. The architecture is enhanced by beautiful gardening, relying on plentiful irrigation. The only substantial building genuinely from the Spanish colonial time is the Mission, which also had to be repaired after the earthquake. There is no doubt that Santa Barbara is one of the most beautiful towns in the US. Unlike similar seaside towns in England, Santa Barbara had a flourishing industrial and commercial life, as well as the large university, and younger people were mostly working.

At some point, maybe later when I became a postdoc in Santa Barbara, Roman visited. When he was in Southern California, Roman always tried to have lunch with Julian Schwinger in Westwood, close to UCLA. Schwinger was on the faculty of UCLA but quite reclusive, and could only be persuaded to meet to discuss physics over lunch, rather than in his office. Roman often took along his more junior associates, and I went to one of these 'lunches with Julian'. Schwinger was the winner of a Nobel Prize, along with Feynman and Tomonaga, but he was somehow still bitter that his contributions to field theory were under-recognised, and his work overshadowed by that of Feynman. The main reason was that the methods he expounded in his magnum opus on quantum field theory were less intuitive than Feynman's diagrams. At the lunch, we talked a little about his work on monopoles, and on electron–positron pair creation, but I don't think he had followed the more recent developments.

After returning to MIT, through contacts with the Harvard theoretical physicists at the joint MIT/Harvard seminar, including my friend Graham Shore who had also been a Ph.D. student in Cambridge U.K., I met Ian Affleck, a Canadian who was then a Harvard Junior Fellow — a prestigious kind of postdoc. He was interested in monopoles and instantons, following in the footsteps of Sidney Coleman. Together, we realised that we could study monopole–antimonopole pair production in a strong magnetic field. This would be dual to the well-known study of electron–positron pair production in a strong electric field by Julian Schwinger. Both are kinds of quantum tunnelling processes, involving an instanton that is circularly symmetric in 4 dimensions.

Simultaneously with monopole pair production, there can also be dyon pair production. The dyon's electric charge arises from the monopole twisting an integer number of times as it goes round the circular loop of the instanton. It is necessary to sum over the contributions of all these twists. The sum converges because dyons are more massive than monopoles, their mass increasing with their electric charge. It was useful, later, to have clarified what the mass of a dyon is. The simple square root formula that emerges from supersymmetry is correct. It has the right quadratic behaviour for modest electric charges, and is consistent with the formula discovered by Bernard Julia and Tony Zee, provided one imposes Dirac quantization of charge. Our conclusion was that a dyon is a quantized version of a monopole, which takes into account the internal twisting of the monopole.

An exciting event in spring 1981 was a visit by Richard Feynman to give a seminar at MIT. Feynman had recently published a paper on pure, quantum Yang–Mills gauge theory, in which he gave arguments why there is no massless particle in the theory. The lowest energy particle would be a positive mass glueball rather than a massless gluon, and one says that the theory has a mass gap. Gluons would exist as particles in a limited sense, but like quarks they would be confined. His paper was mainly about the case of Yang–Mills theory in two space dimensions, where on dimensional grounds, the arguments are simpler than in three space dimensions. There was no rigorous result in the paper, but the ideas were stimulating. Roman knew that Feynman came to Boston fairly frequently to visit a group that was pioneering quantum computers, and persuaded Feynman to take time off to speak about his gauge theory work at MIT. A small group of us went with Roman by car to collect him, and that was an opportunity to discuss his work. He was lively, but we couldn't get him much interested in our research. Roman arranged that Feynman would first give a general seminar open to all, to be followed by a technical session where we would discuss ideas on how to make his results rigorous. The first session was packed out. Feynman included a few technical details, and spoke for well over an hour. At the end, Roman asked non-experts who'd heard enough, and were not interested in further details to leave, but hardly anyone did so. The seminar restarted, and Feynman was asked to go through some of his arguments carefully again. But he felt he had said just about enough, and didn't have a clear idea how to develop his ideas further. So the seminar ended. Even today, there is no proof of Feynman's claim of a mass gap, though the numerical evidence is convincing.

A number of years later, when I was established in Cambridge U.K., I became interested in Skyrmions. It is difficult to solve the Skyrme field equation explicitly, but with Michael Atiyah, I found a construction of Skyrme fields in 3 dimensions using 4-dimensional instanton gauge fields. To study the interactions of two Skyrmions it is very helpful to work with 2-instanton fields. For these, the JNR ansatz of Roman and his collaborators gives all possible instantons (which is not the case for higher instanton charges). JNR had observed that there was a particular shift in the parameters of their ansatz that only had the effect of a gauge transformation. This was understood in a beautiful geometrical way by Hartshorne. A 2-instanton is specified by a circle and an ellipse interior to the circle, lying in some plane in

4-dimensional space. It is necessary for there to exist a triangle with vertices on the circle, having sides tangent to the ellipse. This requirement places one constraint on the position and shape of the ellipse. The JNR instanton data depend on the triangle vertices and the tangent points. Remarkably, once there exists one triangle, it is a result of Poncelet from the 19th century that there is a whole family of triangles with vertices on the circle, having sides tangent to the ellipse. This is called a Poncelet porism. Any point on the circle can be fixed as the first vertex. From here you draw tangents to the ellipse, intersecting the circle again at the second and third vertices. The line joining these vertices is then automatically tangent to the ellipse. In a gauge invariant sense, the instanton doesn't depend on the choice of triangle, but only on the circle/ellipse pair. Changing the triangle infinitesimally, within the porism, is precisely the shift of the JNR parameters that leads to a gauge transformation. Atiyah and I used the JNR ansatz to find concrete Skyrme fields, but it was very helpful to know about the circle/ellipse geometry and the Poncelet porism of triangles. The JNR ansatz has also been useful in the construction of highly symmetric monopoles in hyperbolic 3-space, as I showed with Paul Sutcliffe.

In summary, I am grateful to MIT for appointing me to a postdoc position in 1979–81, and especially thankful to Roman for collaboration, discussions and long-term support, and also for the opportunity to spend time on both the East Coast and West Coast of the US for the first time [2].

References

1. R. Jackiw and N. S. Manton, Symmetries and conservation laws in gauge theories, Ann. Phys. **127**, 257 (1980).
2. N. Manton and P. Sutcliffe, *Topological Solitons* (Cambridge University Press, 2004).

Chapter 8

Recollections of a Most Fruitful and Enjoyable Collaboration: Looking Back at My Work with Roman Jackiw

Claudio Rebbi

Department of Physics, Boston University, Boston, MA 02215, USA

I look back at my collaboration with Roman Jackiw and the very important role it played for my scientific career.

I arrived at MIT in September 1974 with a one-year appointment as visiting associate professor. Prior to that, after completing my studies under the guidance of Professor Sergio Fubini at the University of Torino, I had been a postdoctoral fellow at Caltech, a lecturer at the University of Trieste, and a research associate at CERN. I have always been interested in non-perturbative investigations and while at Caltech I became fascinated by the dual resonance model, which eventually became string theory, as a theory of strong interactions. I did some work on that model, culminating in two papers, published with Goddard and Thorn [1], and with Goddard, Goldstone, and Thorn [2], on the relativistic, massless string, which were well received. After that work and a few further papers, I tried for a while to develop extensions of string theory that would incorporate the full dynamics of quarks attached at the end of the string and that would include currents with acceptable analytical properties. Both attempts were unsuccessful and I remember that when I expressed to Stanley Mandelstam my frustration in being unable to make progress, he told me: "Claudio, not all problems have a solution". That happened in the spring of 1974. I was at Caltech for a four-month visit and had gone to Berkeley to give a seminar. That was a turning point and I moved away from string theory. But what I had learned from the dual resonance model, and from my studies of constrained Hamiltonian systems for attaching quarks at the end of the string, was quite helpful for my further research.

At MIT there was a lot of interest in the so called bag model of hadrons and I was able to put to fruition what I had learned about constrained Hamiltonian systems in two papers on the small oscillations of the hadronic bags [3,4]. My appointment at MIT was renewed for one year and it was in the fall of 1975 that Roman Jackiw came to my office and brought to my attention that the one-dimensional Dirac equation had an interesting zero energy solution when coupled to a soliton field, which effectively acted as a mass term that changed sign across the soliton. Contrary to what normally happens with the solutions of the Dirac equation which come in pairs of opposite energy that can be associated with the anticommuting creation and annihilation of fermions, this solution was isolated and thus could be associated with a single operator a in the expansion of the quantized Fermi field Ψ, while $a^\dagger$, with $\{a^\dagger, a\} = 1$, appeared in the expansion of the conjugate field $\Psi^\dagger$. What was one to make of a? For me, this was reminiscent of something that had fascinated me in the theory of the fermionic string, namely that it described fermions albeit it only contained bosonic fields. The point there was that the string had a zero mode which appeared as a set of gamma matrices in the expansion of string. Thus, although the excitations of the string was described by bosonic creation and annihilation operators, its states had to provide a representation for the algebra of gamma matrices and thus had to be spin one-half fermions. In the case Roman and I were considering, the analogy was that a and $a^\dagger$ where not operators that annihilated or created an excitation, but rather operators which were realized in the algebra of the ground state. But the implication then was that the soliton itself could not be either a fermion or an antifermion, but had to be a state of half-integer fermion number. I brought these considerations to Roman, who may have come to the same conclusions independently, and this was at the origin of the first of several papers I wrote with him and I fondly recall [5].

At that time there was a lot of interest in the scientific literature on soliton-like solutions to gauge theories such as the 't Hooft–Polyakov SU(2) monopole [6,7]. In the continuation of a study also presented in Ref. [5], Roman and I showed that the commingling of spatial and intrinsic degrees of freedom which occurs when one couples a spinless, isospinor field to the SU(2) monopole gives origin to a situation where the intrinsic SU(2) degree of freedom of the isospinor field is converted into spin degrees of freedom [8]. I like to indulge on this study not so much for the results, interesting as they may be, but for a follow-up which speaks highly about Roman's character. We wrote up our work in a paper which we submitted to Physical Review Letters and was accepted for publication. Meanwhile Roman, who was always well informed, had learned that Peter Hasenfratz and Gerard 't Hooft had come up with results similar to ours which they also planned to publish on Physical Review Letters. Displaying a generous attitude, which may be unusual in a world where priority is generally so coveted, Roman suggested that we might request to the editor of PRL to delay publication of our letter by one month so that our letter and the one by Hasenfratz and 't Hooft would appear back to back. I saw eye to eye with Roman in these matters and was very happy to accept his suggestion, and so the two papers did appear one immediately after the other in the same issue of PRL.

In late 1975 a ground breaking paper by Belavin, Polyakov, Schwartz, and Tyup-kin (BPST) presented a solution to the SU(2) Yang–Mills theory in Euclidean four-space [9]. It was originally called a "pseudoparticle" and later an "instanton", because of its feature of representing an event with some degree of localization in

(Euclidean) time. Roman brought it to my attention and in early 1976 we were able to put into evidence that the solution was invariant under an O(5) subgroup of conformal transformations, also establishing a formalism which could be applied to the study of the propagation of fermions in the field of the pseudoparticle [10].

The work of BPST and a subsequent detailed study of the pseudoparticle solutions by 't Hooft [11, 12] suggested that these non-perturbative minima of the action could give CP-violating contributions to the functional integrals which are missed in the ordinary, perturbative vacuum sector. Moreover, the fact that pseudoparticle occurred in Euclidean space-time gave a hint that it could represent some non-perturbative, CP-violating tunneling event. But, if so, between what states did the tunneling occur? Roman and I were intrigued by this question and I remember that in the spring of 1976, as I was driving to MIT from Nahant, where my wife, young daughter and I were living in a beachfront house at the time, I thought of a white handkerchief being picked up by the corners and thus forming a sphere. But then, similarly, if we picked up three-dimensional space by the points at infinity, it became a three-dimensional sphere S^3. The manifold of the SU(2) gauge group is also topologically equivalent to a three-dimensional sphere, thus, very much like a closed loop can wind around a circle any integer number of times, so the SU(2) gauge group could wind an integer number of times around the three-dimensional space closed at infinity. And, in the same way as two loops which wind around a circle a different number of time cannot be continuously deformed into each other, so configurations of the SU(2) gauge group with different numbers of winding could not be continuously deformed into each other. These were the different vacuum states between which the tunneling occurred. Roman must have been thinking along similar lines, because when I mentioned my thoughts to him, there was no discussion and it was clear to us that the tunneling described by the pseudoparticle was between inequivalent sectors of the SU(2) gauge group. But this had major implications. It implied that the quantum theory had a much richer structure than previously thought: rather than having a single vacuum state, it possessed an infinite multiplicity of vacuum states, characterized by field configurations with different winding number. Quantum mechanics then demanded that the actual physical vacuum be a superposition of all the inequivalent vacua, each one occurring with a phase factor $e^{in\theta}$, n being the winding number of the above vacuum states and θ an emerging, new parameter characterizing the gauge theory. Theories built over vacua with different θ, or different θ-vacua with an often used terminology, would likely have different properties (likely because there are situations, e.g. with massless fermions, where they would be equivalent), and a non-zero θ would give origin to CP non-conservation. A lot of theoretical work and suggestions for experiment have followed the introduction of the new vacuum structure.

Roman and I published our work in a letter titled "Vacuum Periodicity in a Yang–Mills Quantum Theory" [13]. Similar results were published at the same time by Callan, Dashen, and Gross in a letter titled "The Structure of the Gauge Theory

Vacuum" [14]. The work on the θ-vacuum was the heyday of my collaboration with Roman. Our letter has received to date over 1400 citations.

After the discovery of the pseudoparticle Roman and I, as certainly many others, were trying to find its multi-particle generalization. Witten beat us to it, finding a multi-particle solution where all the instantons were aligned on an axis. Also, in unpublished work, 't Hooft found a way to build an n-particle solution with $5n$ arbitrary parameters which represented the location and sizes of the instantons. However an analysis of the conformal properties of pseudoparticle configurations done in collaboration with Nohl showed that the most general multi-instanton configuration had yet more than $5n$ degrees of freedom [15]. Roman and I approached the question of the number of degrees of freedom of pseudoparticle configurations by studying their small deformations and were able to show that this number is $8n - 3$ [16]. Our collaboration with Nohl continued in a review paper [17], which was published after I left MIT. In 1977 Roman and I also worked on a spinor formulation of the Yang–Mills theory [18].

With the end of the 1976–77 academic year came also the end of my appointment at MIT. When I had been offered my initial appointment I had been given the choice of a pure research position or a position which entailed also teaching. I love to teach — in my long career found that teaching does not detract from the ability of doing research, if anything it enhances it — and so opted for the latter. I was thus appointed visiting associate professor. My appointment was renewed twice, but it could not be converted into a regular faculty position. In the fall of 1976 I had been offered a tenured scientist position at the Brookhaven National Laboratory, which I was very pleased to accept, and so in June 1977 I moved to BNL. The investigation of semiclassical solutions to quantum field theories and their implications which I had conducted with Roman had been extremely gratifying and I continued to collaborate with him in a few investigations [19, 20]. Still deep behind the study of semiclassical solutions there was also the hope, certainly shared by many, that they could provide a clue to quark confinement and ultimately to the theory of strong interactions. Helping to formulate a theory of strong interactions had been my longstanding interest, dating back to my studies of the dual resonance model and the quantum string, but, as with string theory, the answer to quark confinement was not to be found by semiclassical techniques. In the late 70's, however, supercomputers had become powerful enough to open the possibility that strong interactions could be addressed by computational techniques. So, after moving to BNL, I made the decision to orient my research in that directions. Almost as a warm-up exercise, I calculated with Lawrence Jacobs the interaction energy of superconducting vortices [21], finding numerical evidence that it vanished for a special value of the coupling constant (the so called "Bogomolnyi" limit, a result that was subsequently proven analytically). Two papers written soon after with Creutz and Jacobs [22, 23] paved the way for the computational study of lattice gauge theories, an area of investigation where I focused all my later research and which has by now produced a wealth of impressive results on QCD and other strongly interacting gauge field theories.

In retrospect, I feel privileged to have had Roman as a collaborator and friend. I greatly appreciated his character, his generosity, his sense of humor. Working with him was enjoyable and instructive. I believe that I contributed to our collaboration with some good ideas and good analytical abilities, but he was the one who brought to my attention the problems which we studied together. His mentorship was crucial for the development of my career and I owe him an enormous debt of gratitude.

References

1. P. Goddard, C. Rebbi and C. B. Thorn, "Lorentz covariance and the physical states in dual resonance models," *Nuovo Cim. A* **12**, 425 (1972). doi:10.1007/BF02729555.
2. P. Goddard, J. Goldstone, C. Rebbi and C. B. Thorn, "Quantum dynamics of a massless relativistic string," *Nucl. Phys. B* **56**, 109 (1973). doi:10.1016/0550-3213(73)90223.
3. C. Rebbi, "Nonspherical deformations of hadronic bags," *Phys. Rev. D* **12**, 2407 (1975). doi:10.1103/PhysRevD.12.2407.
4. C. Rebbi, "The spectrum of p wave baryonic excitations in a model with field confinement," *Phys. Rev. D* **14**, 2362 (1976). doi:10.1103/PhysRevD.14.2362.
5. R. Jackiw and C. Rebbi, "Solitons with fermion number 1/2," *Phys. Rev. D* **13**, 3398 (1976). doi:10.1103/PhysRevD.13.3398.
6. G. 't Hooft, "Magnetic monopoles in unified gauge theories," *Nucl. Phys. B* **79**, 276 (1974). doi:10.1016/0550-3213(74)90486-6.
7. A. M. Polyakov, "Particle spectrum in the quantum field theory," *JETP Lett.* **20**, 194 (1974) [*Pisma Zh. Eksp. Teor. Fiz.* **20**, 430 (1974)].
8. R. Jackiw and C. Rebbi, "Spin from isospin in a gauge theory," *Phys. Rev. Lett.* **36**, 1116 (1976). doi:10.1103/PhysRevLett.36.1116.
9. A. A. Belavin, A. M. Polyakov, A. S. Schwartz and Y. S. Tyupkin, "Pseudoparticle solutions of the Yang–Mills equations," *Phys. Lett.* **59B**, 85 (1975). doi:10.1016/0370-2693(75)90163-X.
10. R. Jackiw and C. Rebbi, "Conformal properties of a Yang–Mills pseudoparticle," *Phys. Rev. D* **14**, 517 (1976). doi:10.1103/PhysRevD.14.517.
11. G. 't Hooft, "Symmetry breaking through Bell–Jackiw anomalies," *Phys. Rev. Lett.* **37**, 8 (1976). doi:10.1103/PhysRevLett.37.8.
12. G. 't Hooft, "Computation of the quantum effects due to a four-dimensional pseudoparticle," *Phys. Rev. D* **14**, 3432 (1976). doi:10.1103/PhysRevD.18.2199.3. [Erratum: *ibid* **18**, 2199 (1978). doi:10.1103/PhysRevD.14.3432.]
13. R. Jackiw and C. Rebbi, "Vacuum periodicity in a Yang–Mills quantum theory," *Phys. Rev. Lett.* **37**, 172 (1976). doi:10.1103/PhysRevLett.37.172.
14. C. G. Callan, Jr., R. F. Dashen and D. J. Gross, "The structure of the gauge theory vacuum," *Phys. Lett.* **63B**, 334 (1976). doi:10.1016/0370-2693(76)90277-X.
15. R. Jackiw, C. Nohl and C. Rebbi, "Conformal properties of pseudoparticle configurations," *Phys. Rev. D* **15**, 1642 (1977). doi:10.1103/PhysRevD.15.1642.
16. R. Jackiw and C. Rebbi, "Degrees of freedom in pseudoparticle systems," *Phys. Lett.* **67B**, 189 (1977). doi:10.1016/0370-2693(77)90100-9.
17. R. Jackiw, C. Nohl and C. Rebbi, "Classical and semiclassical solutions of the Yang–Mills theory," MIT-CTP-675.

18. R. Jackiw and C. Rebbi, "Spinor analysis of Yang–Mills theory," *Phys. Rev. D* **16**, 1052 (1977). doi:10.1103/PhysRevD.16.1052.
19. R. Jackiw, I. Muzinich and C. Rebbi, "Coulomb gauge description of large Yang–Mills fields," *Phys. Rev. D* **17**, 1576 (1978). doi:10.1103/PhysRevD.17.1576.
20. R. Jackiw, L. Jacobs and C. Rebbi, "Static Yang–Mills fields with sources," *Phys. Rev. D* **20**, 474 (1979). doi:10.1103/PhysRevD.20.474.
21. L. Jacobs and C. Rebbi, "Interaction energy of superconducting vortices," *Phys. Rev. B* **19**, 4486 (1979). doi:10.1103/PhysRevB.19.4486.
22. M. Creutz, L. Jacobs and C. Rebbi, "Experiments with a gauge invariant Ising system," *Phys. Rev. Lett.* **42**, 1390 (1979). doi:10.1103/PhysRevLett.42.1390.
23. M. Creutz, L. Jacobs and C. Rebbi, "Monte Carlo study of Abelian lattice gauge theories," *Phys. Rev. D* **20**, 1915 (1979). doi:10.1103/PhysRevD.20.1915.

https://doi.org/10.1142/9789811210679_0009

Chapter 9

Roman Jackiw and My MIT Days

Paolo Rossi

Professor of Theoretical Physics, Physics Department
University of Pisa (Italy)

I was a student of Roman long before I met him personally.

In the mid-seventies, when I was a young Italian physics student, every new work by Roman was a source of inspiration for me, enlightening the road for possible future research. Quantum field theory had seemed to be in a very bad shape until just a few years before, but Roman's (and others) papers casted it in a new and exciting perspective. Let me just recall a few titles that were especially relevant for my scientific growth: "New approach to field theory" (1973), "Functional evaluation of the effective potential" (1974), "Effective action for composite operators" (1974) and the whole series of works written in collaboration with Claudio Rebbi.

Then at the end of 1978 I finally had the opportunity to spend a year at MIT, thanks to the Fulbright Student Program, and to meet Roman personally. That was the beginning of one of the most interesting and stimulating periods of my entire scientific career. Working with Roman was not always easy, because his criticisms were very direct and his expectations were very high. After so many years I can still remember his sentence "I'm not impressed", uttered at the end of a long exposition of mine about what I had elaborated on the basis of one of his suggestions. I must admit that he was completely right, even if at the time I was left abased. It was a hard lesson, but it was also a very useful one.

In any case our collaboration, when things were moving in the right direction, gave me also great satisfaction, and I cannot avoid mentioning that one of our works ("Zero modes of the vortex–fermion system", 1981) still keeps being quoted frequently, a success that I had only a few times in my life.

One of the aspects that struck me most about Roman's way of working was his ability to involve people, and especially the younger ones, including myself, in his ideas and his scientific projects, always taking care to value every contribution, even the smallest ones. One of the lessons of professional ethics that I learned from him, and which I have never forgotten, is the duty to always recognize a collaboration in one's own publications, even when for some reason it did not turn out to be particularly useful in achieving the final result. In a scientific context where many times one's results risk not being cited, even if only by distraction, the respect for the work of other people is an important value that I have always seen in Roman.

I must recall an episode that shows quite clearly the deep respect he had of his students, but also his awareness of all the difficulties they had to face. One day he formally introduced to me his doctoral students (some of them are now famous!). After our short meeting, we left the students' room and were strolling along one MIT corridor. Then he told me (more or less) the following words: "Each of the youngsters you have just met would have been first in his field whatever field he had chosen but theoretical physics. As theoretical physicists, most of them will be unhappy and frustrated." That was a shocking sentence, which I could not avoid referring also to myself, even if that was certainly not Roman's intention.

At the end of my one-year stay as a visitor, Roman proposed, without being solicited by me in any way, that I might remain at MIT with a post-doctoral scholarship. So my stay in the US lasted another two years, which were invaluable for my research and my future career. I still remember, as funny as it may seem, that the value of the scholarship was so disproportionate compared to my Italian standards that when faced with the offer I thought that the proposed (monthly) figure referred to the whole year, and in any case I took it seriously in consideration, barely keeping back the leaps of joy when the misunderstanding was clarified.

If this were an autobiographical note I should mention here all the colleagues I had the good fortune to meet in those years, and above all those, apart from Roman, with whom I had the opportunity and the pleasure of collaborating. Fourteen papers in three years are certainly a result far above my previous and subsequent average.

But Roman was also able to involve me in a process of academic socialization that went far beyond simple scientific collaboration. He introduced me personally to Schwinger and Feynman, who were for me two mythical characters, whose scientific achievements I had studied in books without being able to imagine that I would ever meet them. He also gave me the opportunity to meet and appreciate many colleagues who were already, or would soon become, points of reference for theoretical physics in all the world.

I also remember with great pleasure the Californian "expedition" which he organized for me and some of his doctoral students on the occasion of his stay in Santa Barbara. That too was a memorable month, and not only from the point of view of professional experience.

Towards the end of my third year at MIT, it was again Roman who asked me if I was interested in an academic position in a US university, in which case he would have worked to let me get it. But at that time I had an interesting offer from Italy (which for the record did not materialize) and for personal reasons I also felt a strong desire to return to my country of origin. So I declined Roman's generous offer and after a few months, in the summer of 1981 I made my journey back to Italy.

But also on that occasion there was a last important moment of collaboration with Roman. The institution I was returning to (the Scuola Normale Superiore of Pisa) required me to present a report on the work carried out in previous years before a commission. Roman agreed to review my report and to be part of the jury, reaching me in Italy shortly after my return. And he also made the proposal to turn my work into a Physics Report ("Exact results in the theory of non-abelian magnetic monopoles", 1982), another of the few works of mine that have resisted the wear and tear of time and that keep being cited with some frequency.

Unfortunately I did not have many other occasions to meet Roman, because I did no longer cross the Ocean after 1981, but I still often quote some of his teachings, and not only those having a purely scientific content. He has been essential also for my human growth, and that is another reason I have for being extremely grateful to him.

Chapter 10

Roman Jackiw: A Beacon in a Golden Period of Theoretical Physics

Luc Vinet

*Centre de Recherches Mathématiques, Université de Montréal,
P.O. Box 6128, Centre-ville Station, Montréal (Québec), H3C 3J7, Canada*
`vinet@crm.umontreal.ca`

This text offers reminiscences of my personal interactions with Roman Jackiw as a way of looking back at the very fertile period in theoretical physics in the last quarter of the 20th century.

To Roman: A bouquet of recollections as an expression of friendship.

1. Introduction

I owe much to Roman Jackiw: my postdoctoral fellowship at MIT under his supervision has shaped my scientific life and becoming friends with him and So-Young Pi has been a privilege. Looking back at the last decades of the past century gives a sense without undue nostalgia, I think, that those were wonderful years for Theoretical Physics, years that have witnessed the preeminence of gauge field theories, deep interactions with modern geometry and topology, the overwhelming revival of string theory and remarkably fruitful interactions between particle and condensed matter physics as well as cosmology. Roman was a main actor in these developments and to be at his side and benefit from his guidance and insights at that time was most fortunate. Owing to his leadership and immense scholarship, also because he is a great mentor, Roman has always been surrounded by many and has thus generated a splendid network of friends and colleagues. Sometimes, with my own students, I reminisce about how it was in those days; I believe it is useful to keep a memory of the way some important ideas shaped up and were relayed. Hence as a tribute to Roman, I thought of writing the following short account of my personal connections with him in addition to the scientific hommage written in collaboration with my colleagues Nicolas Crampé and Rafael Nepomechie. I fully appreciate that my own little history is of no special interest but I am offering this text as an illustrative testimony of a vibrant intellectual period in the companionship of Roman and of other scientists who like him were larger than life.

2. First encounters with Roman and his work as a graduate student

The Centre de Recherches Mathématiques broadly known as the CRM, was founded in 1968. One only appreciates with hindsight which butterfly wing flaps will have a determining effect on your life. In my case one of these was the Prague spring and its repression which occurred in 1968. As a result, two outstanding Tchech physicists, Jiri Patera and Pavel Winternitz educated in the highest Soviet scientific tradition took up positions at the CRM in the following years. They then gave to Montreal a big research impetus and developed in this city a strong school in mathematical physics. I was fortunate to join the CRM and pursue graduate studies within their group in the mid-70s.

In theoretical physics, the end of the 60s saw the advent of the Weinberg–Salam model [1, 2] unifying the electro-weak interactions, followed by the proof of its renormalizability [3], the development of QCD [4, 5], which had been preceded by the discovery of the Adler–Bell–Jackiw anomaly [6, 7] (beautifully presented in [8]). As a young student, I very much wished to get involved in those striking developments in field theory. With a much enthusiastic postdoctoral fellow then at the CRM, John Harnad, who de facto became my co-supervisor, we started a gauge theory seminar. The importance of classical solutions with topological properties such as magnetic monopoles and instantons in non-perturbative analyses was being revealed at about that time. The review on monopoles by Goddard and Olive [9] as well as Coleman's Erice lectures [10] were extremely formative. In those years, the Theory Division of the Canadian Association of Physicists was organizing summer schools in Banff. This is where I met Roman for the first time in 1977 as he delivered lectures [11] on classical solutions of the Yang–Mills theory that much impressed me.

Two papers on this topic which appeared around that time had a big influence on my Ph.D. work: the first by Roman and his collaborator Rebbi [12] (and reviewed in [11]) where the conformal $SO(5)$ invariance of the BPST one-instanton solution is identified and the second by Witten [13] where a multi-instanton solution is obtained through the dimensional reduction under $SO(3)$ of the self-dual Yang–Mills equations to a Higgs model in curved two-dimensional space. The essential aspect in these studies is that the variance of the Yang–Mills field under the space-time transformation is compensated by a gauge transformation. These publications prompted John Harnad and myself with a geometer, Steve Shnider, then at McGill University to look systematically at the description of gauge fields that are invariant in that sense under space-time transformations. We treated this problem in the global formalism of fiber bundle theory that had gained popularity amongst physicists in part thanks to a paper by Wu and Yang [14]; we classified lifts of (certain) group actions on manifolds to principal bundles over such bases and then characterized the invariant connections under the lifted actions. We then applied these results to obtain solutions to the Yang–Mills equations on compactified Minkowski space. This formed the core of my thesis and the paper on the general framework remains one of my most cited.

At some point in the course of these investigations, Steve made a preliminary presentation [15] at a conference in Lawrence, Kansas in the summer of 1978 which focused on the twistorial approach to instantons. Using this opportunity to take his son Nick as I recall on a road trip across America, Roman was one of the principal speakers [17] at this meeting which was my first occasion to present some of my work to him. Those questions regarding invariant gauge fields were timely then and in fact were addressed simultaneously by Forgacs and Manton [18] in a complementary fashion using infinitesimal methods.

As I was completing these projects, in 1979, I had the occasion to accompany my co-supervisor Pavel Winternitz during his sabbbatical to Saclay. This is how I obtained a Doctorate from the Université Pierre et Marie Curie for work I had done separately on the classification of second-order differential equations in two dimensions that are invariant under subgroups of the conformal group. Upon my return to Montreal, I was wrapping up my Ph.D. and the question came as to where I should go for my postdoctoral fellowship. To explore the possibility of becoming a Research Assistant at MIT, John Harnad extended to Roman an invitation to visit Montreal and this is how a few months later I was moving to Cambridge. We especially remember from that visit that Roman was asking if we knew the three-form whose exterior derivative would give the Chern four-form. He clearly was on his way to developing the Chern–Simons gauge theories, we still regret that at the time we did not have the culture to provide the answer and join him in these investigations.

3. The postdoctoral years at MIT

In some sense Roman is responsible for my meeting my wife. I had become desperate to find an apartment in the Boston area and thought of checking where Roman lived. As it happened I looked in an old phone book that had 808 Memorial Drive as his address. It is essentially where MIT ends and where Harvard begins. As I discovered later, Roman only stayed there briefly, but I found an apartment in that building with a nice view of the Charles. The apartment number was 512. A few months later, a charming girl named Letitia rented apartment 1212. Born in Montreal, she came to take up a research engineer position at MIT's Lincoln Lab. Meeting her has been one of the best things that have happened in my life.

When I arrived in September 1980 at MIT's Center for Theoretical Physics (CTP), Nick Manton had already been there for one year also as a postdoc and by that time had co-authored a paper [19] with Roman applying the theory of invariant gauge fields to the determination of constants of motion in their background. It was great to get to meet Nick, the disappointment came when Roman told me that he had achieved what he wanted to do in relation with my thesis work and was moving on to different problems. Having thus been taught that timing is of importance, I carried on and dove with excitement, delight and some trepidations in the amazing environment that the CTP was providing.

It is not really possible to describe in a few words how vibrant the CTP was. The faculty formed a truly outstanding group. Altogether they received numerous recognitions. During my postdoctoral fellowship alone in 1981, Jeffrey Goldstone obtained the Dannie Heineman Prize (which Roman got in 1995) and Viky Weisskopf was awarded the Wolf Prize. I also recall a meeting in honor of Francis Low who was the Provost of MIT on the occasion of his 60th birthday. This was the time when Alan Guth, a former student of Low who had joined the MIT Faculty in 1980 was developing inflation cosmology [20] and when So-Young Pi was much involved in these developments [21, 22]. And there was Roman who was attracting like a magnet a very large number of highly talented Ph.D. students and postdocs. Many of them are actually contributing to this festschrift with Antti Niemi one of the instigators of this project. It was a privilege to be part of that group which boasted incredible creativity of intensity. Under the leadership of Roman and thanks to his inspiration many important chapters of theoretical physics were developed through various collaborations involving members of that team, among these advances are the studies of the fermion–vortex system and the fermion number fractionization, the development of the Liouville field theory, the topologically massive (Chern–Simons) gauge theories and many more. It was hard to keep track with all these papers "typeset in TeX by Roger L. Gilson". The "brown bag" lunch seminar at the CTP certainly created great occasions for exchanges and the generation of research ideas.

The interactions with Harvard were also highly stimulating. I have much benefitted for instance of the lectures on supergravity given by Steven Weinberg on one of his extended visits. The joint Harvard–MIT theoretical physics seminar in particular was a Thursday ritual not to be missed. Much appreciated also were the dinners with the speaker that followed at which the postdocs were invited. The meals at the Yenching Chinese restaurant near Harvard Square with Sidney Coleman placing the order bring fond memories. I keep saying that this is where I learned to use chopsticks, out of necessity, since I did not manage to eat much because of my poor dexterity the first few times I attended.

In 1981, my second year at the CTP, Eric D'Hoker arrived at MIT for a three-year postdoctoral fellowship and so did Eddie Farhi to take up a Junior Faculty appointment. Eddie and Roman edited a book at that time on dynamical gauge symmetry breaking [23]. Eric and I became close friends and began collaborating on problems that were often prompted by Roman. Thanks to the "Bruno Rossi" exchange program between the INFN and MIT and Sergio Fubini in particular, the CTP entertained close ties with Italian physicists. In a paper published in 1976, de Alfaro, Fubini and Furlan have explored the ramifications of scale invariance in one-dimensional quantum mechanics [24]; building on this work in a more physically motivated context, Roman identified in 1980 the conformal symmetries of the magnetic monopole [25]. At some point Roman passed on to Eric and I manuscript notes from Fubini that led to the paper with Rabinovici developing superconformal quantum mechanics [26] and which stemmed from a lot of on-going interest in supersymmetric theories (see for example [27]). That led us to consider the super-symmetries of monopole systems which we did (see for instance [28]) by extending the work of Roman. We carried on studies [29] that built on the seminal work of Deser, Jackiw and Templeton on topologically massive theories [30]; we did work then on reduction of higher dimensional Chern–Simons theories that for some reason we never published.

Those postdoctoral years at MIT largely thanks to Roman were indeed defining ones. When I left Cambridge at the end of my fellowship, I promised myself to return as soon as possible.

4. Returning to MIT as a young faculty at Université de Montréal

In 1982, holding University Research Fellowship from NSERC, I came back to my hometown to take on an Assistant Professor position at the Université de Montréal. I kept reaping the benefits of my sojourn at MIT: the first postdoc I hired was Haris Panagopoulos whom I had met there; I arranged for Manu Paranjape a fellow Canadian and former student of Goldstone with whom I also had become acquainted at the CTP to come to Montreal as Assistant Professor at UdeM and I kept working with Eric D'Hoker as he moved himself to Columbia University.

In 1987 as I was presenting my dossier for tenure, I held the promise I had made to myself and returned to the CTP as Visiting Researcher for six months at the beginning of the year. I then shared an office with Oscar Eboli, a most charming Brazilian fellow who was working with Roman and So-Young on quantum fields out of thermal equilibrium [31] and with whom I had the pleasure to reconnect in Sao Paulo in the summer of 2018.

At the time So-Young and Roman's son Stefan was two years old.

Shortly after my arrival, Roman introduced me to a brilliant young Italian physi-cist named Roberto Floreanini. The two were working on the functional approach to quantum field theories [32]. It clicked immediately with Roberto. I embarked in that program and co-authored a few articles (e.g. [33]) with Roberto before I returned to Montreal. Once again, because of MIT, I had not only found a superb collaborator but met someone who became a dear friend. And so Roberto and I collaborated intensively until to his disappointment I suppose, I became Provost at McGill University.

As always, it was truly enriching scientifically and personally to be around Ro-man. I was living during that stay at Longfellow Place near So-Young and Roman's home in Beacon Hill. I recall one evening when Roman was giving me a lift in his BMW with the iconic "FFDUAL" plates, the radio was on and he was quite taken by the music saying how beautiful it was. He asked: do you know what it is? I was not completely sure but I thought I had recognized Richard Strauss and said so. A few minutes after I had passed my door, I have a phone call, it was Roman to tell me: you were right it was Ariadne auf Naxos. A cute anecdote to recall how classy Roman is.

At the time in the first half of 1987, Roman was also examining issues connected to Berry potentials [34]. He shared with me the thought that the symmetries of a problem could determine the associated Berry connections. By then, I was back in Montreal. It was quite nice to think that the theory of invariant connections of my Ph.D. days could be brought to bear on this current topic. I sorted this out and wrote a draft. At this point I had not co-authored papers with Roman. I then thought the moment had come to loop the loop and publish something with Roman on the topic that had initially brought me to MIT. In spite of my insistence, Roman declined arguing that I should publish the paper alone [35]. He subsequently wrote an article [36] covering the question in his own way and kindly quoted my publication. Although this left me with the regret of never having had the pleasure of chiseling a text with Roman this turned out to be some kind of blessing for me and (I apologize Roman) a certain curse for him. Indeed as you are all well aware, in any grant application process one needs to suggest prestigious colleagues with whom you have never collaborated (or at least not within a certain period). I have abused of Roman in this respect but magnanimously he always obliged and has been very supportive.

During the academic year 1989–1990, Eric D'Hoker kindly hosted me at UCLA where he had moved. Our first son, Jean-François, was born in LA in July 1990. Roberto Floreanini visited and this is when we launched as we were babysitting what became a vast study of the connections between quantum algebras, q-special functions and their applications.

5. Meeting here and there

The occasion for another prolonged visit at the CTP never presented itself again. Nevertheless life, family and common scientific interests provided opportunities for So-Young, Roman, Letitia and I to get together sometimes with our children and those were always very happy moments. Our friendship built over the years but I think that it certainly strengthened while we were all together in Banff in 1989. I have mentioned before that these summer schools in the Canadian Rockies which provided my first meeting with Roman as a student, were an institution. This one entitled Physics, Geometry and Topology had a stellar group of lecturers including Roman who spoke about planar physics [37]. Although very successful it turned out to be the last school of that nice series because the original sources of funding disappeared. I had enjoyed them so much that I had the idea to revive them under the name *CRM Summer School in Banff* when I became Director of the CRM in the nineties. This turned out to be the precursor of what is today the highly successful Banff International Research Station (BIRS) which is jointly supported by the CONACYT, the NSF and NSERC.

A peculiar encounter with Roman and one that brings laughs in retrospect occurred in Kiev in 1992. There was a conference organized by the Ukrainian Academy of Sciences. Roman was participating to connect with his childhood days I believe, and I was attending because this conference was taking place immediately after a meeting in Alushta, Crimea at which I had been invited by Alexei Morozov and other friends from ITEP in Moscow. Phong who is a member of the Department of Mathematics at Columbia University and a collaborator of Eric D'Hoker was also in Crimea. Based on our understanding of what *ikra* meant in Russian, Phong and I were proudly thinking that we had managed to buy nice caviar. And so I arrived in Kiev with two tin cans of this great finding. My father who had wished to visit Crimea was accompanying me. Once in our room in the hotel of the Academy we noticed that there was no hot water. We got together with Roman at some point to confirm that he did not have hot water either because this time in the summer had been chosen by the hotel management to perform plumbing work throughout the building. Facing this adversity, we told Roman about our caviar and invited him to our room with the hope of indulging in this delicacy only to find after some struggle to open the cans that they contained some dry and highly salted red fish eggs. So much for our refined party!

We have also met in less exotic places like Boston or New York and at times our reunions were prompted by musical reasons. Stefan Jackiw is a magnificent violonist who is unanimously recognized as belonging to an elite group of only a few. My son Jean-François, has done musical performance training as a violist to a high level even though he chose not to pursue a professional career. The 2006 edition of the International Conference on Group Theoretical Methods in Physics (ICGTMP) was held in New York. I took along Jean-François and his younger brother Laurent under the condition that they visit one museum every day, an assignment that they fulfilled. Stefan was already living in New York and So-Young and Roman who were also staying in Manhattan at that moment very kindly arranged for the six of us to have a joyful dinner after Stefan had kindly practiced scales with Jean-François. In 2009, So-Young organized a gathering for Roman's 70th birthday in Boston which was a lovely event and another occasion to celebrate Roman's outstanding impact on science and people. For five years in a row, beginning in 2010, Jean-François participated in the Aspen Music Festival and School. We have always managed to spend some family time in Aspen during those summers. In some of those years Stefan has been a guest artist. I recall in particular being subjugated one evening by an interpretation he gave of the third Brahms sonata. So-Young and Roman were also involved regularly in the Aspen Center for Physics and our stays often overlapped. We never missed the chance to get together and I vividly remember great dinners at the Pine Creek Cookhouse at the base of the Elk Mountains.

6. Roman and the CRM and Montreal

As I bring these reminiscences to a close, I want to stress how generous Roman has been with his visits to Montreal and the connections he has built with this city and its scientific organizations. I shall point at three moments in particular.

In 1988, Yvan Saint-Aubin and I organized the XVIIth ICGTMP in Montreal. This international conference took place against the backdrop of perestroika which created very fortunate circumstances that allowed proeminent Soviet mathematicians and scientists such as Belavin, Faddeev, Fatteev, Manin and Zamolodchikov to attend and lecture at the meeting. For many, this was their first visit to North America. The list of plenary speakers was stunning and we were lucky to have Roman [38] (who also gave an additional talk [41]) and So-Young [39] among them. The recipient of the Wigner medal awarded during the conference was Isadore Singer from MIT whose celebrated index theorems are so intimately connected with Roman's work. This made for a superb program; the conference was very successful and many of the MIT friends (D'Hoker, Eboli, Floreanini, Niemi, Panagopoulos, etc.) attended.

For a while Roman enjoyed smoking little cigars and I wondered if this was not what explained why he appreciated Montreal: cuban cigars could be found in this trendy Canadian city! Indeed whenever in town Roman would make sure to stop at the Davidoff store to stock up. There was thus a time when as a friendly gesture, I would make sure of smuggling some havanas on trips to Boston. For the good of Roman's health and my standing with the US customs officers there is now prescription on these infractions.

In 1993, I was appointed Director of the CRM (for the first time). As a central part of its activities are thematic programs. These concentration periods on topics of special interest bring specialists from all over the world around a number of workshops and conferences; these are planned with significant leadtime. The Aisenstadt Chair is CRM's most prestigious lectureship: it is offered to distinguished scientists upon the recommendation of the CRM International Advisory Committee. The holders of the Chair deliver series of talks that are integrated within the thematic program of the semester or of the year; they are also strongly encouraged to turn their lectures into a book to be published in one of the CRM monograph series. At the beginning of my second term as Director around 1997, together with Philippe Di Francesco, Lisa Jeffrey, André Leclair and Yvan Saint-Aubin we started putting together a theme year in mathematical physics. Little did I know that I would be appointed Provost of McGill University at the beginning of July 1999. Even though I could not enjoy as much as I had intended the deployment of the scientific events that the year entailed, the program was a resounding success with Roman holding the Aisenstadt Chair. He gave his lectures in the framework of two workshops: the first on Strings, Duality and Geometry and the second in Condensed Matter and Non-Equilibrium Physics. The general topic he chose was Fluid Mechanics and as you may imagine Roman offered an original and fascinating view of this broad subject from the perspective of a particle theorist.

Being Provost at McGill led to my becoming Rector (or President) of the Université de Montréal and so I went around the Mont Royal returning to the institution where I had begun my academic career. Time was at a premium in those years but as mentioned above the Jackiws and the Vinets kept meeting and I managed to maintain some research activity. I was determined not to end my term as Rector before ensuring that Roman receives recognition from us for his immense scientific accomplishments and his special relation with Montreal. This happened in 2010 when I had the great pleasure to present him with a Honorary Doctorate from the Université de Montréal. This was a touching celebration that took place within the solemn Ph.D. convocation in a packed amphitheater. Roman with his usual intellectual elegance gave an inspiring acceptance speech and generated a long and enthusiatic applause. I recall that as he was leaving the podium looking quite moved he said to me: Now I know how my son feels!

7. Envoi

Dear Roman:

I see you like a great artist. Writing this short and sketchy chronicle of our encounters over the years gave me the chance to reflect further on your work and its tremendous impact. Your papers and expository texts are like magnificent paintings that reveal subtle and unexpected perspectives. These paintings were much acclaimed when they were first presented and brought you fame. Apprentices came to learn from you and many emulated you from the distance. You generously shared your knowledge and craftmanship. If you had such an influence on me, we can imagine the number of people for whom this has been the case. And then new generations rediscover your work, look at it from different angles, apply it to different phenomena, other artists get inspired by it and create new movements, avatars of your past creations. You engage with that and produce new work making the wheel turn.

Need it be said that we look forward to more of these inspiring pieces and to your advice and views moving ahead. Thank you for the past and thank you for the future. Here is to you from the mind and from the heart.

Acknowledgements

I wish to acknowledge that much of my connections with Roman for which I am so grateful and of which I have given only a poor glimpse here, have been made possible by the Natural Science and Engineering Research Council (NSERC) of Canada. Over the years NSERC has offered me personally postgraduate scholarships, a postdoctoral fellowship and various discovery grants for which Roman relentlessly wrote recommendations. NSERC has also continuously supported this marvellous institute for research in the mathematical sciences that the Centre de

Recherches Mathématiques or CRM is and that I still have the privilege to lead; this has in particular allowed our community to interact so profitably with Roman Jackiw at various occasions. To NSERC and through this Council to the Canadian taxpayers who are making scientific journeys possible: thank you.

References

1. S. Weinberg, A model of leptons, *Phys. Rev. Lett.* **19**, 1264–1266 (1967).
2. A. Salam, Weak and electromagnetic interactions, in *Elementary Particle Theory (Nobel Symposium 8)*, edited by N. Svartholm, (Almqvist & Wiksell, 1968).
3. G. 't Hooft, Renormalizable Lagrangians for massive Yang–Mills fields, *Nucl. Phys. B* **35**, 167–188 (1971).
4. D. Gross, F. Wilczek, Ultraviolet behavior of non-abelian gauge theories, *Phys. Rev. Lett.* **30**, 1343–1346 (1973).
5. H. Politzer, Reliable perturbative results for strong interactions, *Phys. Rev. Lett.* **30**, 1346–1349 (1973).
6. S. Adler, Axial-vector vertex in spinor electrodynamics, *Phys. Rev.* **177**, 2426–2438 (1969).
7. J. Bell, R. Jackiw, A PCAC puzzle: $\eta^0 \to \gamma\gamma$ in the σ-model, *Il Nuovo Cimento A* **60**, 47–61 (1969).
8. R. Jackiw, Field theoretic investigations in current algebra, in *Lectures on Current Algebra and its Applications*, eds. S. B. Treiman, R. Jackiw and D. J. Gross (Princeton University Press, 1972).
9. P. Goddard, D. Olive, Magnetic monopoles in gauge field theories, *Rep. Prog. Phys.* **41**, 1357–1437 (1978).
10. S. Coleman, *Aspects of Symmetry: Selected Erice Lectures of Sidney Coleman* (Cambridge University Press, 1985).
11. R. Jackiw, C. Nohl, C. Rebbi, Classical and semi-classical solutions of the Yang–Mills theory, in *Particles and Fields*, eds. D. H. Boal and A. N. Kamal (Plenum Press, 1978).
12. R. Jackiw, C. Rebbi, Conformal properties of a Yang–Mills pseudoparticle, *Phys. Rev.* **D14**, 517–523 (1976).
13. E. Witten, Some exact multipseudoparticle solutions of classical Yang–Mills theory, *Phys. Rev. Lett.* **38**, 121–124 (1977).
14. T. T. Wu, C. N. Yang, Concept of nonintegrable phase factors and global formulation of gauge fields, *Phys. Rev.* **D12**, 3845–3857 (1975).
15. J. Harnad, S. D. Shnider, L. Vinet, Group actions on principal bundle and invariance conditions for gauge fields, *J. Math. Phys.* **21**, 2719–2733, (1980).
16. J. P. Harnad, S. D. Shnider, L. Vinet, Solutions to Yang–Mills equations on M^4 invariant under subgroups of $O(4,2)$, in *Complex Manifold Techniques in Theoretical Physics*, eds. D. E. Lerner and P. D. Sommers (Pitman Press, 1979).
17. R. Jackiw, Non-linear equations in particle physics, in *Complex Manifold Techniques in Theoretical Physics*, eds. D. E. Lerner and P. D. Sommers (Pitman Press, 1979).
18. P. Forgacs, N. S. Manton, Space-time symmetries in gauge theories, *Commun. Math. Phys.* **72**, 15–35 (1980).
19. R. Jackiw, N. S. Manton, Symmetries and conservation laws in gauge theories, *Ann. Phys.* **127**, 257–273 (1980).
20. A. H. Guth, Inflationary universe: A possible solution to the horizon and flatness problems, *Phys. Rev.* **D23**, 347–356 (1981).

21. A. H. Guth, S.-Y. Pi, Fluctuations in the new inflationary universe, *Phys. Rev. Lett.* **49**, 1110–1113 (1982).
22. S.-Y. Pi, Inflation without tears: A realistic cosmological model, *Phys. Rev. Lett.* **52**, 1725–1728 (1984).
23. E. Farhi, R. Jackiw, *Dynamical Gauge Symmetry Breaking* (World Scientific, 1982).
24. V. de Alfaro, S. Fubini, G. Furlan, Conformal invariance in quantum mechanics, *Nuovo Cimento A* **34**, 569–612 (1976).
25. R. Jackiw, Dynamical symmetry of the magnetic monopole, *Ann. Phys.* **129**, 183–200 (1980).
26. S. Fubini, E. Rabinovici, Superconformal quantum mechanics, *Nucl. Phys.* **B245**, 17–44 (1984).
27. E. Witten, Constraints on supersymmetry breaking, *Nucl. Phys.* **B202**, 253–316 (1982).
28. E. D'Hoker, L. Vinet, Supersymmetry of the Pauli equation in the presence of a magnetic monopole, *Phys. Lett.* **B137**, 72–76 (1984).
29. E. D'Hoker, L. Vinet, Classical solutions to topologically massive Yang–Mills theory, *Ann. Phys.* **162**, 413-440 (1985).
30. S. Deser, R. Jackiw, S. Templeton, Topologically massive gauge theories, *Ann. Phys.* **140**, 372–411 (1982).
31. O. J. P. Eboli, R. Jackiw, S.-Y. Pi, Quantum fields out of thermal equilibrium, *Phys. Rev.* **D37**, 3557–3581 (1988).
32. R. Floreanini, C. T. Hill, R. Jackiw, Functional representation for the isometries of de Sitter space, *Ann. Phys.* **175**, 345–365 (1987).
33. R. Floreanini, L. Vinet, Vacuum state of complex scalar fields in SO(2,1) invariant background, *Phys. Rev.* **D36**, 1731–1739 (1987).
34. R. Jackiw, Berry's phase: Topological ideas from atomic, molecular and optical physics, *Comm. Atom. Mol. Phys.* **21**, 71–82 (1988).
35. L. Vinet, Invariant Berry connections, *Phys. Rev.* **D37**, 2369–2372(R) (1988).
36. R. Jackiw, Three elaborations on Berry's connection, curvature and phase, *Int. J. Mod. Phys. A* **3**, 285–297 (1988).
37. R. Jackiw, Topics in planar physics, in *Physics, Geometry and Topology*, ed. H. C. Lee, NATO ASI Series, Vol. 238 (Springer, 1990), pp. 191–239.
38. R. Jackiw, Field theoretic results in the Schrödinger representation, in *Proceedings of the XVIIth International Colloquium on Group Theoretical Methods in Physics*, eds. Y. Saint-Aubin and L. Vinet (World Scientific, 1989), pp. 90–92.
39. S.-Y. Pi, Inflation and non-equilibrium dynamics, in *Proceedings of the XVIIth International Colloquium on Group Theoretical Methods in Physics*, eds. Y. Saint-Aubin and L. Vinet (World Scientific, 1989), pp. 127–136.
40. R. Jackiw, Quantum gravity in flatland, in *Proceedings of the XVIIth International Colloquium on Group Theoretical Methods in Physics*, eds. Y. Saint-Aubin and L. Vinet (World Scientific, 1989), pp. 331–336.
41. R. Jackiw, *Lectures on Fluid Dynamics: A Particle Theorist's View of Supersymmetric, Non-Abelian, Noncommutative Fluid Mechanics and d-Branes*, CRM Series in Mathematical Physics (Springer, 2002).

Chapter 11

Roman Jackiw

Steven Weinberg

Theoretical physicists are well aware of the many important contributions of Roman Jackiw to the foundations of quantum theory of fields, as shown by his numerous honors. His paper with Bell on anomalies, which saved the program of using broken chiral symmetry in strong interaction physics, was only the most famous of these brilliant works. But I have as well my own personal memories of the pleasure that his ebullient personality gave in our friendship. Here is one example: Roman and I were at some conference or other in San Francisco. Somehow we and our wives found ourselves together downtown. It was getting on toward midnight, and we were about to part reluctantly, when Roman said that we couldn't say goodnight because we had to go to the Ritz Old Poodle Dog. What is that? Why, a restaurant of course. But why do we have to go? Well, of course, because it was a great San Francisco eatery, and the King Crab was in. What was that about the King Crab? Why, it only comes to San Francisco in June. But Roman, it is nearly midnight. Of course, that's when you have to go have King Crab at the Ritz Old Poodle Dog. Convinced by this logic, we arrived at a magnificent establishment, and the waiter said in reverent tones "You have come for the King Crab." There we were at midnight, the four of us, having the time of our lives. Not to mention a couple of hours of great conversation, and some fabulous crabs on ice. With champagne — as Roman said, "of course." Who could forget times like this with Roman Jackiw?

Chapter 12

Tribute to a Mentor

L. C. R. Wijewardhana

Dept. of Physics, University of Cincinnati,
Cincinnati, OH 45221, USA

Roman Jackiw is an outstanding physicist who has made seminal contributions to many areas of theoretical physics. He has also been an exceptional educator and mentored a large number of Ph.D. students. In this note I describe my experiences in working with him and explain how his influence shaped my career as a researcher, teacher and a mentor of graduate and undergraduate students.

I had the privilege of conducting my thesis research under the direction of Professor Roman Jackiw from 1980–1984. A few days after arriving at MIT in the summer of 1980, I went around looking for a research advisor. A professor at the CTP suggested I speak with Professor Jackiw. I was told he had a collection of starter problems used to test if new students could perform research in theoretical physics. Unlike my classmates who had arrived to do theory, I had had no prior research experience. After a few minutes of talking to me Roman realized my lack of experience and told me I could complete a problem he had started with a senior student. This involved computing the next-to-leading order in large N contribution to the vacuum energy of a large N matrix quantum mechanics model. He suggested that I use the collective field method and also predicted the type of answer I should get. I found this method too hard, but I figured out how to compute what he wanted using the next-to-leading order WKB approximation. I also found an answer that did not look at all like what he thought it would be. I was apprehensive to report what I had found, thinking that he would not be pleased. After procrastinating for a while I went back and told him what my results were. He was very nice, helpful, and asked me to draft a short paper explaining my results. Those days' initial drafts were hand written, and after making numerous corrections he asked me to get it typed. He spent a quite a bit of time correcting my grammar. When I stated "I will" do something, he took a red pen and changed it to "I shall". In the acknowledgements I stated "I would like to thank" him and the senior students for their help, he called the senior students to his office, had me thank the two of them in person, then promptly took his red pen and scratched out "would like to". With that, his corrections were complete. I had the paper typed, submitted it to

Physical Review, and after a couple of iterations it was published. That was how my journey into theoretical physics research began.

From that point on he did not assign me any more problems to work on. He encouraged all his students to work with other students or visitors. I wrote a number of papers with my fellow students Antti Niemi, Bruce McClain, Cyrus Taylor, Manu Paranjape, A. Taormina, Norman Redlich, Steve Blau, and Eduardo Guendelman. I have continued to collaborate with Gordon Semenoff, who I met when he was visiting Roman's research group in the early 80's. Roman was an excellent advisor: demanding, but fair. He cared about the welfare of his graduate students. He insisted his students quickly acquire the background necessary for research in theoretical physics. Through his influence I regularly attended seminars and colloquia, read preprints and review articles about current research topics, developed an interest in other researchers' work, and acquired a broad understanding of contemporary theoretical research. Some of the work I did with Niemi, McClain, and Taylor was on field theories coupled to random external sources. Roman encouraged me to explore the statistical mechanical aspects of this work, which led to my collaborating and writing a PRL paper with condensed matter theorists David Andelman and Henri Orland on the lower critical dimension of the random field Ising model. Roman was also helpful to students working with other professors. One of my office mates, who was working with a Mathematics professor on infrared divergences in three-dimensional gauge theories, benefitted immensely by discussing his research work with Roman. When it was time for me to graduate, Roman went out of his way to help me find a post-doctoral research position. He has always made it a point to help his students succeed in their research, find research positions, and eventually permanent employment.

Here I briefly mention three research problems I have worked on over the years which are connected to what I learned as a graduate student in Roman's research group. The first is the analysis of the thermal effective action of field theories with non-zero densities for chiral fermions. In a number of seminal papers written in the early eighties, Roman studied the properties of planar gauge and gravity theories, with Chern–Simons forms playing the roles of gauge-invariant mass terms.[1] Norman Redlich and I worked on deriving such terms starting from a four-dimensional gauge theory at finite temperature using a one-loop computation. We found that in an $SU(2)$ gauge theory with an even number of left handed fermions, which is free of all anomalies, a Chern–Simons term would be generated in the high temperature expansion.[2] Initially the motivation was to use such a term, which would be a mass term for the magnetic part of the gauge theory, to screen the magnetic field. The term we found corresponded to an imaginary mass, signifying an instability. Non-Abelian Chern–Simons terms require the quantization of the Chern–Simons coefficient to maintain global gauge invariance. The term we found had a temperature dependent coefficient and could not be quantized. Later work by Niemi[3] and Rutherford[4] clarified the issue of global gauge invariance. The CP violating part of the gauge theory's effective action is proportional to the eta invariant of a Dirac operator, which contains the Chern–Simons term plus other terms. The eta invariant is gauge invariant under all gauge transformations, and the gauge non-invariance of the Chern–Simons term is cancelled by the gauge variance of the other terms of the eta invariant. Such effective Chern–Simons terms can also be generated in U(1) gauge theories. For example a U(1) gauge theory containing two left handed fermions with charge assignment e and $-e$, selected for anomaly cancellation, would, when the fermions are integrated out, have a Chern–Simons term of the form mentioned here. Such terms generated in the early universe, which signifies an instability, have been invoked in the study of baryogenesis[5] and as a means of generating cosmic magnetic fields.[6] In recent times the instability generated by the effective Chern–Simons term is termed the chiral magnetic effect[7] or chiral magnetic anomaly.[8]

The second problem was mapping the phase structure of non-compact QED in 2+1 dimensions (QED3). I started working on QED3 during my graduate student days, and continued with it during my stint as a post-doctoral fellow at Yale. QED3 with $2N$ two-component, mass zero, Dirac fermions has a U(2N) global symmetry. The question is whether this symmetry is manifest in the Wigner–Weyl mode or the Nambu–Goldstone mode. Gauge dynamics could break this symmetry down to U(N) × U(N) giving rise to a set of $2N^2$ Goldstone bosons.[9] T. Appelquist, D. Nash, and I showed that beyond a critical fermion number N_{cr}, the symmetry is realized in the Wigner–Weyl mode.[10] Below N_{cr} the symmetry is spontaneously broken. The actual value of the critical fermion number is yet to be determined with a high degree of accuracy. Values ranging from four all the way down to zero have been obtained[11; 12; 13] using different techniques. A review of various attempts to determine N_{cr} can be found in Refs. AWReview and Ghukov. If this number is greater than zero, at the critical point QED3 undergoes a structural phase transition in theory space. When the number of fermion flavors is greater than the critical value, QED3 is in its conformal window. This theory with $N = 2$ has been utilized as an effective continuum theory for the 2D quantum antiferromagnet.[16; 17] The observed Neel ordering of this system corresponds to dynamical fermion mass generation.[18; 19] Therefore the accurate determination of the actual N_{cr} is of physical interest.

The third and most recent research problem is analyzing the properties of axion stars. Axions are pseudo-Goldstone bosons which arise in a set of models postulated to solve the strong CP problem in QCD. They become pseudo-Goldstone particles due to the presence of the axial anomaly. Roman was one of the first to realize that quantization of chiral gauge theories leads to the anomalous non-conservation of the axial current. My collaborators and I have worked on developing a formalism to study the stability of axion stars, which are gravitationally bound configurations of axion particles. We have shown that beyond a certain maximum mass, axion stars become unstable and collapse,[20] emitting relativistic axion particles along the way,[21] and generating what is termed a "bosenova" event.[22; 23] We have also worked on developing a low energy effective field theory to study axion condensates,[24] and to analyze if collapsing axion stars reach more compact configurations, so-called "dense axion stars".[25] We find that only dilute axion star configurations live long enough to survive until the current cosmological epoch.

I have mentored a few graduate and undergraduate students during my years as a faculty member. As a mentor I have strived to have them work hard, encouraged them to develop a strong work ethic, ensured they receive support for travel to conferences, and helped them to develop as well-rounded professionals. All these actions, which are expected from a caring and dedicated mentor, are inspired by the example Professor Roman Jackiw set when he was a mentor to me. I take this opportunity to thank him for being such a dedicated and truly exemplary mentor, and I wish him a healthy, long life.

References

1. S. Deser, R. Jackiw, and S. Templeton, Three-dimensional massive gauge theories, *Phys. Rev. Lett.* **48**, 975–978 (1982).
2. A. N. Redlich and L. C. R. Wijewardhana, Induced Chern–Simons terms at high temperatures and finite densities, *Phys. Rev. Lett.* **54**, 970 (1985).
3. A. J. Niemi, Topological terms induced by finite temperature and density fluctuations, *Phys. Rev. Lett.* **57**, 1102–1105 (1986).
4. A. R. Rutherford, Spectral asymmetry and gauge theories at finite fermion density, *Phys. Lett. B* **182**, 187–192 (1986).
5. M. E. Shaposhnikov, Structure of the high temperature gauge ground state and electroweak production of the baryon asymmetry, *Nucl. Phys. B* **299**, 797–817 (1988).
6. M. Joyce and M. E. Shaposhnikov, Primordial magnetic fields, right-handed electrons, and the Abelian anomaly, *Phys. Rev. Lett.* **79**, 1193–1196 (1997).
7. D. E. Kharzeev, The chiral magnetic effect and anomaly-induced transport, *Prog. Part. Nucl. Phys.* **75**, 133–151 (2014).
8. Y. Kojima and Y. Miura. The growth of chiral magnetic instability in a large-scale magnetic field, *PTEP* **2019**, 043E01 (2019).
9. R. D. Pisarski, Chiral symmetry breaking in three-dimensional electrodynamics, *Phys. Rev. D* **29**, 2423 (1984).
10. T. Appelquist, D. Nash, and L. C. R. Wijewardhana, Critical behavior in (2+1)-dimensional QED, *Phys. Rev. Lett.* **60**, 2575 (1988).
11. D. Nash, Higher order corrections in (2+1)-dimensional QED, *Phys. Rev. Lett.* **62**, 3024 (1989).
12. T. Appelquist, A. G. Cohen and M. Schmaltz, A new constraint on strongly coupled gauge theories, *Phys. Rev. D* **60**, 045003 (1999).
13. N. Karthik and R. Narayanan, Scale-invariance of parity-invariant three-dimensional QED, *Phys. Rev. D* **94**, 065026 (2016).
14. T. Appelquist and L. C. R. Wijewardhana, Phase structure of noncompact QED3 and the Abelian Higgs model, in *Quantum Theory and Symmetries, Proceedings of the 3rd International Symposium*, Cincinnati, USA, 10–14 September 2003 (World Scientific, 2004), pp.177–191.
15. S. Gukov, RG flows and bifurcations, *Nucl. Phys. B* **919**, 583–638 (2017).

16. J. B. Marston and I. Affleck, Large-N limit of the Hubbard–Heisenberg model, *Phys. Rev. B* **39**, 11538 (1989).
17. M. C. Diamantini, P. Sodano, E. Langmann, and G. W. Semenoff, SU(N) antiferromagnets and the phase structure of QED in the strong coupling limit, *Nucl. Phys. B* **406**, 595–630 (1993).
18. J. B. Marston, U(1) gauge theory of the Heisenberg antiferromagnet, *Phys. Rev. Lett.* **61**, 1914–1917 (1988).
19. D. H. Kim and P. A. Lee, Theory of spin excitations in undoped and underdoped cuprates, *Annals Phys.* **272**, 130–164 (1999).
20. J. Eby, P. Suranyi, C. Vaz, and L. C. R. Wijewardhana, Axion stars in the infrared limit, *JHEP* **1503**, 080 (2015).
21. J. Eby, P. Suranyi, and L. C. R. Wijewardhana, The lifetime of axion stars, *Mod. Phys. Lett.* **A31**, 1650090 (2016).
22. J. Eby, M. Leembruggen, P. Suranyi, and L. C. R. Wijewardhana, Collapse of axion stars, *JHEP* **1612**, 066P (2016).
23. M. H. P. M. van Putten, Pair condensates produced in bosenovae, *Phys. Lett.* **A374** 3346 (2010).
24. J. Eby, P. Suranyi, and L. C. R. Wijewardhana, Expansion in higher harmonics of boson stars using a generalized Ruffini–Bonazzola approach, Part 1: Bound states, *JCAP* **1804**(04), 1 (2018).
25. J. Eby, M. Leembruggen, L. Street, P. Suranyi, and L. C. R. Wijewardhana, Global view of QCD axion stars *Phys. Rev. D* **100**, 063002 (2019).

Part 2

Scientific Contributions

Chapter 13

Entanglement in Fermionic Chains and Bispectrality

Nicolas Crampé[1], Rafael I. Nepomechie[2] and Luc Vinet[3]

[1] *Institut Denis-Poisson CNRS/UMR 7013 - Université de Tours
- Université d'Orléans, Parc de Grandmont, 37200 Tours, France*
`crampe1977@gmail.com`
[2] *Physics Department, P.O. Box 248046, University of Miami,
Coral Gables, FL 33124 USA*
`nepomechie@miami.edu`
[3] *Centre de Recherches Mathématiques, Université de Montréal, P.O. Box 6128,
Centre-ville Station, Montréal (Québec), H3C 3J7, Canada*
`vinet@crm.umontreal.ca`

Entanglement in finite and semi-infinite free Fermionic chains is studied. A parallel is drawn with the analysis of time and band limiting in signal processing. It is shown that a tridiagonal matrix commuting with the entanglement Hamiltonian can be found using the algebraic Heun operator construct in instances when there is an underlying bispectral problem. Cases corresponding to the Lie algebras $\mathfrak{su}(2)$ and $\mathfrak{su}(1,1)$ as well as to the q-deformed algebra $\mathfrak{so}_q(3)$ at q a root of unity are presented.

This paper is dedicated to Roman Jackiw with admiration and gratitude on the occasion of his 80th birthday.

1. Introduction

Throughout his career Roman Jackiw has achieved a number of important scientific advances and in the process he has brought many modern geometrical, topological and representation theoretic results to bear on the elaboration and understanding of physical theories. He has hence much contributed to increasing the level of interactions between physicists and mathematicians. We here wish to thankfully pay tribute to him by discussing how symmetry and algebraic considerations can contribute to entanglement studies in light of a parallel with long-studied issues in signal processing. We hope that this report will hence capture some of the bridge building spirit of Roman's insightful and inspiring papers.

A fundamental feature of quantum theories, entanglement enables correlations and is a key resource in applications to information. It is therefore relevant to obtain quantitative evaluation of this property and this is being much explored using the notion of entropy. This paper belongs to that class of studies and focuses on systems consisting of free fermionic chains that have been much looked at [1, 2] because of their simplicity.

Basically, for these systems, the entanglement entropy is determined by the eigenvalues of a truncated correlation matrix. However a significant difficulty in carrying out their computation arises for large chains because the spectra of these correlation matrices usually accumulate near certain points thereby rendering the numerical analysis problematic. As a matter of fact this entanglement problem proves closely analogous [1–4] to the classical question of the time and band limiting of signals where the corresponding calculation difficulty is circumvented thanks to the discovery made by Slepian *et al.* [5, 6, 17] of an operator easy to handle numerically that commutes with the limiting operator.

The main goal of the present paper is to explain how this efficient processing of signals can be adapted in certain cases to the entropy analysis of fermionic chains.

Although this is still not fully understood, the circumstances for the existence of the commuting operator are perceived to stem from bispectrality situations [7, 8] where the functions involved depend on two variables and satisfy a pair of eigenvalue problems such that in the first equation the operator acts on one variable and the eigenvalue is a function of the other variable and vice versa in the second equation. The hypergeometric polynomials of the Askey scheme [9, 10] offer examples of such bispectral problems: they are eigenfunctions of a differential or difference operator in the variable with the eigenvalues depending on the degree and, their orthogonality requires that they satisfy a recurrence relation which is viewed as an eigenvalue equation for an operator acting on the degree taken as a discrete variable with the eigenvalue in this case solely depending on the standard variable. One notes that there are two pictures for the pair of operators: the variable picture and the degree picture much like the coordinate and momentum representations in quantum mechanics.

That bispectrality has something to do with the existence of a commuting operator in problems of the time and band limiting class was revisited recently in Ref. [11]. Assume that the limiting takes place by restricting the range of the two variables associated to the problem. A first observation is that the bilinear combinations of the two bispectral operators provide generalizations of the Heun operator which itself actually arises in the particular case of the Jacobi polynomials. The reader will recall that the usual Heun operator defines the Fuchsian second-order differential equation with four regular singularities. To each bispectral problem is thus associated what has been called an algebraic Heun operator. Once this is recognized, it is easy to determine how these generalized Heun operators should be specialized so as to commute with the projectors on the restrained domains and as a consequence with the limiting operator.

Basically, determining the entanglement of fermionic chains amounts to: (i) taking the chain in some state which we will assume to be the ground state,

(ii) dividing the system into two parts, and (iii) examining how these two parts are coupled in the chosen state. The analogy with time and band limiting arises as energy is gapped by the Fermi sea filling and space is chopped through the partitioning of the chain. In cases where the fermionic chain Hamiltonian exhibits bispectral features, we shall show how algebraic Heun operators provide matrices that commute with the entanglement Hamiltonian and have nice properties from the point of view of numerical analysis.

This paper enlarges and complements our recent article [12] on this topic where the emphasis in the characterization of the chains and their properties was put on the associated orthogonal polynomials. Here the focus is on bispectrality and algebras. The parallel with time and band limiting will be explained with the help of a review of the classic results in this field and the connection with algebraic Heun operators will be illustrated in this context first. Supplementing the set of chains considered in [12], we shall discuss a semi-infinite chain as well as a finite one based on a representation of a q-deformed algebra at q a root of unity that has as special case the uniform chain treated in [2] and [12].

The presentation will proceed along the following lines. The free fermionic Hamiltonians and their diagonalization are described in Sec. 2 that will also establish notation. Section 3 introduces the restricted correlation matrix as the central quantity for the study of entanglement. Section 4 momentarily leaves the topic of fermionic chains to offer a short overview of the classical problem of limiting in time a signal which is banded in frequency. It shall explain how the Heun operator associated to the Fourier bispectral problem leads to the second-order differential operator that commutes with the integral operator that effects the limiting in this case. Section 5 returns to fermionic chains in light of this understanding and discusses generally when the Hamiltonians are characterized by a bispectral problem. For finite chains this will involve Leonard pairs which are known to be in correspondence with the families of orthogonal polynomials of the terminating branch of the Askey scheme. Section 6 derives the tridiagonal matrices that commute with the chopped correlation matrix from the algebraic Heun operators attached to Hamiltonians with bispectral underpinnings. Special bispectral situations that will be considered as examples shall be arising from the representation theory of Lie and q-algebras. This will be the contents of Secs. 7–9. Section 7 will reproduce results from [12] by discussing the chain based on $\mathfrak{su}(2)$. Section 8 will treat the case of the semi-infinite chain associated to $\mathfrak{su}(1,1)$. Section 9 will focus on the chain whose couplings are given by the representation matrices of the non-standard deformation $\mathfrak{so}_q(3)$ of $\mathfrak{so}(3)$ at q root of unity. This will have as a special case the uniform chain treated in Refs. [2] and [12]. Section 10 shall bring the paper to a close with concluding remarks.

2. Free-fermion Hamiltonian and its diagonalization

We consider the following open quadratic free-fermion inhomogeneous Hamiltonian with nearest neighbour interactions and with magnetic fields

$$\widehat{\mathcal{H}} = \sum_{n=0}^{N-1} (J_n c_n^\dagger c_{n+1} + J_n^* c_{n+1}^\dagger c_n) - \sum_{n=0}^{N} B_n c_n^\dagger c_n, \tag{1}$$

where B_n (resp. J_n) are real (resp. complex) parameters, J_n^* is the complex conjugate of J_n and $\{c_m^\dagger, c_n\} = \delta_{m,n}$. For the sake of simplicity of the following computations, we enumerate the sites of the lattice from 0 to N. We can also consider the case $N \to +\infty$ which corresponds to a semi-infinite chain (see Sec. 8 for an example).

In order to diagonalize $\widehat{\mathcal{H}}$, it is convenient to rewrite it as follows

$$\widehat{\mathcal{H}} = (c_0^\dagger, \dots, c_N^\dagger)\, \widehat{H} \begin{pmatrix} c_0 \\ \vdots \\ c_N \end{pmatrix}. \tag{2}$$

The $(N+1) \times (N+1)$ matrix $\widehat{H}$ is an Hermitian tridiagonal matrix given by

$$\widehat{H} = \sum_{n=0}^{N} \left(J_{n-1}|n-1\rangle\langle n| - B_n|n\rangle\langle n| + J_n^*|n+1\rangle\langle n| \right), \tag{3}$$

with the convention $J_N = J_{-1} = 0$. The set $\{|0\rangle, |1\rangle, \dots, |N\rangle\}$ of elements in $\mathbb{C}^{N+1}$ denotes the canonical orthonormal basis and will be called the position basis. The spectral problem for $\widehat{H}$ reads

$$\widehat{H}|\omega_k\rangle = \omega_k|\omega_k\rangle, \tag{4}$$

where

$$|\omega_k\rangle = \sum_{n=0}^{N} \phi_n(\omega_k)|n\rangle. \tag{5}$$

We order the $N+1$ eigenvalues $\omega_0, \omega_1, \dots, \omega_N$ so that $\omega_k < \omega_{k+1}$. We also normalize the eigenvectors $|\omega_0\rangle, |\omega_1\rangle, \dots, |\omega_N\rangle$ so that they form an orthonormal basis of $\mathbb{C}^{N+1}$, to be called the momentum basis. Having diagonalized $\widehat{H}$, we see that the Hamiltonian $\widehat{\mathcal{H}}$ (1) can be rewritten as

$$\widehat{\mathcal{H}} = \sum_{k=0}^{N} \omega_k \tilde{c}_k^\dagger \tilde{c}_k, \tag{6}$$

where the annihilation operators $\tilde{c}_k$ are defined by

$$\tilde{c}_k = \sum_{n=0}^{N} \phi_n^*(\omega_k)\, c_n, \tag{7}$$

and the corresponding formulas for the creation operators $\tilde{c}_k^\dagger$ are given by the Hermitian conjugation of (7). These operators obey the anticommutation relations

$$\{\tilde{c}_k^\dagger, \tilde{c}_p\} = \delta_{k,p}, \qquad \{\tilde{c}_k^\dagger, \tilde{c}_p^\dagger\} = \{\tilde{c}_k, \tilde{c}_p\} = 0. \tag{8}$$

One can invert relation (7) to get

$$c_n = \sum_{k=0}^{N} \phi_n(\omega_k)\, \tilde{c}_k. \tag{9}$$

The eigenvectors of $\widehat{\mathcal{H}}$ are therefore given by

$$|\Psi\rangle\!\rangle = \tilde{c}_{k_1}^\dagger \ldots \tilde{c}_{k_r}^\dagger |0\rangle\!\rangle, \tag{10}$$

where $k_1 < \ldots < k_r \in \{0, \ldots, N\}$, and the vacuum state $|0\rangle\!\rangle$ is annihilated by all the annihilation operators

$$\tilde{c}_k |0\rangle\!\rangle = 0, \qquad k = 0, \ldots, N. \tag{11}$$

The corresponding energy eigenvalues of $\widehat{\mathcal{H}}$ are simply given by

$$E = \sum_{i=1}^{r} \omega_{k_i}. \tag{12}$$

3. Correlations and the entanglement Hamiltonian

For the sake of concreteness, we shall consider entanglement in the ground state described below. We shall further review how the reduced density matrix for the first $\ell + 1$ sites of the chain is determined by the 1-particle correlation matrix and equivalently by the entanglement Hamiltonian.

3.1. *Defining the ground state*

The fact that the ground state is constructed by filling the Fermi sea leads to a restriction in energy. Indeed, the ground state $|\Psi_0\rangle\!\rangle$ of the Hamiltonian (1) is given by

$$|\Psi_0\rangle\!\rangle = \tilde{c}_0^\dagger \ldots \tilde{c}_K^\dagger |0\rangle\!\rangle, \tag{13}$$

where $K \in \{0, 1, \ldots, N\}$ is the greatest integer below the Fermi momentum, such that

$$\omega_K < 0, \qquad \omega_{K+1} > 0. \tag{14}$$

Let us remark that K can be modified by adding a constant term to the external magnetic fields B_n. We shall in fact choose this constant magnetic field so as to ensure that $\omega_k \neq 0$ for any k in order to avoid dealing with a degenerate ground state.

The correlation matrix $\widehat{C}$ in the ground state is an $(N+1) \times (N+1)$ matrix with the following entries

$$\widehat{C}_{mn} = \langle\!\langle \Psi_0 | c_m^\dagger c_n | \Psi_0 \rangle\!\rangle. \tag{15}$$

Expressing everything in terms of annihilation and creation operators using (9) and (10), and then using the anticommutation relations (8) and the property (11) of the vacuum state, we obtain

$$\widehat{C}_{mn} = \sum_{k=0}^{K} \phi_m^*(\omega_k)\phi_n(\omega_k)\,, \qquad 0 \le n, m \le N\,. \tag{16}$$

It is then manifest that

$$\widehat{C} = \sum_{k=0}^{K} |\omega_k\rangle\langle\omega_k|\,, \tag{17}$$

namely, that $\widehat{C}$ is the projector onto the subspace of $\mathbb{C}^{N+1}$ spanned by the vectors $|\omega_k\rangle$ with $k = 0, ..., K$ running over the labels of the excitations in the ground state.

3.2. *Entanglement entropy*

In order to examine entanglement, we must first define a bipartition of our free fermionic chain. As subsystem (part 1) we shall take the first $\ell + 1$ consecutive sites, and shall find how it is intertwined with the rest of the chain in the ground state $|\Psi_0\rangle\rangle$. To that end, we need the reduced density matrix

$$\rho_1 = \text{tr}_2 |\Psi_0\rangle\rangle\langle\langle\Psi_0|\,, \tag{18}$$

where part 2, the complement of part 1, is comprised of the sites $\{\ell+1, \ell+2, ..., N\}$; from this quantity one can compute for instance the von Neumann entropy

$$S_1 = -\text{tr}(\rho_1 \log \rho_1)\,. \tag{19}$$

The explicit computations of this entanglement entropy amounts to finding the eigenvalues of ρ_1.

It has been observed that this reduced density matrix ρ_1 is determined by the spatially "chopped" correlation matrix C, which is the following $(\ell + 1) \times (\ell + 1)$ submatrix of $\widehat{C}$:

$$C = |\widehat{C}_{mn}|_{0 \le m, n \le \ell}\,. \tag{20}$$

The argument which we take from Ref. [13] (see also Ref. [14]) goes as follows. Because the ground state of the Hamiltonian $\widehat{\mathcal{H}}$ is a Slater determinant, all correlations can be expressed in terms of the one-particle functions, i.e. in terms of the matrix elements of $\widehat{C}$. Restricting to observables A associated to part 1, since the expectation value of A is given by $\langle A \rangle = \text{tr}(\rho_1 A)$, the factorization property will hold according to Wick's theorem if ρ_1 is of the form

$$\rho_1 = \kappa \, \exp(-\mathcal{H})\,, \tag{21}$$

with the entanglement Hamiltonian $\mathcal{H}$ given by

$$\mathcal{H} = \sum_{m,n=0}^{\ell} h_{mn} \, c_m^\dagger c_n\,. \tag{22}$$

The hopping matrix $h = |h_{mn}|_{0 \le m,n \le \ell}$ is defined so that

$$C_{mn} = \text{tr}(\rho_1 \, c_m^\dagger c_n), \qquad m, n \in \{0, 1, \ldots, \ell\}, \tag{23}$$

holds, and one finds through diagonalization that

$$h = \log[(1 - C)/C]. \tag{24}$$

We thus see that the $2^{(\ell+1)} \times 2^{(\ell+1)}$ matrix ρ_1 is obtained from the $(\ell+1) \times (\ell+1)$ matrix C or equivalently, from the entanglement Hamiltonian $\mathcal{H}$.

Introducing the projectors

$$\pi_1 = \sum_{n=0}^{\ell} |n\rangle\langle n| \quad \text{and} \quad \pi_2 = \sum_{k=0}^{K} |\omega_k\rangle\langle\omega_k| = \widehat{C}, \tag{25}$$

the chopped correlation matrix can be written as (see for instance Refs. [15, 16])

$$C = \pi_1 \pi_2 \pi_1 \,. \tag{26}$$

To calculate the entanglement entropies one therefore has to compute the eigenvalues of C. As explained in Ref. [1], this is not easy to do numerically because the eigenvalues of that matrix are exponentially close to 0 and 1. We shall show in the following how to go about this problem by drawing on methods developed in signal processing.

4. A review of time and band limiting

We here digress to underscore that the treatment of time and band limiting problems is of relevance for the characterization of entanglement in fermionic chains. To make that clear, we shall review aspects of the classic problem of optimizing the concentration in time of a band-limited signal. In the first part of this section we shall show that the limiting integral operator can also be expressed in terms of projectors exactly as in (26). The diagonalization of this operator that would give the optimization solution is also plagued by computational difficulties. In the second part of the section we shall indicate how the underlying bispectrality provides a way to overcome this numerical analysis challenge by allowing to identify a differential operator that commutes with the limiting one.

Let $f(t)$ be a signal limited to the band of frequencies $[-W, W]$:

$$f(t) = \frac{1}{\sqrt{2\pi}} \int_{-W}^{W} dp \, e^{ipt} F(p) \in B_W, \tag{27}$$

and call B_W the space of such functions taken to be real. It is natural to want a signal of finite duration, that is to ask that $f(t)$ vanishes outside the interval $-T < t < T$:

$$f \ne 0 \quad \text{only for} \quad -T < t < T. \tag{28}$$

It is however readily realized that this is impossible: since $f(t) \in B_W$, it is entire in complex t-plane; therefore if $f(t) = 0$ for any interval, it follows that $f(t)$ is identically zero ($f(t) \equiv 0$). In the 1960s and 1970s Slepian, Landau, Pollak from Bell labs (see the reviews [5, 6]) considered how to approximate the situation wished for and asked the question: Which band-limited signal $\in B_W$ is best concentrated in the time interval $-T < t < T$, second best concentrated etc.? In other words which functions $f(t) \in B_W$ are maximizing

$$
\alpha^2(T) = \frac{\displaystyle\int_{-T}^{T} f^2(t)dt}{\displaystyle\int_{-\infty}^{\infty} f^2(t)dt}
$$

$$
= 2\,\frac{\displaystyle\int_{-W}^{W} dp' \int_{-W}^{W} dp'' \left[\frac{\sin((p'-p'')T)}{(p'-p'')}\right] F(p'')F^*(p')}{\displaystyle\int_{-W}^{W} dp'\, F(p')F^*(p')}. \tag{29}
$$

As is well known from the calculus of variations, the answer to that question is provided by the solutions of

$$
GF(p) = \lambda F(p), \tag{30}
$$

where the integral operator G is defined by

$$
GF(p) = \int_{-W}^{W} dp'\, K(p-p')F(p'), \tag{31}
$$

with $K(p-p')$ the sinc kernel

$$
K(p-p') = \frac{\sin((p-p')T)}{\pi(p-p')}. \tag{32}
$$

Let us remark that $GF(p)$ is zero if p is not between $-W$ and W as the functions $F(p)$ we start with.

In principle this should settle the concentration problem. However the spectrum of G accumulates sharply at the origin and this makes the numerical computations intractable. Slepian, Landau, Pollak [5, 6, 17] quite remarkably found a way out by showing that there exists a second-order differential operator D that commutes with the integral operator G. This is important because D has common eigenfunctions with G and second-order differential operators are typically well behaved numerically. It is interesting to mention that D actually arises in separating the Laplacian in prolate spheroidal coordinates. Let us indicate how this commuting operator is obtained using the bispectral framework of Fourier transform.

We shall first note that G can be written in a form similar to that given in (26) for the chopped correlation matrix. Consider projectors on an interval:

$$
\pi_L^x f(x) = \begin{cases} f(x) & -L < x < L \\ 0 & \text{otherwise} \end{cases}
$$

$$
= [\Theta(x+L) - \Theta(x-L)]f(x), \tag{33}
$$

with $\Theta(x)$ the step function. Let $\mathcal{F} : f(t) \mapsto F(p)$ denote the Fourier transform and $\mathcal{F}^{-1}$ its inverse. Take the following projectors in Fourier (or band) space:

$$\pi_W^p \qquad \text{and the Fourier transformed} \qquad \hat{\pi}_T^p = \mathcal{F}\pi_T^t\mathcal{F}^{-1}. \tag{34}$$

It is straightforward to see that

$$G = \pi_W^p \hat{\pi}_T^p \pi_W^p. \tag{35}$$

Operators X and Y form a bispectral pair if they have common eigenfunctions $\psi(x, n)$ such that

$$X\psi(x, n) = \omega(x)\psi(x, n), \tag{36}$$
$$Y\psi(x, n) = \lambda(n)\psi(x, n), \tag{37}$$

with X acting on the variable n and Y on the variable x. When forming products of these operators X and Y, we shall understand that they are both taken in the same representation "n" or "x". The functions $\psi(t, p) = e^{ipt}$ in Fourier transforms satisfy

$$-\frac{d^2}{dt^2}\psi(t, p) = p^2\psi(t, p), \qquad -\frac{d^2}{dp^2}\psi(t, p) = t^2\psi(t, p), \tag{38}$$

and are thus associated to the simplest bispectral problem: the functions $\psi(t, p)$ are eigenfunctions of an operator acting on t with eigenvalues depending on p and vice versa. All the orthogonal polynomials of the Askey scheme are solutions of bispectral problems defined by the differential/difference equation and the recurrence relation.

How does this help find the differential operator that commutes with the limiting operator G?

To each bispectral problem, one can attach an *Algebraic Heun Operator* [18] defined as the most general operator W_H which is bilinear in the bispectral operators X and Y:

$$W_H = \tau_1\{X, Y\} + \tau_2[X, Y] + \tau_3 X + \tau_4 Y + \tau_0, \tag{39}$$

with $\tau_i, i = 0, 1, \dots, 4$, constants and $\{X, Y\} = XY + YX$. The name comes from the fact that the standard Heun operator results when this construct is applied to the bispectral operators of the Jacobi polynomials, namely the hypergeometric operator and multiplication by the variable x. We claim that the commuting operators belong to that class of operators. Let us return to the Fourier case where in the "frequency" representation

$$X = -\frac{d^2}{dp^2}, \qquad Y = p^2. \tag{40}$$

In this representation, taking $\tau_0 = 0$ and $\tau_1 = -1/2$, the algebraic Heun operator which we will now denote by D takes the form:

$$D = \frac{1}{2}\{\frac{d^2}{dp^2}, p^2\} + \tau[-\frac{d^2}{dp^2}, p^2] - \mu\frac{d^2}{dp^2} + \nu p^2 \tag{41}$$

$$= (p^2 - \mu)\frac{d^2}{dp^2} + (2 - 4\tau)p\frac{d}{dp} + \nu p^2 - 2\tau + 1. \tag{42}$$

Given (35), such an operator will commute with G if it commutes with both $\tilde{\pi}_W^p$ and $\tilde{\pi}_T^p$. Consider a general second-order differential operator written as

$$\mathcal{D} = A(p)\frac{d^2}{dp^2} + B(p)\frac{d}{dp} + C(p) . \qquad (43)$$

Let us look first at the projector onto the semi-infinite interval $[W, \infty)$

$$\tilde{\pi}_W^p = \Theta(p - W) . \qquad (44)$$

It is easy to see that $[\mathcal{D}, \tilde{\pi}_W^p] = 2A(p)\delta(p - W)\frac{d}{dp} + (-A'(p) + B(p))\delta(p - W) = 0$ if $A(W) = 0$ and $A'(W) = B(W)$. Now recall that

$$\pi_W^p = \Theta(p + W) - \Theta(p - W) .$$

In this case $[\mathcal{D}, \pi_W^p] = 0$ is satisfied if

$$A(\pm W) = 0 \qquad \text{and} \qquad A'(\pm W) = B(\pm W) . \qquad (45)$$

Applying these conditions to D as given by (42) is readily seen to imply that

$$\mu = W^2 \qquad \text{and} \qquad \tau = 0. \qquad (46)$$

Now if in addition $[D, \hat{\pi}_T^p] = 0$, we would have $[D, G] = 0$. Clearly $[D, \hat{\pi}_T^p] = [D, \mathcal{F}\pi_T^t \mathcal{F}^{-1}] = 0$ is tantamount to $[\mathcal{F}^{-1}D\mathcal{F}, \pi_T^t] = 0$, namely to the condition that the Fourier transform $\tilde{D} = \mathcal{F}^{-1}D\mathcal{F}$ of D commutes with a projector in t with parameter T that is similar to π_W^p. Under the Fourier transform: $p^2 \to -\frac{d^2}{dt^2}, -\frac{d^2}{dp^2} \leftrightarrow t^2$ and $\tilde{D}$ is obtained from D by exchanging p and t as well as μ and ν and by taking τ into $-\tau$. It is then obvious that the condition $[\tilde{D}, \pi_T^t] = 0$ is satisfied by taking

$$\nu = T^2 \qquad \text{and again} \qquad \tau = 0. \qquad (47)$$

It thus follows that the second-order differential operator that commutes with the limiting integral operator is simply obtained from the algebraic Heun operator (42) by imposing the conditions $\tau = 0$, $\mu = W^2$ and $\nu = T^2$ on the parameters.

The parallel with the study of entanglement in fermionic chains is quite clear. Taking the chain in its ground state (or in any other reference state) involves restricting the energies and corresponds to band limiting. Associated to that is the projector π_2 in (25). Establishing the bipartition truncates space and this is akin to time limiting. Attached to this is the projector π_1 in (25). The task is to solve the eigenvalue problem for the chopped correlation matrix $C = \pi_1\pi_2\pi_1$ which looks very much like the limiting operator G as given in (35) (the picture is actually the dual one here). On the basis of this similarity, we may therefore hope that there could be a tridiagonal matrix — the discrete analog of a second-order differential operator — that would commute with both π_1 and π_2 and hence with C so as to ease the numerical analysis. Recalling that the existence of the commuting operator was predicated on the fact that there was an underlying bispectral problem, we shall discuss next what this requirement entails for the specifications of the fermionic chains that shall henceforth be considered.

5. A bispectral framework for fermionic chains

In order to identify fermionic chains that are based on bispectral problems, let us recall that two natural bases, the position basis $\{|n\rangle\}$ and the momentum basis $\{|\omega_k\rangle\}$, are associated to the chains. The $(N+1) \times (N+1)$ matrix $\widehat{H}$ (3) that defines the Hamiltonian is irreducible tridiagonal in the first of these bases and diagonal in the second. (By irreducible it is understood that there are no zeros on the sub- and super-diagonals.) From (4) we have

$$\langle n|\widehat{H}|\omega_k\rangle = \omega_k \langle n|\omega_k\rangle \tag{48}$$

and thus in view of (3) the wavefunctions $\phi_n(\omega_k) = \langle n|\omega_k\rangle$ satisfy the eigenvalue equation

$$\omega_k \phi_n(\omega_k) = J_n \phi_{n+1}(\omega_k) - B_n \phi_n(\omega_k) + J_{n-1}\phi_{n-1}(\omega_k), \qquad 0 \leq n \leq N. \tag{49}$$

We wish the functions $\phi_n(\omega_k)$ to be solutions of a bispectral problem. To that end we need to adjoin to $\widehat{H}$ a companion operator $\widehat{X}$ with the property of being diagonal in the basis $\{|n\rangle\}$ and irreducible tridiagonal in the basis $\{|\omega_k\rangle\}$. In other words we need an $\widehat{X}$ such that

$$\widehat{X} = \sum_{n=0}^{N} \lambda_n |n\rangle\langle n|, \tag{50}$$

and

$$\widehat{X} = \sum_{k=0}^{N} \left(\overline{J}_{k-1}|\omega_{k-1}\rangle\langle\omega_k| - \overline{B}_k|\omega_k\rangle\langle\omega_k| + \overline{J}_k^*|\omega_{k+1}\rangle\langle\omega_k| \right), \tag{51}$$

with the convention $\overline{J}_{-1} = \overline{J}_{N+1} = 0$. It then follows that

$$\langle n|\widehat{X}|\omega_k\rangle = \lambda_n \langle n|\omega_k\rangle \tag{52}$$

becomes the difference equation

$$\lambda_n \phi_n(\omega_k) = \overline{J}_k^* \phi_n(\omega_{k+1}) - \overline{B}_k \phi_n(\omega_k) + \overline{J}_{k-1}\phi_n(\omega_{k-1}), \qquad 0 \leq k \leq N. \tag{53}$$

Equations (49) and (53) provide a bispectral problem for $\phi_n(\omega_k)$ which is a discrete version of the bispectral problem (38) at the root of the previous section.

When N is finite, the couple of operators $\widehat{H}$ and $\widehat{X}$ form by definition a Leonard pair [19]. One can deduce that the eigenvalues $\{\omega_k\}$ of $\widehat{H}$ are pairwise distinct and similarly for the eigenvalues $\{\lambda_n\}$ of $\widehat{X}$ (see Lemma 1.3. in [20]). Leonard pairs have been classified [19] and shown to be in correspondence with the orthogonal polynomial families of the truncating part of the Askey tableau. As a matter of fact, all discrete hypergeometric polynomials of that scheme, not only the finite classes, provide admissible $\widehat{H}$ and $\widehat{X}$ through their recurrence relation and difference equation.

Summing up, the fermionic chains susceptible of admitting a commuting tridiagonal matrix are those whose specifications are dictated by a duo of operators $\widehat{H}$ and $\widehat{X}$ with the special properties described above. Operators that would qualify are for instance two generators of the Askey–Wilson algebra or, for $q = 1$, of the Racah algebra; these are quadratic algebras which respectively describe the bispectral properties of the polynomials sitting at the top of the Askey scheme. As particular and simpler cases, a moment's thought will make one realize that two generators of rank-one Lie or q-deformed Lie algebras will meet the requirement that one of these elements will be represented by an irreducible tridiagonal matrix in the eigenbasis of the other and vice versa. These are the situations on which we will focus in Secs. 7, 8 and 9.

Given such bispectral contexts, the time and band-limiting experience has taught us that nice commuting operators can be simply obtained from the associated algebraic Heun operator. This is what we will explain in the next section before we come to examples.

6. Algebraic Heun operators and commuting matrices

Looking for a tridiagonal matrix T that commutes with C, in the spirit of Sec. 4, we introduce the "discrete - discrete" version of the algebraic Heun operator (42). As per (39), we take this operator to be [18] the following bilinear combination of the two operators that define the bispectral problem:

$$\widehat{T} = \{\widehat{X}, \widehat{H}\} + \tau[\widehat{X}, \widehat{H}] + \mu\widehat{X} + \nu\widehat{H} \,. \tag{54}$$

At this point the parameters τ, μ, ν are free. (Note that allowing for redefinition by an irrelevant overall factor, the coefficient of $\{\widehat{X}, \widehat{H}\}$ has been set equal to 1.) It is immediate to see that $\widehat{T}$ is tridiagonal in both the position basis

$$\begin{aligned}
\widehat{T}|n\rangle = {}&J_{n-1}\left(\lambda_{n-1}(1+\tau) + \lambda_n(1-\tau) + \nu\right)|n-1\rangle \\
&+ (\mu\lambda_n - 2B_n\lambda_n - \nu B_n)|n\rangle \\
&+ J_n\left(\lambda_n(1-\tau) + \lambda_{n+1}(1+\tau) + \nu\right)|n+1\rangle \,,
\end{aligned} \tag{55}$$

and the momentum basis

$$\begin{aligned}
\widehat{T}|\omega_k\rangle = {}&\overline{J}_{k-1}(\omega_{k-1}(1-\tau) + \omega_k(1+\tau) + \mu)|\omega_{k-1}\rangle \\
&+ (\nu\omega_k - 2\overline{B}_k\omega_k - \mu\overline{B}_k)|\omega_k\rangle \\
&+ \overline{J}_k(\omega_k(1+\tau) + \omega_{k+1}(1-\tau) + \mu)|\omega_{k+1}\rangle \,.
\end{aligned} \tag{56}$$

As a matter of fact, it has been shown in Ref. [21] that $\widehat{T}$ is the most general operator which is tridiagonal in both bases in finite-dimensional situations.

Let $\widehat{T}_{mn} = \langle m|\widehat{T}|n\rangle$ and define the "chopped" matrix T by

$$T = |\widehat{T}_{mn}|_{0 \leq m,n \leq \ell} \,. \tag{57}$$

Following the results of Refs. [18] and [22], we know that T and C will commute,

$$[T, C] = 0 \,, \tag{58}$$

if the parameters in $\widehat{T}$ (54) are given by

$$\tau = 0 \,, \quad \mu = -(\omega_K + \omega_{K+1}) \quad \text{and} \quad \nu = -(\lambda_\ell + \lambda_{\ell+1}) \,. \tag{59}$$

Indeed, with the particular value of ν given by (59), we see that the matrix $\widehat{T}$ leaves the subspace $\{|n\rangle , n = 0, 1, \ldots, \ell\}$ invariant. Therefore, T commutes with π_1. Similarly, with μ specified by (59), $\widehat{T}$ leaves the subspace $\{|\omega_k\rangle, k = 0, 1, \ldots, K\}$ invariant and T commutes with π_2. Finally, in view of (26), it is easy to see that (58) holds.

The main result of this section is that the tridiagonal matrix T (57) i.e.

$$
T = \begin{pmatrix}
d_0 & t_0 & & & & \\
t_0 & d_1 & t_1 & & & \\
& t_1 & d_2 & t_2 & & \\
& & \ddots & \ddots & \ddots & \\
& & & t_{\ell-2} & d_{\ell-1} & t_{\ell-1} \\
& & & & t_{\ell-1} & d_\ell
\end{pmatrix},
\tag{60}
$$

whose nonzero matrix elements are given by (see (55))

$$
t_n = J_n(\lambda_n + \lambda_{n+1} - \lambda_\ell - \lambda_{\ell+1}),
\tag{61}
$$

$$
d_n = -B_n(2\lambda_n - \lambda_\ell - \lambda_{\ell+1}) - \lambda_n(\omega_K + \omega_{K+1})
\tag{62}
$$

commutes with the correlation matrix (58). A key ingredient obviously is the operator $\widehat{X}$ defined in (50). In the following sections, we apply this construction to examples of both finite and semi-infinite free fermionic chains.

If $t_n \neq 0$ (which is the case in the examples below), T is non-degenerate (see e.g. Lemma 3.1 in Ref. [19]) and the commuting matrices T and C have a unique set of common eigenvectors. Since T is tridiagonal, its eigenvectors can be readily computed numerically. By acting with C on these eigenvectors, the eigenvalues of C can be easily obtained. The eigenvalues of the entanglement Hamiltonian $\mathcal{H}$, and therefore the entanglement entropy of the model, can then also be straightforwardly determined.

7. The chain based on $\mathfrak{su}(2)$

In this section, and the subsequent ones, we use unitary representations of Lie and q-deformed algebras to identify appropriate pairs of bispectral Hermitian operators $\widehat{H}$ and $\widehat{X}$. We then construct the Heun operator to obtain explicit examples of matrices T that commute with the respective entanglement Hamiltonians.

We begin with the simplest case, that is, $\mathfrak{su}(2)$. The spin s $(s \in \mathbb{Z}/2)$ representation of $\mathfrak{su}(2)$ is given by

$$
s^x = \frac{1}{2} \sum_{n=0}^{2s} \sqrt{(n+1)(2s-n)} \left(|n\rangle\langle n+1| + |n+1\rangle\langle n| \right),
\tag{63}
$$

$$
s^y = -\frac{i}{2} \sum_{n=0}^{2s} \sqrt{(n+1)(2s-n)} \left(|n\rangle\langle n+1| - |n+1\rangle\langle n| \right),
\tag{64}
$$

$$
s^z = -\sum_{n=0}^{2s} (n-s)|n\rangle\langle n|.
\tag{65}
$$

We choose

$$\widehat{H} = \cos(\theta)s^z - \sin(\theta)s^x - b, \tag{66}$$

where b and θ are real constants. In view of (3), we study a chain with $N = 2s$ and with parameters (see (1)) given by

$$B_n = \cos(\theta)(n - s) + b, \qquad J_n = -\frac{1}{2}\sin(\theta)\sqrt{(n+1)(2s-n)}. \tag{67}$$

To diagonalize $\widehat{H}$, we observe that $\widehat{H} = U(s^z - b)U^\dagger$ with $U = e^{i\theta s^y}$, and hence $\widehat{H}|\omega_k\rangle = \omega_k|\omega_k\rangle$ with

$$|\omega_k\rangle = U|2s - k\rangle \qquad \text{and} \qquad \omega_k = k - s - b, \tag{68}$$

where $k = 0, 1, \ldots, 2s$. The integer K in (14) is the unique integer satisfying[a] $s + b - 1 \leq K < s + b$. Let us mention that $\phi_n(\omega_k) = \langle n|\omega_k\rangle = \langle n|U|k\rangle$ are given in terms of the Krawtchouk polynomials in this case.

The operator $\widehat{X}$ (50) can be chosen as $\widehat{X} = s^z$, which is diagonal in the position basis with $\lambda_n = s - n$. We observe that

$$s^z = U\left(\cos(\theta)s^z + \sin(\theta)s^x\right)U^\dagger. \tag{69}$$

Hence, in the momentum basis, in light of the first equation in (68), $\widehat{X}$ is given by

$$\widehat{X} = \cos(\theta)\sum_{k=0}^{2s}(k-s)|\omega_k\rangle\langle\omega_k|$$
$$+ \frac{1}{2}\sin(\theta)\sum_{k=0}^{2s}\sqrt{(k+1)(2s-k)}\left(|\omega_k\rangle\langle\omega_{k+1}| + |\omega_{k+1}\rangle\langle\omega_k|\right). \tag{70}$$

Comparing with the general form (51) for $\widehat{X}$ in the momentum basis, we have

$$\overline{B}_k = -\cos(\theta)(k - s), \qquad \overline{J}_k = \frac{1}{2}\sin(\theta)\sqrt{(k+1)(2s-k)}. \tag{71}$$

The Heun operator associated to the Lie algebra $\mathfrak{su}(2)$ has been studied previously in [23]. We conclude that the matrix T is given by (60) with

$$t_n = -\sin(\theta)(n - \ell)\sqrt{(n+1)(2s-n)}, \tag{72}$$
$$d_n = [\cos(\theta)(n - s) + b](2n - 2\ell - 1) + (s - n)(2s - 2K + 2b - 1). \tag{73}$$

---

[a] We choose b such that $K \in \{0, 1, \ldots, N\}$. The other case $K < 0$ (resp. $K > N$) corresponds to an empty (resp. full) ground state which is not interesting from the point of view of the entanglement entropy.

8. The chain based on $\mathfrak{su}(1,1)$

In this section, we focus on the irreducible discrete series unitary representation of the Lie algebra $\mathfrak{su}(1,1)$ given by (see e.g. [24])

$$\sigma^x = \frac{1}{2} \sum_{n=0}^{\infty} \sqrt{(n+1)(\kappa+n)} \Big(|n\rangle\langle n+1| + |n+1\rangle\langle n| \Big), \tag{74}$$

$$\sigma^y = \frac{i}{2} \sum_{n=0}^{\infty} \sqrt{(n+1)(\kappa+n)} \Big(|n\rangle\langle n+1| - |n+1\rangle\langle n| \Big), \tag{75}$$

$$\sigma^z = \sum_{n=0}^{\infty} \left(n + \frac{\kappa}{2} \right) |n\rangle\langle n|, \tag{76}$$

where κ is a real positive parameter. Indeed, one can show that

$$[\sigma^x, \sigma^y] = -i\sigma^z, \qquad [\sigma^z, \sigma^x] = i\sigma^y, \qquad [\sigma^z, \sigma^y] = -i\sigma^x. \tag{77}$$

We choose for $\widehat{H}$

$$\widehat{H}^{ell} = \cosh(\theta)\sigma^z - \sinh(\theta)\sigma^x + b, \tag{78}$$

where b and θ are real. The superscript "ell" stands for elliptic. To justify this name, we recall that a rotation by an element of the group $SU(1,1)$ of a generic element $l_x\sigma^x + l_y\sigma^y + l_z\sigma^z$ preserves the non-definite form $l_x^2 + l_y^2 - l_z^2$. For the Lie element $\cosh(\theta)\sigma^z - \sinh(\theta)\sigma^x$ in (78), this non-definite form is negative with the element thus belonging to the elliptic orbit.

We are therefore studying in this section a chain with an infinite number of sites. In view of (3), the parameters of the Hamiltonian $\widehat{H}$ defined by (1) are given by

$$B_n^{ell} = -\cosh(\theta)\left(n + \frac{\kappa}{2} \right) - b, \qquad J_n^{ell} = -\frac{1}{2}\sinh(\theta)\sqrt{(n+1)(\kappa+n)}. \tag{79}$$

To obtain the eigenvalues and eigenvectors of $\widehat{H}^{ell}$ (78), we note here that $\widehat{H}^{ell} = U(\sigma^z + b)U^\dagger$ with $U = e^{i\theta\sigma^y}$, and find that $\widehat{H}^{ell}|w_k\rangle = w_k|w_k\rangle$ with

$$|w_k\rangle = U|k\rangle \qquad \text{and} \qquad w_k = k + \frac{\kappa}{2} + b, \tag{80}$$

for $k = 0, 1, \dots$ Let us mention that the wavefunctions $\phi_n(w_k)$ are expressed in terms of the Meixner polynomials in this case.

The operator $\widehat{X}$ is taken to be $\widehat{X} = \sigma^z$, and is diagonal in the position basis with $\lambda_n = n + \frac{\kappa}{2}$. We observe that

$$\widehat{X}^{ell} = U\left(\cosh(\theta)\sigma^z + \sinh(\theta)\sigma^x \right) U^\dagger. \tag{81}$$

Proceeding as for the $\mathfrak{su}(2)$ model and referring to (51), we observe that the expression of $\widehat{X}$ in the momentum basis involves the following coefficients:

$$\overline{B}_k^{ell} = -\cosh(\theta)\left(k + \frac{\kappa}{2} \right), \qquad \overline{J}_k^{ell} = \frac{1}{2}\sinh(\theta)\sqrt{(k+1)(k+\kappa)}. \tag{82}$$

The Heun operator associated to the Lie algebra $\mathfrak{su}(1,1)$ has been studied previously in [23]. We conclude that the matrix T in this case is given by (60) with

$$t_n = -\sinh(\theta)(n - \ell)\sqrt{(n+1)(\kappa+n)}, \tag{83}$$

$$d_n = \left[\cosh(\theta)\left(n + \frac{\kappa}{2} \right) + b \right](2n - 2\ell - 1) - \left(n + \frac{\kappa}{2} \right)(\kappa + 2K + 2b + 1). \tag{84}$$

9. The chain based on $\mathfrak{so}_q(3)$ at q root of unity

In this section, we offer a final explicit example based on an irreducible unitary representation of the q-deformed Lie algebra $\mathfrak{so}_q(3)$ at q root of unity. Let N be a positive integer and $d = 1, 2, \ldots N - 1$. There is a $(d+1) \times (d+1)$ irreducible representation of $\mathfrak{so}_q(3)$ with $q = \exp(2i\pi/N)$ given by [25]

$$K_1 = -\frac{1}{2} \sum_{n=0}^{d-1} \sqrt{\frac{\sin\left(\frac{\pi(n+1)}{N}\right) \sin\left(\frac{\pi(d-n)}{N}\right)}{\cos\left(\frac{\pi(d-2n-2)}{2N}\right) \cos\left(\frac{\pi(d-2n)}{2N}\right)}} \left(|n\rangle\langle n+1| + |n+1\rangle\langle n| \right), \quad (85)$$

$$K_0 = \sum_{n=0}^{d} \sin\left(\frac{\pi(2n-d)}{2N}\right) |n\rangle\langle n|. \quad (86)$$

We define $K_2 = e^{i\pi/(2N)} K_0 K_1 - e^{-i\pi/(2N)} K_1 K_0$. Then, one gets

$$e^{i\pi/(2N)} K_1 K_2 - e^{-i\pi/(2N)} K_2 K_1 = -\sin^2\left(\frac{\pi}{N}\right) K_0, \quad (87)$$

$$e^{i\pi/(2N)} K_2 K_0 - e^{-i\pi/(2N)} K_0 K_2 = -\sin^2\left(\frac{\pi}{N}\right) K_1, \quad (88)$$

thus realizing the defining relations of $\mathfrak{so}_q(3)$ (we have changed the normalisation of the generators K_i for later convenience).

We take for $\widehat{H}$

$$\widehat{H}^{\mathfrak{so}} = K_1 + b, \quad (89)$$

where b is a real constant. This defines a chain with $d+1$ sites. In view of (3), the couplings of the Hamiltonian $\widehat{H}$ defined by (1) are in this case given by

$$B_n^{\mathfrak{so}} = -b, \quad J_n^{\mathfrak{so}} = -\frac{1}{2} \sqrt{\frac{\sin\left(\frac{\pi(n+1)}{N}\right) \sin\left(\frac{\pi(d-n)}{N}\right)}{\cos\left(\frac{\pi(d-2n-2)}{2N}\right) \cos\left(\frac{\pi(d-2n)}{2N}\right)}}. \quad (90)$$

Let us remark that when the number of sites is related to the order of the unity root, i.e. when $d = N - 2$, these reduce to

$$B_n^{\mathfrak{so}} = -b, \quad J_n^{\mathfrak{so}} = -\frac{1}{2}. \quad (91)$$

Hence, the model treated here generalizes the homogeneous chain studied in [12].

Note that the q-commutation relations (88) of $\mathfrak{so}_q(3)$ are symmetric under the exchange $K_0 \leftrightarrow K_1$; hence, in the present representation where this permutation is unitarily realized, K_1 has the same spectrum as K_0 [25]. Therefore, $\widehat{H}$ given by (89) is diagonalized as follows: for $k = 0, 1, \ldots, d$,

$$\widehat{H}^{\mathfrak{so}}|\omega_k\rangle = \omega_k|\omega_k\rangle, \quad \omega_k = \sin\left(\frac{\pi(2k-d)}{2N}\right) + b. \quad (92)$$

Let us mention that the wavefunctions $\phi_n(\omega_k)$ involve the q-ultraspherical polynomials at q a root of unity. It is interesting to realize that the finite Chebychev polynomials that occur in the uniform chain are a special case of these q-polynomials.

The operator $\widehat{X}$ can be chosen as $\widehat{X}^{so} = K_0$, which is diagonal in the position basis with $\lambda_n = \sin\left(\frac{\pi(2n-d)}{2N}\right)$. In the momentum basis, this operator $\widehat{X}$ is also tridiagonal and reads

$$\overline{B}_k^{so} = 0, \qquad \overline{J}_k^{so} = \frac{1}{2}\sqrt{\frac{\sin\left(\frac{\pi(k+1)}{N}\right)\sin\left(\frac{\pi(d-k)}{N}\right)}{\cos\left(\frac{\pi(d-2k-2)}{2N}\right)\cos\left(\frac{\pi(d-2k)}{2N}\right)}}. \tag{93}$$

We conclude that the matrix T is given by (60) with

$$t_n = 2\cos\left(\frac{\pi}{2N}\right)\sin\left(\frac{\pi(\ell-n)}{2N}\right)\cos\left(\frac{\pi(\ell+n-d+1)}{2N}\right)$$

$$\times \sqrt{\frac{\sin\left(\frac{\pi(n+1)}{N}\right)\sin\left(\frac{\pi(d-n)}{N}\right)}{\cos\left(\frac{\pi(d-2n-2)}{2N}\right)\cos\left(\frac{\pi(d-2n)}{2N}\right)}}, \tag{94}$$

$$d_n = -2\cos\left(\frac{\pi}{2N}\right)\left[b\sin\left(\frac{\pi(2\ell-d+1)}{2N}\right)\right.$$

$$\left. + \sin\left(\frac{\pi(2n-d)}{2N}\right)\sin\left(\frac{\pi(2K-d+1)}{2N}\right)\right]. \tag{95}$$

This coincides with the matrix found in [2] and [12] when $d = N - 2$.

10. Concluding remarks

This paper has discussed entanglement in free fermionic chains and focused in particular on the challenges associated to the diagonalization of the entanglement Hamiltonian. It has underscored in this respect the connection that these studies bear with the classic treatment of time and band limiting in signal processing. This article has illustrated how the methods developed in the latter context can be usefully imported in the entanglement analyses of fermionic chains. The key feature that has thus been adapted is the existence of a second-order differential (or difference) operator that commutes with the non-local limiting operator. In time this remarkable fact has been understood to arise from an underlying bispectral situation, and recently [18] the related algebraic Heun operator construct was seen to lead to these commuting operators. This was reviewed here and was seen to be transposable to the entanglement of fermionic chains.

The specifications of chains which have a bispectral underpinning have been characterized. Involved are two operators ($\widehat{H}$ and $\widehat{X}$) which are diagonal in the momentum and position bases respectively and tridiagonal in the other. They define the bispectral problem that the wavefunctions satisfy. Attached to chains of that type are algebraic Heun operators that readily yield a tridiagonal matrix that commutes with the restricted correlation matrix which is the fundamental operator that needs to be diagonalized. It was pointed out that the bispectral operators generate algebraic structures of interest and are connected to orthogonal polynomials. With that perspective, three pairs of bispectral operators were identified from representations of the Lie and q-deformed algebras $\mathfrak{su}(2)$, $\mathfrak{su}(1,1)$ and $\mathfrak{so}_q(3)$. The corresponding free fermionic chains were introduced and the commuting matrices presented. The first model gave an example of a finite chain, the second of a semi-infinite one and the third based on representations of $\mathfrak{so}_q(3)$ at q a root of unity offered a one-parameter generalization of the chain with uniform couplings.

A number of interesting questions are pending and deserve further investigations. In all our considerations, the bipartition of the chains has been defined by considering one part as the subset of sites consisting of consecutive nodes starting with the first one. It would obviously be of relevance to extend the approach to other space limiting. Studies of entanglement of fermions (and bosons) on different graphs have been undertaken [26, 27]. We plan on examining how the considerations developed in this paper could extend in that context. It would also be nice to carry this out in field theory especially in the Schrödinger representation (see in particular [28]) that Roman Jackiw has at times advocated [29, 30].

Acknowledgements

The authors are grateful to the editors for the invitation to contribute to this volume in honour of Roman Jackiw. N. Crampé warmly thanks the Centre de Recherches Mathématiques (CRM) for hospitality and support during his visit to Montreal in the course of this investigation. The research of L. Vinet is supported in part by a Discovery Grant from the Natural Science and Engineering Research Council (NSERC) of Canada.

References

[1] I. Peschel, On the reduced density matrix for a chain of free electrons, *J. Stat. Mech.* **2004**(6), 06004 (2004). doi: 10.1088/1742-5468/2004/06/P06004.
[2] V. Eisler and I. Peschel, Properties of the entanglement Hamiltonian for finite free-fermion chains, *J. Stat. Mech.* **2018**(10), 104001 (oct, 2018).
[3] D. Gioev and I. Klich, Entanglement entropy of fermions in any dimension and the widom conjecture, *Phys Rev Lett.* 96(10):100503 (Mar, 2006). doi: 10.1103/PhysRevLett.96.100503.
[4] V. Eisler and I. Peschel, Free-fermion entanglement and spheroidal functions, *J. Stat. Mech.* **2013**(4), 04028 (2013). doi: 10.1088/1742-5468/2013/04/P04028.
[5] D. Slepian, Some comments on Fourier analysis, uncertainty and modeling, *SIAM Rev.* **25**(3), 379–393 (1983). doi: 10.1137/1025078.
[6] H. Landau, "An Overview of Time and Frequency Limiting", in ed. J. F. Price, *Fourier Techniques and Applications*, (Springer US, Boston, MA, 1985), pp. 201–220. doi: 10.1007/978-1-4613-2525-3_12.
[7] J. J. Duistermaat and F. A. Grünbaum, Differential equations in the spectral parameter, *Commun. Math. Phys.* **103**(2), 177–240 (1986). doi: 10.1007/BF01206937.

[8] F. A. Grünbaum, Time-band limiting and the bispectral problem, *Commun. Pur. Appl. Math.* **47**(3), 307–328 (1994). doi: 10.1002/cpa.3160470305.

[9] R. Koekoek and R. F. Swarttouw. The Askey-scheme of hypergeometric orthogonal polynomials and its q-analogue. Technical Report 98-17, Delft University of Technology (1998). `https://homepage.tudelft.nl/11r49/askey.html`.

[10] R. Koekoek, P. A. Lesky, and R. F. Swarttouw, *Hypergeometric Orthogonal Polynomials and Their q-Analogues.* (Springer-Verlag, 2010). doi: 10.1007/978-3-642-05014-5.

[11] F. A. Grünbaum, L. Vinet, and A. Zhedanov, Tridiagonalization and the Heun equation, *J. Math. Phys.* **58**(3):031703 (2017). doi: 10.1063/1.4977828.

[12] N. Crampé, R. I. Nepomechie, and L. Vinet, Free-fermion entanglement and orthogonal polynomials, *J. Stat. Mech.* **2019**(9), 093101 (Sep, 2019). doi: 10.1088/1742-5468/ab3787.

[13] I. Peschel, Letter to the Editor: Calculation of reduced density matrices from correlation functions, *J. Phys. A.* **36**(14), L205–L208 (2003). doi: 10.1088/0305-4470/36/14/101.

[14] I. Peschel and V. Eisler, Reduced density matrices and entanglement entropy in free lattice models, *J. Phys. A.* **42**(50), 504003 (2009). doi: 10.1088/1751-8113/42/50/504003.

[15] C. H. Lee, P. Ye, and X.-L. Qi, Position–momentum duality in the entanglement spectrum of free fermions, *J. Stat. Mech.* **1410**(10), P10023 (2014). doi: 10.1088/1742-5468/2014/10/P10023.

[16] Z. Huang and D. P. Arovas, Entanglement spectrum and Wannier center flow of the Hofstadter problem, *Phys. Rev. B.* **86**(24), 245109 (2012). doi: 10.1103/PhysRevB.86.245109.

[17] D. Slepian and H. O. Pollak, Prolate spheroidal wave functions, Fourier analysis and uncertainty. I, *Bell Syst. Tech. J.* **40**(1), 43–63 (1961).

[18] F. A. Grünbaum, L. Vinet, and A. Zhedanov, Algebraic Heun operator and band-time limiting, *Commun. Math. Phys.* **364**(3), 1041–1068 (Dec, 2018). doi: 10.1007/s00220-018-3190-0.

[19] P. Terwilliger, Two linear transformations each tridiagonal with respect to an eigenbasis of the other, *Linear Algebra Appl.* **330**(1-3), 149–203 (2001). doi: 10.1016/S0024-3795(01)00242-7.

[20] P. Terwilliger, Two linear transformations each tridiagonal with respect to an eigenbasis of the other: Comments on the split decomposition, *J. Comput. Appl. Math.* **178**, 437–452 (Jun, 2005).

[21] K. Nomura and P. Terwilliger, Linear transformations that are tridiagonal with respect to both eigenbases of a Leonard pair, *Linear Algebra Appl.* **420**(1), 198–207 (2007). doi: 10.1016/j.laa.2006.07.004.

[22] R. K. Perline, Discrete time-band limiting operators and commuting tridiagonal matrices, *SIAM J. Algebraic Discrete Methods.* **8**(2), 192–195 (Apr., 1987). doi: 10.1137/0608016.

[23] N. Crampé, L. Vinet, and A. Zhedanov, Heun algebras of Lie type, *Proc. Amer. Math. Soc.* (2019). doi: 10.1090/proc/14788.

[24] B. G. Wybourne, *Classical Groups for Physicists* (Wiley, 1974).

[25] V. Spiridonov and A. Zhedanov, q-ultraspherical polynomials for q a root of unity, *Lett. Math. Phys.* **37**(2), 173–180 (Jun, 1996). doi: 10.1007/BF00416020.

[26] M. A. Jafarizadeh, F. Eghbalifam, and S. Nami, Entanglement entropy of free fermions on directed graphs, *Eur. Phys. J. Plus.* **132**(12), 539 (2017). doi: 10.1140/epjp/i2017-11805-1.

[27] M. A. Jafarizadeh, F. Eghbalifam, and S. Nami, Entanglement entropy in the spinless

free fermion model and its application to the graph isomorphism problem, *J. Phys. A.* **51**(7), 075304 (jan, 2018). doi: 10.1088/1751-8121/aaa1da.

[28] C. G. Callan, Jr. and F. Wilczek, On geometric entropy, *Phys. Lett.* **B333**, 55–61 (1994). doi: 10.1016/0370-2693(94)91007-3.

[29] R. Floreanini, C. T. Hill, and R. Jackiw, Functional representation for the isometries of de Sitter space, *Annals Phys.* **175**, 345 (1987). doi: 10.1016/0003-4916(87)90213-2.

[30] R. Floreanini and R. Jackiw, Functional representation for fermionic quantum fields, *Phys. Rev.* **D37**, 2206 (1988). doi: 10.1103/PhysRevD.37.2206.

Chapter 14

Gravitational Wilson Lines in AdS$_3$*

Eric D'Hoker and Per Kraus

Mani L. Bhaumik Institute for Theoretical Physics
Department of Physics and Astronomy
University of California, Los Angeles, CA 90095, USA
dhoker@physics.ucla.edu, pkraus@ucla.edu

The construction of gravitational Wilson lines in the Chern–Simons formulation of AdS$_3$ gravity in terms of composite operators in the dual boundary conformal field theory is reviewed. New evidence is presented that the Wilson line, dimensionally regularized and suitably renormalized, behaves as a bi-local operator of two conformal primaries whose dimension is predicted by $SL(2, \mathbb{R})$ current algebra.

1. Introduction

The deep impact of Roman Jackiw's contributions to quantum field theory extends from particle physics and condensed matter physics to mathematics. The fundamental role played by the Adler–Bell–Jackiw anomaly [1, 2] and its generalizations in the renormalizability of Yang–Mills theory [3, 4], the non-perturbative dynamics of gauge theories [5], and the consistency of string theory [6, 7] is well known. Conformal symmetry [8], Chern–Simons field theory [9, 10], and three-dimensional gravity [11] are but a few other subjects to which Roman has contributed brilliantly, and which provide the arena for the present work on quantum observables in AdS$_3$, along with Liouville theory [12] in collaboration with the first author, and two-dimensional gravity [13, 14].

While three-dimensional gravity does not support propagating gravitational waves or gravitons, it does have black hole solutions of finite mass, Bekenstein–Hawking entropy and temperature [15]. Accounting for the entropy of these black holes in terms of a precise counting of quantum micro-states should provide a simplified but informative warm-up for the study of the quantum behavior of physical four-dimensional black holes.

Gravity in a space-time which is asymptotically AdS$_3$ may be reformulated in terms of Chern–Simons theory for gauge group $SL(2, \mathbb{R}) \times SL(2, \mathbb{R})$ [16, 17]. Clas-

---

*Dedicated to Professor Roman Jackiw on the occasion of his 80th birthday.

sical solutions, such as thermal AdS and BTZ black holes, then correspond to flat connections, and AdS$_3$ gravity is now power-counting renormalizible [17]. Furthermore, AdS$_3$ gravity is holographically dual to two-dimensional conformal field theory, where powerful methods are available for counting states and evaluating correlation functions. In fact, an early clue to the existence of AdS/CFT duality was the discovery of the Virasoro asymptotic symmetry of AdS$_3$ with a central charge proportional to the radius of AdS$_3$ [18]. Whereas the large central charge behavior is semi-classical, the regime of finite central charge corresponds to fully quantized gravity in AdS$_3$. There is increasing evidence that AdS$_3$ may well provide the first example where the AdS/CFT duality can be proven [19].

Fundamental observables in Chern–Simons theory are the Wilson line and the Wilson loop, namely path-ordered integrals of the gauge connection in diverse representations of the gauge group. Given the flatness of the connection on physical AdS$_3$ solutions, the (closed) Wilson loop measures the holonomy of the connection; for instance its value when wrapping the horizon of a BTZ black hole yields the Bekenstein–Hawking entropy [20,21]. Another interesting observable is the open Wilson line anchored by two points on the conformal boundary of AdS$_3$. It is invariant under gauge transformations and diffeomorphisms that vanish at the boundary. In gravitational language, its expectation value is related to the propagation of a massive point particle, including the effects of gravitational self-interaction which renormalize its mass [22]. The possibility of understanding such quantum gravity effects in a controlled setting provides one motivation for studying the quantum properties of the open Wilson line.

The Wilson line in the bulk of AdS$_3$ has a natural counterpart in the boundary, given by a Wilson line for a composite gauge field built out of the stress tensor of the CFT. This observable arises naturally by using the flatness of the connection in the bulk to push the Wilson line from the bulk onto the boundary. As reviewed below, classical considerations (i.e. large c) indeed confirm that this Wilson line between two points z_1, z_2 on the boundary corresponds to a bi-local observable in the CFT which transforms under conformal transformations as a product of two conformal primaries of identical dimensions located at the points z_1 and z_2. More precisely, it yields the Virasoro vacuum OPE block, capturing all operators built out of the stress tensor that appear in the operator product expansion of the two primaries. The composite nature of the corresponding quantum Wilson line operator requires subtle regularization and renormalization. Bi-locality and the conformal properties of the Wilson line are obscured during the regularization and renormalization process. References that develop the role of Wilson lines, and networks of Wilson lines, as representing conformal blocks include [22–30]. We also note [31] which foreshadows some of these developments.

In the present paper we shall begin by reviewing earlier work [24,26,30] where an expansion in powers of $1/c$ was used to establish agreement between the quantum dimension of the Wilson line predicted from general considerations of $SL(2, \mathbb{R})$

current algebra and explicit calculations in perturbation theory in $1/c$ to order $1/c^3$ included. We shall then extend these results by showing that the correlator between a single stress tensor and the Wilson line is as predicted by conformal invariance to order $1/c$ and that the correlators between an arbitrary number of stress tensors and the Wilson line agree with predictions from conformal field theory to leading order in $1/c$, thereby providing further evidence that the Wilson line operator behaves as a bi-local operator of conformal primaries.

2. AdS3 Chern–Simons

In this section we shall review the formulation of AdS3 gravity in terms of $SL(2, \mathbb{R}) \times SL(2, \mathbb{R})$ Chern–Simons theory.

2.1. *Chern–Simons*

We consider an oriented 3-dimensional manifold M which is locally asymptotic to AdS3 and whose conformal boundary is a Riemann surface Σ. In terms of a metric g on M the standard Einstein–Hilbert action with negative cosmological constant is given by

$$I[g] = -\frac{1}{16\pi G} \int_M d^3x \sqrt{g} \left(R + \frac{2}{\ell^2} \right) \tag{1}$$

where G is the three-dimensional Newton constant, R is the Ricci scalar curvature, and $\ell > 0$ is the radius of the AdS3 vacuum solution to Einstein's equations. Equivalently, the action may be recast in terms of the frame e^a and connection ω^a one-forms with $a = 1, 2, 3$ and, as a special feature of three dimensions, without reference to the inverse frame. A convenient way to package this data is in terms of two gauge fields $A, \tilde{A}$ obtained as a linear combination of the frame e^a and the connection ω^a,

$$A = A^a t_a = (\omega^a + \lambda e^a) t_a \qquad\qquad \tilde{A} = \tilde{A}^a t_a = (\omega^a - \lambda e^a) \tilde{t}_a \tag{2}$$

whose dynamics is governed by a Chern–Simons action,

$$S[A, \tilde{A}] = -\frac{k}{4\pi} \int_M \mathrm{tr} \left(AdA + \tfrac{2}{3}A^3 \right) + \frac{k}{4\pi} \int_M \mathrm{tr} \left(\tilde{A}d\tilde{A} + \tfrac{2}{3}\tilde{A}^3 \right) + S_{\partial M}[A, \tilde{A}] \tag{3}$$

Here $S_{\partial M}$ is the boundary action required for the variational principle, as discussed in Ref. [36], but whose explicit form will not be needed here.

AdS3 with Minkowski signature has isometry group $SO(2, 2) = SL(2, \mathbb{R}) \times SL(2, \mathbb{R})/\mathbb{Z}_2$. The gauge fields A and $\tilde{A}$ are independent of one another and take values in the Lie algebra of $SL(2, \mathbb{R}) \times SL(2, \mathbb{R})$. [a] The parameters k and λ are real

---

[a] The generators t_a and $\tilde{t}_a$ of the Lie algebra of $SL(2, \mathbb{R}) \times SL(2, \mathbb{R})$ may be chosen real and normalized by $\mathrm{tr}(t_a t_b) = \mathrm{tr}(\tilde{t}_a \tilde{t}_b) = \tfrac{1}{2}\eta_{ab}$ for $\eta = \mathrm{diag}(+ + -)$ with $[t_a, t_b] = \varepsilon_{abc}\eta^{cd}t_d$ and $[\tilde{t}_a, \tilde{t}_b] = \varepsilon_{abc}\eta^{cd}\tilde{t}_d$ for $\varepsilon_{123} = 1$. For example, in the defining representation we may choose $t_1 = \tfrac{1}{2}\sigma_3$, $t_2 = \tfrac{1}{2}\sigma_1$, $t_3 = -\tfrac{i}{2}\sigma_2$ in terms of Pauli matrices. We shall often prefer to use a Cartan basis of generators $L_0, L_{\pm 1}$ whose structure relations are $[L_m, L_n] = (m - n)L_{m+n}$ and which are related to the t_a generators by $L_0 = t_1$ and $L_{\pm 1} = t_3 \pm t_2$.

and related to ℓ by $4Gk = \ell$ and $\lambda = 1/\ell$, and the level k is quantized in integer values.

AdS$_3$ with Euclidean signature is, however, the natural framework for the holographic correspondence with two-dimensional CFT on a compact conformal boundary Riemann surface Σ. The isometry group $SO(3,1)$ is no longer the product of rank one Lie groups. To recover the Chern–Simons formulation, we complexify the isometry group $SO(3,1)$ to $SO(4,\mathbb{C}) = SL(2,\mathbb{C}) \times SL(2,\mathbb{C})/\mathbb{Z}_2$ and complexify the frame e^a and the connection ω^a. The gauge fields $A, \tilde{A}$ are still given by (2) but now with imaginary λ,

$$\lambda = \frac{i}{\ell} \qquad\qquad k = \frac{\ell}{4G} \qquad\qquad (4)$$

The generators t_a and $\tilde{t}_a$ of the Lie algebra of $SL(2,\mathbb{C}) \times SL(2,\mathbb{C})$ may be chosen to coincide with those of $SL(2,\mathbb{R}) \times SL(2,\mathbb{R})$ as given in the footnote, but the Lie algebra is now over $\mathbb{C}$. In this complexified formulation, the fields A and $\tilde{A}$ are independent of one another, just as was the case with Minkowski AdS$_3$. Physical solutions, for which e^a and ω^a must be real, are obtained by imposing the condition $\tilde{A}^a = -(A^a)^*$. For the holographic correspondence with the boundary CFT it will be convenient to work with the complexified formulation to obtain chiral conformal blocks, and use the reality condition only to construct a Hermitian pairing of left and right-moving conformal blocks.

The field equations in the bulk express the flatness of both connections,

$$F = dA + A \wedge A = 0 \qquad\qquad \tilde{F} = d\tilde{A} + \tilde{A} \wedge \tilde{A} = 0 \qquad (5)$$

which are solved locally by $A = U^{-1}dU$ and $\tilde{A} = \tilde{U}^{-1}d\tilde{U}$, for $U, \tilde{U} \in SL(2,\mathbb{C})$.

2.2. AdS$_3$ asymptotics

Metrics on the three-dimensional space-time M which are asymptotically AdS$_3$ may be parametrized by Fefferman–Graham coordinates (r, x^μ) for $\mu = 1, 2$. The coordinate r is transverse to the conformal boundary Σ (which is reached in the limit $r \to \infty$) and x^μ are local coordinates parallel to the boundary. The metric ds^2 then takes the form

$$\frac{1}{\ell^2} ds^2 = \frac{dr^2}{r^2} + r^2 \gamma_{\mu\nu}(r,x)dx^\mu dx^\nu \qquad\qquad (6)$$

The transverse metric $\gamma_{\mu\nu}$ admits an expansion in powers of r^2 for large r given by

$$\gamma_{\mu\nu}(r,x) = \gamma_{\mu\nu}^{(0)}(x) + \frac{1}{r^2}\gamma_{\mu\nu}^{(2)}(x) + \mathcal{O}(r^{-4}) \qquad\qquad (7)$$

With a suitable gauge choice for the radial component of the gauge fields,

$$A_r(r,x) = \frac{1}{r}L_0 \qquad\qquad \tilde{A}_r(r,x) = -\frac{1}{r}\tilde{L}_0 \qquad\qquad (8)$$

Fefferman–Graham coordinates may also be used to express the gauge fields,

$$A(r,x) = +\frac{dr}{r}L_0 + rA^{(0)}(x) + A^{(1)}(x) + \mathcal{O}(r^{-1})$$

$$\tilde{A}(r,x) = -\frac{dr}{r}\tilde{L}_0 + r\tilde{A}^{(0)}(x) + \tilde{A}^{(1)}(x) + \mathcal{O}(r^{-1}) \qquad\qquad (9)$$

where $A^{(0)}, \tilde{A}^{(0)}, A^{(1)}, \tilde{A}^{(1)}$ have vanishing components along the differential dr. According to the standard AdS/CFT dictionary, $\gamma^{(0)}, A^{(0)}$, and $\tilde{A}^{(0)}$ are the sources to the bulk fields, while $\gamma^{(2)}, A^{(1)}$, and $\tilde{A}^{(1)}$ are the expectation values of the dual CFT operators; see, e.g., [32, 36].

We shall be interested in pure gravity solutions with AdS3 asymptotics. Locally, we can set $\gamma^{(0)}$ equal to the flat metric on Σ. The metric endows Σ with a complex structure, and we choose local complex coordinates $z, \bar{z}$ in terms of which the metric is given by

$$\gamma^{(0)}_{\mu\nu} dx^{\mu} dx^{\nu} = dz\, d\bar{z} \tag{10}$$

The boundary conditions on A and $\tilde{A}$ depend on the complex structure, and are as follows,

$$A^{(0)}(x) = A^{(0)}_z(z)dz \qquad\qquad \tilde{A}^{(0)}(x) = \tilde{A}^{(0)}_{\bar{z}}(\bar{z})d\bar{z} \tag{11}$$

where the flatness of the connections implies that $A^{(0)}_z(z)$ is holomorphic in z and $\tilde{A}^{(0)}_{\bar{z}}(\bar{z})$ is holomorphic in $\bar{z}$. The $\gamma^{(2)}_{\mu\nu}$ part of the metric is related to the expectation value of the stress tensor T which, in the above complex coordinates, is given as follows,

$$\gamma^{(2)}_{\mu\nu} dx^{\mu} dx^{\nu} = \frac{6}{c}\left(T(z)dz^2 + \tilde{T}(\bar{z})d\bar{z}^2\right) \tag{12}$$

Since the boundary theory is conformal, the trace part of the stress tensor is absent [32]. The asymptotic form of the metric (6) to order $\mathcal{O}(r^{-2})$ is invariant under infinitesimal conformal transformations $\delta z = \varepsilon(z)$ (accompanied by $2\delta\rho = -\partial_z\varepsilon$ and $2r^2\delta\bar{z} = -\partial_z^2\varepsilon$) provided T transforms as the stress tensor of a CFT of central charge c, whose value in terms of G and ℓ is given by the Brown–Henneaux formula [18],

$$\delta T = \varepsilon\partial_z T + 2(\partial_z\varepsilon)T - \frac{c}{12}\partial_z^3\varepsilon \qquad\qquad c = \frac{3\ell}{2G} \tag{13}$$

2.3. The general solutions asymptotic to AdS3

The general solution takes the form

$$A = b(r)^{-1}\left(d + a(z)\right)b(r) \qquad\qquad a = dz\, L_1 + \frac{6}{c}T(z)dz\, L_{-1}$$

$$\tilde{A} = \tilde{b}(r)\left(d + \tilde{a}(\bar{z})\right)\tilde{b}(r)^{-1} \qquad\qquad \tilde{a} = d\bar{z}\, \tilde{L}_1 + \frac{6}{c}\tilde{T}(\bar{z})d\bar{z}\, \tilde{L}_{-1} \tag{14}$$

where $b(r) = r^{L_0}$ and $\tilde{b}(r) = r^{\tilde{L}_0}$. The corresponding metric is given by

$$\frac{1}{\ell^2}ds^2 = \frac{dr^2}{r^2} + r^2|dz|^2 + \frac{6}{c}\left(T(z)dz^2 + \tilde{T}(\bar{z})d\bar{z}^2\right) + \frac{36}{c^2 r^2}T(z)\tilde{T}(\bar{z})|dz|^2 \tag{15}$$

The solution is exact for any holomorphic $T(z)$ in z and $\tilde{T}(\bar{z})$ in $\bar{z}$. The real solution is obtained by setting $\tilde{A}^a = -(A^a)^*$ and $\tilde{T}(\bar{z}) = T(z)^*$.

3. Gravitational Wilson lines

In this section, we shall review gravitational Wilson lines and their role as bi-local conformal primary fields in the holographic dual conformal field theory.

3.1. *The classical Wilson line*

We begin by considering the Wilson line in the background gauge field of a general classical solution given in the preceding section. Henceforth, we shall restrict attention to the chiral sector of the theory, which is governed by the chiral stress tensor T and the chiral gauge field A which takes values in a representation of the Lie algebra of $SL(2,\mathbb{C})$ labelled by its spin j. As proposed in Refs. [22,24] it suffices to consider the finite-dimensional representations of $SL(2,\mathbb{C})$ for which j is a positive half-integer. The classical Wilson line $\mathcal{W}[Z_2, Z_1]$ between two arbitrary points $Z_i = (r_i, z_i)$ for $i = 1, 2$ in the bulk, is defined as follows,

$$\mathcal{W}_A[Z_2, Z_1] = P \exp \int_{Z_1}^{Z_2} A \tag{16}$$

Path ordering is required even classically because A takes values in the non-Abelian Lie algebra of $SL(2,\mathbb{C})$. The Wilson line $\mathcal{W}_A$ takes values in the representation of the group $SL(2,\mathbb{C})$ of spin j. Under a gauge transformation $U \in SL(2,\mathbb{C})$ with $A \to U^{-1}(d + A)U$ the Wilson line transforms as follows,

$$\mathcal{W}_A[Z_2, Z_1] \to U(Z_2)^{-1} \mathcal{W}_A[Z_2, Z_1] U(Z_1) \tag{17}$$

Since all r-dependence of A arises in (14) through a gauge transformation b of an r-independent gauge field a, we may extract all r-dependence of the Wilson line,

$$\mathcal{W}_A[Z_2, Z_1] = b(r)^{-1} \mathcal{W}_a[z_2, z_1] b(r) \qquad \mathcal{W}_a[z_2, z_1] = P \exp \int_{z_1}^{z_2} a \tag{18}$$

In particular, the r-dependence of the matrix element of $\mathcal{W}_A$ between highest and lowest weight states $|j, \pm j\rangle$ (which satisfy $b(r)|j, \pm j\rangle = r^{\pm j}|j, \pm j\rangle$) is given by

$$\langle j, -j|\mathcal{W}_A[Z_2, Z_1]|j, j\rangle = r^{-2h} \langle j, -j|\mathcal{W}_a[z_2, z_1]|j, j\rangle \tag{19}$$

which indicates that the classical Wilson line has dimension $h = -j$.

Having factored out the r-dependence, the central object of study is the remaining matrix element for given j which we shall denote by

$$W[z_2, z_1] = \langle j, -j|P \exp \int_{z_1}^{z_2} dz \left(L_1 + \frac{6}{c}T(z)L_{-1} \right)|j, j\rangle \tag{20}$$

The classical Wilson line W transforms under local conformal transformations as a bi-local primary field of dimension $h = -j$ at both points z_1 and z_2. To see this, we start with $T = 0$, in which case the path-ordered integral reduces to an ordinary integral and we have

$$W[z_2, z_1]\Big|_{T=0} = \langle j, -j|e^{(z_2-z_1)L_1}|j, j\rangle = (z_2 - z_1)^{-2h} \tag{21}$$

Under an arbitrary local conformal transformation $z \to f(z)$, the function T transforms to a non-zero value, given by the Schwarzian of f,

$$T_f(z) = \frac{c}{12} \left(\frac{f'''(z)}{f'(z)} - \frac{3}{2} \left(\frac{f''(z)}{f'(z)} \right)^2 \right) \tag{22}$$

as may be established by integrating the infinitesimal transformation law of (13). The Wilson line for this value of $T = T_f$ is found to be

$$W[z_2, z_1] \Big|_{T=T_f} = \frac{[f'(z_2)f'(z_1)]^h}{[f(z_2) - f(z_1)]^{2h}} \tag{23}$$

where we again have $h = -j$. Although the path-ordered integral depends on the values of $T(z)$ along the integration path from z_1 to z_2, the value of the above matrix element of the Wilson line depends only on the end-points, so that the Wilson line indeed behaves as a bi-local conformal primary field of dimension h at the points z_1 and z_2. Bi-locality may be traced back, of course, directly to the fact that the Chern–Simons field equations express the flatness of the connection A in the bulk of AdS₃, so that the value of the Wilson line between any two points is independent of the path between the points.

3.2. *The quantum Wilson line*

The classical Wilson line $W[z_2, z_1]$ defined in (20) may be promoted to a quantum operator in the boundary CFT by promoting T to the stress tensor operator of the CFT. This construction is formal as short-distance singularities in the OPE of stress tensor operators,

$$T(z)T(w) = \frac{c/2}{(z-w)^4} + \frac{2T(w)}{(z-w)^2} + \frac{\partial_w T(w)}{z-w} + \text{regular} \tag{24}$$

arise in the path-ordered exponential when two or more stress tensor operators collide. The presence of these short-distance singularities will require regularization and renormalization. For finite values of c the quantum Wilson line, if it can be suitably renormalized, will provide quantum observables of AdS₃ gravity at finite coupling, whence the importance of the construction of these operators. In the limit of large c, the correlators of the quantum Wilson line are expected to tends to the values predicted by the classical Wilson line.

The central question of the present investigations is whether a renormalized quantum Wilson line operator $W_R[z_2, z_1]$ can be constructed with the following properties.

(1) The renormalized Wilson line $W_R[z_2, z_1]$ is equivalent to a bi-local operator $\mathcal{O}(z_2)\mathcal{O}(z_1)$, in the sense that their correlators with an arbitrary number p of stress tensors coincide,

$$\langle T(w_1) \cdots T(w_p) W_R[z_2, z_1] \rangle = \langle T(w_1) \cdots T(w_p) \mathcal{O}(z_2)\mathcal{O}(z_1) \rangle \tag{25}$$

These identities mean that the Virasoro vacuum OPE block may be generated by the operator $W_R[z_2, z_1]$, so that the renormalized Wilson line operator $W_R[z_2, z_1]$ captures all terms in the OPE of $\mathcal{O}(z_2)\mathcal{O}(z_1)$ which involve only the stress tensor.

(2) The Wilson line W_R and the operator $\mathcal{O}$ have chiral conformal dimension h which coincides with the predictions from $SL(2, \mathbb{R})$ current algebra [35],

$$h(j) = -j + \frac{m+1}{m} j(j+1) \qquad\qquad c = 1 - \frac{6}{m(m+1)} \qquad (26)$$

If a renormalized quantum operator W_R with these properties can be constructed, then its vacuum expectation value must be given by

$$\langle W_R[z_2, z_1] \rangle = (z_2 - z_1)^{-2h(j)} \qquad (27)$$

while its correlator with one stress tensor must be

$$\langle T(w) W_R[z_2, z_1] \rangle = \frac{h(j)(z_2 - z_1)^2}{(z_2 - w)^2 (z_1 - w)^2} \langle W_R[z_2, z_1] \rangle \qquad (28)$$

We note that the exact order 2 of the poles in w at z_1 and z_2 of this correlator signifies that $T(w)$ suffers no discontinuity as it is moved across the line of integration from z_1 to z_2, thereby confirming the bi-local nature of the operator W_R. The remainder of this paper, starting in the subsequent section, will be devoted to developing evidence for the existence of a renormalized Wilson line operator W_R with these properties.

3.3. Wilson line as solution to Virasoro Ward identity

In this section we show how to arrive at the Wilson line from the perspective of Virasoro Ward identities, which are equivalent to imposing condition (1) in the previous subsection. We show that the unrenormalized Wilson line formally solves these Ward identities. The solution is formal because the unrenormalized Wilson line is singular as a quantum operator. However, UV divergences only set in at subleading order in $1/c$, and so the following argument will establish the validity of the Wilson line, in the sense of obeying conditions (1) and (2), at leading order in $1/c$.

We begin with a definition of Virasoro OPE blocks [34]. Let $\mathcal{O}_i(z)$ denote a Virasoro primary field of (chiral) scaling dimension h_i, where we suppress dependence on anti-holomorphic data. Given two such primary fields we can act on the CFT vacuum state to obtain the state $\mathcal{O}_1(z_1)\mathcal{O}_2(z_2)|0\rangle$. This state may be decomposed into irreducible representations of the Virasoro algebra, where as usual we are defining the Hilbert space in terms of radial quantization around the origin $z = 0$. Each such representation is labelled by a corresponding primary field $\mathcal{O}_p$. Letting P_p denote the projector onto the representation labelled by $\mathcal{O}_p$, we write the decomposition as

$$\mathcal{O}_1(z_1)\mathcal{O}_2(z_2)|0\rangle = \sum_p C_{12p} \frac{1}{C_{12p}} P_p \mathcal{O}_1(z_1)\mathcal{O}_2(z_2)|0\rangle \qquad (29)$$

where C_{12p} is the primary three-point coefficient, $\mathcal{O}_1(z)\mathcal{O}_2(0) \sim C_{12p}z^{h_p-h_1-h_2}\mathcal{O}_p(0) + \ldots$. Each term

$$\Psi_{12p}[z_2, z_1] = \frac{1}{C_{12p}} P_p \mathcal{O}_1(z_1)\mathcal{O}_2(z_2)|0\rangle \tag{30}$$

is referred to as an "OPE block", since it corresponds to using the OPE to express $\mathcal{O}_1(z_1)\mathcal{O}_2(z_2)$ in terms of local operators at the origin, and keeping only those operators in the conformal family of $\mathcal{O}_p$. Since we have pulled out C_{12p} from its definition, the OPE block is a universal object, completely fixed by Virasoro symmetry. Its precise form is most efficiently extracted as the solution to Ward identities, as we now discuss.

The CFT stress tensor has the mode expansion,

$$T(z) = \sum_{n=-\infty}^{\infty} l_n z^{-n+2} \tag{31}$$

where the l_n obey the Virasoro algebra,

$$[l_m, l_n] = (m - n)l_{m+n} + \frac{c}{12}(m^3 - m)\delta_{m,-n} \tag{32}$$

For a primary operator $\mathcal{O}(z)$ of dimension h we have

$$[l_n, \mathcal{O}(z)] = -\mathcal{L}_n^{(z)}\mathcal{O}(z) \tag{33}$$

where the differential operators,

$$\mathcal{L}_n^{(z)} = -z^{n+1}\partial_z - (n+1)hz^n \tag{34}$$

obey

$$[\mathcal{L}_m^{(z)}, \mathcal{L}_n^{(z)}] = (m - n)\mathcal{L}_{m+n}^{(z)} \tag{35}$$

Using $[l_n, P_p] = 0$, along with $l_n|0\rangle = 0$ for $n \geq -1$, and the trivial identity $[l_n, \mathcal{O}_1(z_1)\mathcal{O}_2(z_2)] = [l_n, \mathcal{O}_1(z_1)]\mathcal{O}_2(z_2) + \mathcal{O}_1(z_1)[l_n, \mathcal{O}_2(z_2)]$, we have

$$\left(l_n + \mathcal{L}_n^{(z_2)} + \mathcal{L}_n^{(z_1)}\right)\Psi_{12p}[z_2, z_1] = 0, \quad n \geq -1 \tag{36}$$

We refer to this system of equations as the Virasoro Ward identity.

One approach to determining $\Psi_{12p}[z_2, z_1]$ is to solve (36) order by order in the level expansion. In particular, take $z_1 = 0$, $z_2 = z$ and expand

$$\Psi_{12p}[z, 0] = z^{h_p-h_1-h_2}\left(1 + c_1 z l_{-1} + c_{11} z^2 l_{-1} l_{-1} + c_{22} z^2 l_{-2} + \ldots\right)\mathcal{O}_p(0)|0\rangle \tag{37}$$

The Ward identity turns into recursion relations for the expansion coefficients. An efficient method for solving these recursion relations is to use an oscillator representation of the l_n, as explained in [33].

We now focus on the "vacuum OPE block", corresponding to taking $\mathcal{O}_p$ to be the identity operator I. We further take $\mathcal{O}_1 = \mathcal{O}_2 = \mathcal{O}$, since $C_{12I} = 0$ unless $\mathcal{O}_1 = \mathcal{O}_2$. One way to think about the vacuum OPE block is that it captures all information about correlation functions of $\mathcal{O}_1(z_1)\mathcal{O}_2(z_2)$ with any number of stress tensor insertions,

$$\langle 0|T(w_1)\dots T(w_n)\mathcal{O}_1(z_1)\mathcal{O}_2(z_2)|0\rangle = \langle 0|T(w_1)\dots T(w_n)\Psi_{120}[z_2,z_1] \quad (38)$$

In the remainder of this section we establish that the unrenormalized Wilson line provides a formal solution to the Ward identity and hence can be identified with the vacuum OPE block,

$$\Psi_{120}[z_2,z_1] = W[z_2,z_1]|0\rangle \quad (39)$$

As already noted, the solution will be formal in the sense that we will ignore the existence of UV divergences, which is only correct at leading order in $1/c$. For the same reason, the conformal dimension appearing in the Ward identity will be $h = -j$, corresponding to the large-c limit of the general result in (26).

Our objective is to show

$$\left(l_n + \mathcal{L}_n^{(z_2)} + \mathcal{L}_n^{(z_1)}\right)\langle j,-j|\left[P\exp\int_{z_1}^{z_2}a\right]|j,j\rangle|0\rangle = 0 , \quad n \geq -1 \quad (40)$$

with

$$a(z) = L_1 + \frac{6}{c}T(z)L_{-1} \quad (41)$$

The general strategy is to use the relation between infinitesimal $SL(2,\mathbb{C})$ gauge transformations and conformal transformations. Under an infinitesimal $SL(2,\mathbb{C})$ gauge transformations we have $\delta_\lambda a = d\lambda + [\lambda,a]$. We consider

$$\lambda_n(z) = \epsilon_n(z)L_1 + \partial\epsilon_n(z)L_0 + \left(\frac{1}{2}\partial^2\epsilon_n(z) + \frac{6}{c}T(z)\epsilon_n(z)\right)L_{-1} \quad (42)$$

with

$$\epsilon_n(z) = z^{n+1} \quad (43)$$

Note that in (42) $T(z)$ is the stress tensor *operator*, so that $\lambda_n(y)$ defines an operator-valued gauge transformation. The form of $\lambda_n(z)$ is chosen so as to preserve the form of $a(z)$ in (41), with $T(z)$ transforming as a stress tensor under the infinitesimal conformal transformation $z \to z + \epsilon_n(z)$,

$$\delta_\lambda T = \epsilon_n\partial T + 2\partial\epsilon_n T + \frac{c}{12}\partial^3\epsilon_n \quad (44)$$

Since the Virasoro algebra implies $\delta_{\lambda_n}T(z) = [l_n,T(z)]$, it follows that

$$\delta_{\lambda_n}a(z) = [l_n,a(z)] . \quad (45)$$

The gauge covariance of the Wilson line (17) then yields the relation

$$\left[l_n,P\exp\int_{z_1}^{z_2}a\right] = \lambda(z_2)\left(P\exp\int_{z_1}^{z_2}a\right) - \left(P\exp\int_{z_1}^{z_2}a\right)\lambda(z_1) \quad (46)$$

Next, using $L_{-1}|j,j\rangle = \langle j,-j|L_{-1} = 0$, we have

$$\partial_{z_2}\langle j,-j|\left(P\exp\int_{z_1}^{z_2}a\right)|j,j\rangle = \langle j,-j|L_1\left(P\exp\int_{z_1}^{z_2}a\right)|j,j\rangle$$

$$\partial_{z_1}\langle j,-j| \left(P\exp \int_{z_1}^{z_2} a \right) |j,j\rangle = -\langle j,-j| \left(P\exp \int_{z_1}^{z_2} a \right) L_1|j,j\rangle . \qquad (47)$$

We then find

$$\mathcal{L}_n^{(z_2)}\langle j,-j| \left(P\exp \int_{z_1}^{z_2} a \right) |j,j\rangle = -\langle j,-j|\lambda_n(z_2) \left(P\exp \int_{z_1}^{z_2} a \right) |j,j\rangle$$

$$\mathcal{L}_n^{(z_1)}\langle j,-j| \left(P\exp \int_{z_1}^{z_2} a \right) |j,j\rangle = \langle j,-j| \left(P\exp \int_{z_1}^{z_2} a \right) \lambda_n(z_1)|j,j\rangle \qquad (48)$$

with $\mathcal{L}_n^{(z_1,2)}$ given by the differential operators in (34) with $h = -j$. Finally, combining (46) and (48) implies the desired Ward identity (40).

This analysis shows that the Wilson line correctly yields the vacuum Virasoro OPE block in the large-c limit. In evaluating correlators of the Wilson line with some number of external stress tensors, the leading large-c contribution will not involve any OPE singularities among stress tensors on the Wilson line, and hence no associated divergences, so we have shown that the Wilson line correctly reproduces these leading large-c contributions. On the other hand, at subleading orders in $1/c$ the Wilson line needs to be renormalized, and the preceding argument does not hold in the presence of a UV regulator. On general grounds, we expect that once the regulator is taken away, the renormalized Wilson line will obey the Ward identity but with the renormalized dimension (26); evidence for this claim appears in [22, 25–27, 30] as well as in the discussion that follows.

4. Renormalization of Wilson line correlators

In this section, we shall review the regularization and renormalization of the vacuum expectation value of the Wilson line operator in an expansion in powers of $1/c$. We shall summarize the results of Ref. [30] which prove that this correlator indeed obeys the form predicted for a bi-local operator of dimension $h(j)$ given in (27) to order $1/c^3$. We shall then present new results for the correlator of the Wilson line with one stress tensor insertion and show that it obeys (28) to order $1/c$. Finally, we present arguments that the equivalence of (25) holds to leading order in $1/c$ for an arbitrary number of stress tensor insertions.

4.1. *Regularization*

No regularization of short-distance singularities which preserves the full conformal symmetry in two-dimensional space-time is known to exist. However, dimensional regularization from two dimensions to $d = 2 - \varepsilon$ dimensions preserves dilation symmetry of stress tensor correlators in d dimensions for all values of d where the integrals of the individual Feynman graphs are absolutely convergent. For this reason, dimensional regularization and subsequent analytic continuation in ε appear better-suited for regularizing correlators in scale-invariant theories than other schemes.[b]

As should be expected in $d \neq 2$, the Ward identity (24), by which all stress tensor correlators can be computed on the two-dimensional plane, no longer holds and cannot be used to this end. To evaluate stress tensor correlators in the absence of the Ward identities a concrete quantum field theory model is needed which is valid in arbitrary dimension d. Such a model is provided by taking c integer and considering c scalar fields ϕ^γ with $\gamma = 1, \cdots, c$ in d space-time dimensions. After renormalization and continuation to two dimensions, it should be expected that all correlators become independent of the specific model used. The fact that c takes integer values is immaterial in the large-c expansion used here.

The free-field correlators in this model are standard. Parametrizing space-time $\mathbb{R}^d$ by coordinates $(z, \bar{z}, \mathbf{z})$ where $z, \bar{z}$ are the complex coordinates for $\mathbb{C}$ and $\mathbf{z} \in \mathbb{R}^{d-2}$, we readily evaluate the normalized two-point function of the field $\partial_z \phi^\gamma$,

$$\left\langle \partial_z \phi^\gamma(z) \partial_w \phi^{\gamma'}(w) \right\rangle = \frac{-\delta^{\gamma\gamma'}(\bar{z} - \bar{w})^2}{\left(|z - w|^2 + (\mathbf{z} - \mathbf{w})^2\right)^{\frac{d}{2}+1}} \tag{49}$$

The overall normalization may be absorbed by a renormalization function in the Wilson line and is immaterial, as we shall see below. For two points in the complex plane we have $\mathbf{z} = \mathbf{w} = 0$, and for two points on the real line the correlator in $d = 2 - \varepsilon$ dimensions simplifies to the following formula we shall use throughout,

$$\left\langle \partial_w \phi^\gamma(z) \partial_z \phi^{\gamma'}(w) \right\rangle = \frac{-\delta^{\gamma\gamma'}}{|z - w|^{2-\varepsilon}} \tag{50}$$

In this model, the holomorphic stress tensor $T(z)$ for $z \in \mathbb{C}$ is defined as the T_{zz} component of the d-dimensional traceless stress tensor for the free field ϕ^γ, which is given by

$$T(z) = -\frac{1}{2} \sum_{\gamma=1}^{c} \; : \partial_z \phi^\gamma(z) \partial_z \phi^\gamma(z) : \tag{51}$$

where the normal ordering symbol :: instructs us to omit all self-contractions.

4.2. Renormalization

Following the discussion of the preceding subsection, we shall adopt the model of c free scalar fields in dimension $d = 2 - \varepsilon$ as a systematic regularization, and evaluate all stress tensor correlators with the help of (49). We shall translate to $z_1 = 0$, rotate to $z_2 = z \in \mathbb{R}^+$, and choose the path of integration along $\mathbb{R}^+$. Additional stress tensors will be inserted at points $w_1, \cdots, w_p \in \mathbb{R}^-$ in (25), and subsequently analytically continued to the complex plane. Therefore, all correlators may effectively be evaluated by using (50).

---

[b]The use of short-distance regulators other than dimensional regularization was attempted in Ref. [22] with a dimensionful cut-off and in Sec. 6 of Ref. [30] with a strictly two-dimensional regularization of the operator. Both lead to inconsistency with the conformal Ward identities at sufficiently high order in $1/c$.

We know of no systematic procedure to renormalize a highly composite operator such as the Wilson line and to guarantee that its renormalized correlators satisfy the two-dimensional conformal Ward identities. The proposal made in Ref. [30] for the renormalized Wilson line is to include two renormalization functions: a function N multiplying the Wilson line operator, and a function α multiplying $1/c$,

$$W_R[z,0] = N\langle j, -j| P\exp\left\{\int_0^z dy\left(L_1 + \frac{6\alpha}{c}T(y)L_{-1}\right)\right\}|j,j\rangle \tag{52}$$

The functions N and α depend on the regulator ε as well as on c and j. Below we shall summarize the results of Ref. [30] which show that, in an expansion in powers of $1/c$, these two renormalization functions suffice to the vacuum expectation value $\langle W_R[z,0]\rangle$ to order $1/c^3$ included. Whether N and α suffice to renormalize W_R to all orders in $1/c$, or non-perturbatively in c remains an open question.

4.3. *Expansion in powers of* $1/c$

To carry out a systematic expansion in powers of $1/c$, it is useful to recast the Wilson line operator in the "interaction picture" by using the following identity,

$$P\exp\left\{\int_0^z dy\left(L_1 + \frac{6\alpha}{c}T(y)L_{-1}\right)\right\} = e^{zL_1}P\exp\left\{\frac{6\alpha}{c}\int_0^z dy\,X(y)T(y)\right\} \tag{53}$$

where $X(y)$ is given by

$$X(y) = e^{-yL_1}L_{-1}e^{yL_1} = L_{-1} - 2yL_0 + y^2L_1 \tag{54}$$

The Wilson line operator thus takes the form

$$W_R[z,0] = N\langle j, -j|e^{zL_1}P\exp\left\{\frac{6\alpha}{c}\int_0^z dy\,X(y)T(y)\right\}|j,j\rangle \tag{55}$$

The expansion in powers of $1/c$ is obtained by first expanding in powers of α/c, then constructing N and α in a power series in $1/c$, and finally extracting from this combination the systematic expansion in $1/c$. The expansion in α/c of a general correlator is given by

$$\langle T(w_1)\cdots T(w_p)W_R[0,z]\rangle = Nz^{2j-2p}\sum_{n=0}^{\infty}\frac{\alpha^n}{c^n}z^{(n+p)\varepsilon}T^pW_n \tag{56}$$

where the coefficients T^pW_n are given by

$$T^pW_n = \frac{6^n\,z^{2p}}{z^{(n+p)\varepsilon}}\int_0^z dy_n\cdots\int_0^{y_2} dy_1 F_n(z;y_n,\cdots,y_1)\langle T(w_1)\cdots T(w_p)T(y_1)\cdots T(y_n)\rangle \tag{57}$$

and the functions F_n are defined by

$$F_n(z;y_n,\cdots,y_1) = \langle j,-j|X(y_n)\cdots X(y_1)|j,j\rangle \tag{58}$$

The function $F_n(z;y_n,\cdots,y_1)$ is a homogeneous polynomial of degree n in its variables $z, y_n, \cdots, y_1$. The factor $z^{2p-(n+p)\varepsilon}$ is included in the coefficients T^pW_n to render it scale-invariant and thus independent of z when expressed in terms of the dimensionless variables $v_1 = w_1/z, \cdots, v_p = w_p/z$ and $x_1 = y_1/z, \cdots, x_n = y_n/z$, so that we have

$$T^p W_n = 6^n \int_0^1 dx_n \cdots \int_0^{x_2} dx_1 F_n(1; x_n, \cdots, x_1) \langle T(v_1) \cdots T(v_p) T(x_1) \cdots T(x_n) \rangle \quad (59)$$

The functions F_n obey a recursion relation, which allows for their ready evaluation [30].

The renormalization functions N and α may similarly be expanded in a power series in $1/c$,

$$N = 1 + \sum_{n=1}^{\infty} \frac{N_n}{c^n} \qquad\qquad \alpha = 1 + \sum_{n=1}^{\infty} \frac{\alpha_n}{c^n} \qquad (60)$$

where the coefficients N_n and α_n depend on j and have a Laurent expansion in powers of ε. They are expected to be independent of the external stress tensors, and thus independent of the number p. The most singular power of ε in N_n and α_n is expected to be $1/\varepsilon^n$.

4.4. *Regularized stress tensor correlators*

The stress tensor correlators are evaluated in the theory of c free scalar fields. We shall place the points $w_1, \cdots, w_p$ on $\mathbb{R}^-$ so that the stress tensor correlators may be computed with two-point function (50). The combinatorics of these correlators was given in detail in Ref. [30], and proceeds as follows. Evidently, we have $\langle T(y) \rangle = 0$. The Feynman diagrams for a correlator $\langle T(y_1) \cdots T(y_m) \rangle$ for $m \geq 2$ may be distinguished by the number of connected one-loop sub-graphs. Each sub-graph may be labelled by a *partition P into cycles* of the set of points $\{y_1, \cdots, y_m\}$, with each cycle containing at least two points. Two partitions are equivalent if they are related by cyclic permutations and/or reversal of orientation of the points in each cycle, and under permutations of the cycles. This partitioning of a Feynman graph into inequivalent cycles is unique.

We shall denote a cycle of ordered points $y_{i_1}, \cdots, y_{i_\mu}$ by a square bracket $[i_1, \cdots, i_\mu]$ and the value of the corresponding one-loop diagram along this cycle by

$$\langle T^2 \rangle_{[i_1, i_2]} = \frac{c/2}{|y_{i_1} - y_{i_2}|^{4-2\varepsilon}}$$

$$\langle T^\mu \rangle_{[i_1, \cdots, i_\mu]} = \frac{c}{|y_{i_1} - y_{i_2}|^{2-\varepsilon} |y_{i_2} - y_{i_3}|^{2-\varepsilon} \cdots |y_{i_\mu} - y_{i_1}|^{2-\varepsilon}}, \qquad \mu \geq 3 \ (61)$$

The correlator is given by a sum over all possible inequivalent partitions $P = C_1 \cup C_2 \cup \cdots \cup C_\nu$ into ν cycles, with $C_s \cap C_{s'} = \emptyset$ for $s' \neq s$, of the set $\{y_1, \cdots, y_m\}$,

$$\langle T(y_1) \cdots T(y_m) \rangle = \sum_P \langle T^m \rangle_P , \qquad\qquad \langle T^m \rangle_P = \prod_{s=1}^{\nu} \langle T^{\mu_s} \rangle_{C_s} \quad (62)$$

For example, when $m = 4$ six partitions contribute,

$$\langle T(y_1) \cdots T(y_4) \rangle = \langle T^4 \rangle_{[12][34]} + \langle T^4 \rangle_{[13][24]} + \langle T^4 \rangle_{[14][23]}$$
$$+ \langle T^4 \rangle_{[1234]} + \langle T^4 \rangle_{[1342]} + \langle T^4 \rangle_{[1324]} \tag{63}$$

When additional stress tensors $T(w_1) \cdots T(w_p)$ are present, the above combinatorics should be applied to the set of all $m = n + p$ stress tensors.

4.5. *Renormalization of* $\langle W_R[z,0] \rangle$

It was shown in Ref. [30] that the functions N and α suffice to renormalize $\langle W_R[z,0] \rangle$, produce the scaling in z of (27), with chiral dimension of (26) to order $1/c^3$,

$$h(j) = -j - \frac{6}{c} j(j+1) - \frac{78}{c^2} j(j+1) - \frac{1230}{c^3} j(j+1) + \mathcal{O}(c^{-4}) \tag{64}$$

provided the renormalization function α is chosen as follows,

$$\alpha = 1 + \frac{1}{c} \left(\frac{6}{\varepsilon} + 3 \right) + \frac{1}{c^2} \left(\frac{30}{\varepsilon^2} + \frac{55}{\varepsilon} + \frac{185}{3} - \frac{16\pi^2}{5} (3j(j+1) - 1) \right) + \mathcal{O}(c^{-3})(65)$$

Terms of order $1/c^3$ are not needed to evaluate $h(j)$ to the order given in (64). The expansion of the function N to order $1/c^3$ included is required to evaluate (64) but is significantly more involved than α, and we shall give its expression to order $1/c^2$ only,

$$N = 1 - \frac{6J^2}{c\varepsilon} - \frac{1}{c}(10J^2 - 6j) + \frac{6J^2}{c^2 \varepsilon^2}(3J^2 - 2) + \frac{3J^2}{c^2 \varepsilon}(20J^2 - 16j - 17)$$
$$+ \frac{1}{c^2} \left((50j^2 + 2j - 160)J^2 + 42j + 8\pi^2 J^2 \right) + \mathcal{O}(c^{-3}) \tag{66}$$

where $J^2 = j(j+1)$ is the quadratic Casimir value in the representation with spin j.

5. Renormalization and bi-locality of $\langle T(w) W_R[z,0] \rangle$

If a renormalized Wilson line operator can be constructed which satisfies the criteria of bi-locality and conformal behavior spelled out in Sec. 3.2, then its correlator with one stress tensor insertion must behave according to (28),

$$\langle T(w) W_R[z,0] \rangle = \frac{h(j) z^2}{w^2 (w-z)^2} \langle W_R[z,0] \rangle \tag{67}$$

In the absence of complete arguments for the existence of such an operator we construct its correlators in a power expansion in $1/c$, and verify that they satisfy the criteria spelled out in Sec. 3.2. In this section, we shall do so for the correlator $\langle T(w) W_R[z,0] \rangle$ to order $1/c$, which is the lowest order at which non-trivial renormalization is required.

5.1. Expansion in $1/c$

To compute $\langle T(w)W_R[z,0]\rangle$ we use the formalism of Sec. 4.3 for $p=1$ which gives

$$\langle T(w)W_R[z,0]\rangle = N z^{2j-2} \sum_{n=0}^{\infty} \frac{\alpha^n}{c^n} z^{(n+1)\varepsilon} TW_n \tag{68}$$

where the coefficients TW_n are given in terms of the scaling variable $v = w/z \in \mathbb{R}^-$ by

$$TW_n = 6^n \int_0^1 dx_n \cdots \int_0^{x_2} dx_1 F_n(1;x_n,\cdots,x_1)\langle T(v)T(x_1)\cdots T(x_n)\rangle \tag{69}$$

The renormalization functions N and α are identical to those derived in (66) and (65) for the renormalization of $\langle W_R[z,0]\rangle$ which, to this order, are given by their contributions to order $1/c$ included. We shall study the correlator to order $1/c$ and thus retain contributions from $n=1,2,3$ (since $\langle T(v)\rangle = 0$), which requires the following stress tensor correlators,

$$\langle T(v)T(x_1)\rangle = \frac{c/2}{(x_1-v)^{4-2\varepsilon}}$$

$$\langle T(v)T(x_1)T(x_2)\rangle = \frac{c}{(x_1-v)^{2-\varepsilon}(x_2-v)^{2-\varepsilon}(x_2-x_1)^{2-\varepsilon}}$$

$$\langle T(v)T(x_1)T(x_2)T(x_3)\rangle = \frac{c^2/4}{(x_1-v)^{4-2\varepsilon}(x_3-x_2)^{4-2\varepsilon}} + 2 \text{ perms} + \mathcal{O}(c) \tag{70}$$

The $\mathcal{O}(c)$ terms in the four-T correlator will not contribute to order $1/c$. The contributions from $n=1,2$ are given by the integrals,

$$TW_1 = 3 \int_0^1 dx_1 \frac{F_1(1;x_1)}{(x_1-v)^{4-2\varepsilon}}$$

$$TW_2 = \frac{36}{c} \int_0^1 dx_2 \int_0^{x_2} dx_1 \frac{F_2(1;x_2,x_1)}{(x_1-v)^{2-\varepsilon}(x_2-v)^{2-\varepsilon}(x_2-x_1)^{2-\varepsilon}} \tag{71}$$

The contribution from $n=3$ is the sum of three terms given by the three terms in the corresponding four-T correlator of (81),

$$TW_3 = TW_3^{(1)} + TW_3^{(2)} + TW_3^{(3)} \tag{72}$$

with each term given by

$$TW_1^{(1)} = \frac{54}{c} \int_0^1 dx_3 \int_0^{x_3} dx_2 \int_0^{x_2} dx_1 \frac{F_3(1;x_3,x_2,x_1)}{(x_1-v)^{4-2\varepsilon}(x_3-x_2)^{4-2\varepsilon}}$$

$$TW_2^{(2)} = \frac{54}{c} \int_0^1 dx_3 \int_0^{x_3} dx_2 \int_0^{x_2} dx_1 \frac{F_3(1;x_3,x_2,x_1)}{(x_2-v)^{4-2\varepsilon}(x_3-x_1)^{4-2\varepsilon}}$$

$$TW_3^{(3)} = \frac{54}{c} \int_0^1 dx_3 \int_0^{x_3} dx_2 \int_0^{x_2} dx_1 \frac{F_3(1;x_3,x_2,x_1)}{(x_3-v)^{4-2\varepsilon}(x_2-x_1)^{4-2\varepsilon}} \tag{73}$$

The functions F_n are given by

$$F_1(1; x_1) = 2jx_1(x_1 - 1) \tag{74}$$

$$F_2(1; x_2, x_1) = 2jx_1(x_2 - 1)\big((2j-1)x_1x_2 - 2jx_2 + x_1\big)$$

$$F_3(1; x_3, x_2, x_1) = 4jx_1(x_3 - 1)\Big((2j^2 - 3j + 1)x_1x_2^2x_3 - (2j^2 - j)x_2^2x_3 + (2j-1)x_1x_2^2$$

$$-jx_2^2 - (2j^2 - 3j + 1)x_1x_2x_3 + 2j^2x_2x_3 - (j-1)x_1x_2 - jx_1x_3\Big)$$

To evaluate the contributions of order c^0 and c^{-1} to $\langle T(w)W_R[z,0]\rangle$, we need TW_1 to orders ε^0 and ε^1 since the renormalization $N\alpha$, which multiplies this contribution, will pick up order ε^0 terms that are of order $1/c$. The remaining terms are needed to orders ε^{-1} and ε^0.

5.2. Calculation of the coefficients TW_n

The integral for TW_1 is straightforward and we find

$$TW_1 = \frac{-j}{v^2(v-1)^2} - 2j\varepsilon \left(\frac{1+2v}{v^2}\ln(-v) + \frac{3-2v}{(v-1)^2}\ln(1-v)\right.$$

$$\left. + \frac{2}{v(1-v)} + \frac{5}{6v^2(v-1)^2}\right) + \mathcal{O}(\varepsilon^2) \tag{75}$$

The expression is symmetric under $v \to 1 - v$.

To evaluate TW_2 we set $x_2 = x$ and change variables from x_1 to $x_1 = x - u$ with $0 \le u \le x$,

$$TW_2 = \frac{36}{c}\int_0^1 dx \int_0^x du \frac{F_2(1; x, x-u)}{(x-u-v)^{2-\varepsilon}(x-v)^{2-\varepsilon}u^{2-\varepsilon}} \tag{76}$$

Next, we define the function

$$R_\varepsilon = \frac{1}{(x-u-v)^{2-\varepsilon}} - \frac{1}{(x-v)^{2-\varepsilon}} - \frac{(2-\varepsilon)u}{(x-v)^{3-\varepsilon}} \tag{77}$$

which is constructed so as to vanish to second order in u at $u = 0$ for all values of ε. In terms of R_ε, we may recast TW_2 as follows,

$$TW_2 = \frac{36}{c}\int_0^1 dx \int_0^x du \frac{F_2(1; x, x-u)}{(x-v)^{2-\varepsilon}u^{2-\varepsilon}}\left(\frac{1}{(x-v)^{2-\varepsilon}} + \frac{(2-\varepsilon)u}{(x-v)^{3-\varepsilon}} + R_\varepsilon\right) \tag{78}$$

Since R_ε vanishes as u^2, the integral over R_ε converges for $\varepsilon = 0$, so that we have

$$TW_2 = \frac{36}{c}\int_0^1 dx \int_0^x du \frac{F_2(1; x, x-u)}{(x-v)^{5-2\varepsilon}u^{2-\varepsilon}}(x-v+(2-\varepsilon)u)$$

$$+ \frac{36}{c}\int_0^1 dx \int_0^x du \frac{F_2(1; x, x-u)}{(x-v)^2u^2}R_0 + \mathcal{O}(\varepsilon) \tag{79}$$

Since F_2 is polynomial in its arguments, each integral is elementary and may now be performed using MAPLE: the result is quite involved and will not be exhibited here.

To evaluate $TW_3^{(i)}$ for $i = 1, 2, 3$, we rearrange the integrations so as to carry out the integrals over the corresponding x_i last. To this end, we relabel $(x_1, x_2, x_3) \to (x_3, x_2, x_1)$ in $TW_3^{(1)}$, and $(x_1, x_2, x_3) \to (x_1, x_3, x_2)$ in $TW_2^{(2)}$, leaving the last integral unchanged. Next, we also change variables from x_1 to $u = x_2 - x_1$, so that we find

$$TW_1^{(1)} = \frac{54}{c} \int_0^1 dx_3 \int_{x_3}^1 dx_2 \int_0^{x_2 - x_3} du \, \frac{F_3(1; x_2, x_2 - u, x_3)}{(x_3 - v)^{4 - 2\varepsilon} u^{4 - 2\varepsilon}}$$

$$TW_2^{(2)} = \frac{54}{c} \int_0^1 dx_3 \int_{x_3}^1 dx_2 \int_{x_3 - x_2}^{x_2} du \, \frac{F_3(1; x_2, x_3, x_2 - u)}{(x_3 - v)^{4 - 2\varepsilon} u^{4 - 2\varepsilon}}$$

$$TW_3^{(3)} = \frac{54}{c} \int_0^1 dx_3 \int_0^{x_3} dx_2 \int_0^{x_2} du \, \frac{F_3(1; x_3, x_2, x_2 - u)}{(x_3 - v)^{4 - 2\varepsilon} u^{4 - 2\varepsilon}} \qquad (80)$$

Interchanging the integrations over x_2 and u reveals the convenient fact that the integrals over x_2 are of a polynomial in x_2 and may be performed easily, whereafter the integrals over u and x_3 may be performed in MAPLE by elementary methods as well. Again, the results are quite involved and will not be presented individually here.

Assembling all contributions, inserting the renormalization functions N and α, and using MAPLE to simplify the final results, we find that all logarithmic contributions $\ln(-v) = \ln(-w/z)$ and $\ln(1 - v) = \ln(1 - w/z)$ cancel exactly to this order, and the resulting correlator is governed by the relation (67) expected under the assumption of conformal invariance with the operator dimension $h(j)$ given by (64) to order $\mathcal{O}(1/c)$ included.

6. Higher order correlators $\langle T(w_1) \cdots T(w_p) W_R[z, 0] \rangle$

Assuming the existence of a bi-local conformal primary renormalized Wilson line operator $W_R[z, 0]$, its correlator with p stress tensor insertions can be derived completely from the conformal Ward identities. In turn, showing that such correlators obey the relations expected from conformal invariance and bi-locality will provide further evidence for the existence of a renormalized Wilson line operator satisfying the criteria of Sec. 3.2.

In this last section, we shall show that even to leading order in $1/c$, where no renormalization effects are required, individual contributions to the correlator do not respect conformal properties, and in particular exhibit logarithmic singularities when z approaches one of the insertion points $w_1, \cdots, w_p$. For the correlator with $p = 1$, this issue did not arise as the only contribution was given by the coefficient TW_1 whose conformal behavior is manifest. We shall begin by examining the simplest case, namely for $p = 2$, and show by explicit calculation that the combined contributions to the correlator $\langle T(w_1) T(w_2) W_R[z, 0] \rangle$ see all their logarithmic contributions at $z \sim w_1, w_2$ cancelled to order c^0. We shall then present general arguments why this result extends to all values of p at order c^0.

6.1. *The correlator* $\langle T(w_1)T(w_2)W_R[z,0]\rangle$

The correlator is given by (56) and (59) for the case $p = 2$. To order c^0, no short-distance singularities appear, we may set $N = \alpha = 1$, and $\varepsilon = 0$. The contribution from $n = 0$ is given by a disconnected correlator which is of order c and trivial. The remaining connected contributions of order c^0 are given by $n = 1, 2$. Working with rescaled variables $v = w/z, x = y/z$, we need the three-T stress tensor correlators,

$$\langle T(v_1)T(v_2)T(x_1)\rangle = \frac{c}{(x_1 - v_1)^2(x_1 - v_2)^2(v_2 - v_1)^2} \tag{81}$$

and the four-T connected correlators,

$$\langle T(v_1)T(v_2)T(x_1)T(x_2)\rangle_c = \frac{c^2/4}{(x_1 - v_1)^4(x_2 - v_2)^4} + \frac{c^2/4}{(x_1 - v_2)^4(x_2 - v_1)^4} \tag{82}$$

The corresponding contributions T^2W_1 and T^2W_2, to order c^0, are given by

$$T^2W_1 = \int_0^1 dx_1 \frac{6F_1(1;x_1)}{(x_1 - v_1)^2(x_1 - v_2)^2(v_2 - v_1)^2}$$

$$T^2W_2 = \int_0^1 dx_2 \int_0^{x_2} dx_1 \left(\frac{9F_2(1;x_2,x_1)}{(x_1 - v_1)^4(x_2 - v_2)^4} + \frac{9F_2(1;x_2,x_1)}{(x_1 - v_2)^4(x_2 - v_1)^4} \right) \tag{83}$$

Evaluation of the integrals is elementary, and we find,

$$T^2W_1 = \frac{24j}{(v_1 - v_2)^4} + \frac{12j(2v_1v_2 - v_1 - v_2)}{(v_1 - v_2)^5} \ln\left(\frac{v_1(1 - v_2)}{v_2(1 - v_1)} \right) \tag{84}$$

The contribution T^2W_2 similarly has logarithmic singularities which, however, all cancel in the sum with T^2W_1, and we find

$$T^2W_1 + T^2W_2 = \frac{j^2}{v_1^2(1 - v_1)^2v_2^2(1 - v_2)^2} + \frac{j}{v_1(1 - v_1)v_2(1 - v_2)(v_1 - v_2)^2} \tag{85}$$

Assembling all contributions and reverting to the variables w and z, we find

$$\langle T(w_1)T(w_2)W_R[z,0]\rangle = \frac{j^2z^{2j+4}}{w_1^2(z - w_1)^2w_2^2(z\ w_2)^2} + \frac{jz^{2j+2}}{w_1(z - w_1)w_2(z - w_2)(w_1 - w_2)^2} \tag{86}$$

in agreement with the predictions from conformal Ward identities.

6.2. *The correlator* $\langle T(w_1)\ldots T(w_n)W_R[z,0]\rangle$ *at leading order in* $1/c$

In Sec. 3.3 we derived the Ward identity

$$\left(l_n + \mathcal{L}_n^{(z_2)} + \mathcal{L}_n^{(z_1)} \right) W[z_2, z_1]|0\rangle = 0, \quad n \geq -1 \tag{87}$$

at leading order in the $1/c$ expansion where renormalization is not necessary. We conclude by showing how the Ward identity fixes correlators of the Wilson line with any number of external stress tensor insertions. We proceed recursively. Given $\langle T(w_1)\ldots T(w_n)W[z_2, z_1]\rangle$ we wish to insert an additional stress tensor. We work in the operator formulation. Using the general structure of the Virasoro algebra we have $[l_n, T] \sim T + I$, so we can write

$$
\begin{aligned}
\langle T(u)T(w_1)\dots T(w_n)W[z_2, z_1]\rangle &= \sum_{m=2}^{\infty} \frac{1}{u^{m+2}} \langle l_m T(w_1)\dots T(w_n)W[z_2, z_1]\rangle \\
&= \sum_{m=2}^{\infty} \frac{1}{u^{m+2}} \langle T(w_1)\dots T(w_n)l_m W[z_2, z_1]\rangle + \text{(fewer)} \\
&= -\sum_{m=2}^{\infty} \frac{1}{u^{m+2}} \left(\mathcal{L}_m^{(z_2)} + \mathcal{L}_m^{(z_1)} \right) \langle T(w_1)\dots T(w_n)W[z_2, z_1]\rangle \\
&\quad + \text{(fewer)}
\end{aligned}
$$

(88)

where "fewer" stands for correlators with n or $n-1$ stress tensor insertions, which are assumed to be known. These recursion relations determine all correlators of the Wilson line with stress tensors in terms of $\langle W[z_2, z_1]\rangle$.

Our expectation is that the quantum Wilson line, renormalized in the way we have discussed, will continue to obey (87) but with the renormalized value of h appearing in $\mathcal{L}_n$. If so, this will fix all correlators with any number of stress tensor insertions in terms of $\langle W_R[z_2, z_1]\rangle$.

Acknowledgments

We thank Mert Besken and Ashwin Hegde for collaboration on the work reviewed here. Research supported in part by the National Science Foundation under grant PHY-19-14412.

References

1. J. S. Bell and R. Jackiw, "A PCAC puzzle: $\pi^0 \to \gamma\gamma$ in the σ model," *Nuovo Cim. A* **60**, 47 (1969).
2. S. L. Adler, "Axial vector vertex in spinor electrodynamics," *Phys. Rev.* **177**, 2426 (1969).
3. D. J. Gross and R. Jackiw, "Effect of anomalies on quasi-renormalizable theories," *Phys. Rev. D* **6**, 477 (1972).
4. C. Bouchiat, J. Iliopoulos and P. Meyer, "An anomaly free version of Weinberg's model," *Phys. Lett.* **38B**, 519 (1972).
5. G. 't Hooft, Cargese Summer Institute Lecture Notes (1979), in *Dynamical Gauge Symmetry Breaking*, E. Farhi and R. Jackiw, eds. (World Scientific, Singapore, 1982).
6. L. Alvarez-Gaume and E. Witten, "Gravitational anomalies," *Nucl. Phys. B* **234**, 269 (1984).
7. M. B. Green and J. H. Schwarz, "Anomaly cancellation in supersymmetric $D = 10$ gauge theory and superstring theory," *Phys. Lett.* **149B**, 117 (1984).
8. C. G. Callan, Jr., S. R. Coleman and R. Jackiw, "A new improved energy-momentum tensor," *Annals Phys.* **59**, 42 (1970).
9. S. Deser, R. Jackiw and S. Templeton, "Topologically massive gauge theories," *Annals Phys.* **140**, 372 (1982) [Erratum: *ibid.* **185**, 406 (1988)].

10. R. Jackiw and S. Y. Pi, "Classical and quantal nonrelativistic Chern–Simons theory," *Phys. Rev. D* **42**, 3500 (1990) [Erratum: *ibid.* **48**, 3929 (1993)].

11. S. Deser and R. Jackiw, "Three-dimensional cosmological gravity: Dynamics of constant curvature," *Annals Phys.* **153**, 405 (1984).

12. E. D'Hoker and R. Jackiw, "Liouville field theory," *Phys. Rev. D* **26**, 3517 (1982);
E. D'Hoker and R. Jackiw, "Space translation breaking and compactification in the Liouville theory," *Phys. Rev. Lett.* **50**, 1719 (1983);
E. D'Hoker, D. Z. Freedman and R. Jackiw, "SO(2,1) invariant quantization of the Liouville theory," *Phys. Rev. D* **28**, 2583 (1983).

13. R. Jackiw, "Lower dimensional gravity," *Nucl. Phys. B* **252**, 343 (1985).

14. C. Teitelboim, "Gravitation and Hamiltonian structure in two space-time dimensions," *Phys. Lett.* **126B**, 41 (1983).

15. M. Banados, C. Teitelboim and J. Zanelli, "The black hole in three-dimensional space-time," *Phys. Rev. Lett.* **69**, 1849 (1992)

16. A. Achucarro and P. K. Townsend, "A Chern–Simons action for three-dimensional anti-de Sitter supergravity theories," *Phys. Lett. B* **180**, 89 (1986).

17. E. Witten, "(2+1)-dimensional gravity as an exactly soluble system," *Nucl. Phys. B* **311**, 46 (1988).

18. J. D. Brown and M. Henneaux, "Central charges in the canonical realization of asymptotic symmetries: An example from three-dimensional gravity," *Commun. Math. Phys.* **104**, 207 (1986).

19. L. Eberhardt, M. R. Gaberdiel and R. Gopakumar, "Deriving the AdS$_3$/CFT$_2$ correspondence," arXiv:1911.00378 [hep-th].

20. J. de Boer and J. I. Jottar, "Thermodynamics of higher spin black holes in AdS$_3$," *JHEP* **1401**, 023 (2014) [arXiv:1302.0816 [hep-th]].

21. M. Ammon, A. Castro and N. Iqbal, "Wilson lines and entanglement entropy in higher spin gravity," *JHEP* **1310**, 110 (2013) [arXiv:1306.4338 [hep-th]].

22. M. Besken, A. Hegde and P. Kraus, "Anomalous dimensions from quantum Wilson lines," arXiv:1702.06640 [hep-th].

23. A. Bhatta, P. Raman and N. V. Suryanarayana, *JHEP* **1606**, 119 (2016) [arXiv:1602.02962 [hep-th]].

24. M. Besken, A. Hegde, E. Hijano and P. Kraus, "Holographic conformal blocks from interacting Wilson lines," *JHEP* **1608**, 099 (2016) [arXiv:1603.07317 [hep-th]].

25. A. L. Fitzpatrick, J. Kaplan, D. Li and J. Wang, "Exact Virasoro blocks from Wilson lines and background-independent operators," *JHEP* **1707**, 092 (2017) [arXiv:1612.06385 [hep-th]].

26. Y. Hikida and T. Uetoko, "Conformal blocks from Wilson lines with loop corrections," *Phys. Rev. D* **97**, 086014 (2018) [arXiv:1801.08549 [hep-th]].

27. Y. Hikida and T. Uetoko, "Correlators in higher-spin AdS$_3$ holography from Wilson lines with loop corrections," *PTEP* **2017**, 113B03 (2017) [arXiv:1708.08657 [hep-th]].

28. N. Anand, H. Chen, A. L. Fitzpatrick, J. Kaplan and D. Li, "An exact operator that knows its location," *JHEP* **1802**, 012 (2018) [arXiv:1708.04246 [hep-th]].

29. P. Kraus, A. Sivaramakrishnan and R. Snively, "Late time Wilson lines," *JHEP* **1904**, 026 (2019) [arXiv:1810.01439 [hep-th]].

30. M. Besken, E. D'Hoker, A. Hegde and P. Kraus, "Renormalization of gravitational Wilson lines," *JHEP* **1906**, 020 (2019) [arXiv:1810.00766 [hep-th]].

31. H. L. Verlinde, "Conformal field theory, 2-D quantum gravity and quantization of Teichmuller space," *Nucl. Phys. B* **337**, 652 (1990).

32. P. Kraus, "Lectures on black holes and the AdS(3) / CFT(2) correspondence," *Lect. Notes Phys.* **755**, 193 (2008) [hep-th/0609074].

33. M. Beşken, S. Datta and P. Kraus, "Quantum thermalization and Virasoro symmetry," arXiv:1907.06661 [hep-th].

34. B. Czech, L. Lamprou, S. McCandlish, B. Mosk and J. Sully, "A stereoscopic look into the bulk," *JHEP* **1607**, 129 (2016) [arXiv:1604.03110 [hep-th]].

35. M. Bershadsky and H. Ooguri, "Hidden SL(n) symmetry in conformal field theories," *Commun. Math. Phys.* **126**, 49 (1989).

36. A. Campoleoni, S. Fredenhagen, S. Pfenninger and S. Theisen, "Asymptotic symmetries of three-dimensional gravity coupled to higher-spin fields," *JHEP* **1011**, 007 (2010) [arXiv:1008.4744 [hep-th]].

Chapter 15

Bions and Instantons in Triple-Well and Multi-Well Potentials

Gerald V. Dunne, Tin Sulejmanpasic and Mithat Ünsal

*Department of Physics, University of Connecticut,
Storrs, CT 06269-3046, USA
Department of Mathematical Sciences, Durham University,
Durham, DH1 3LE, United Kingdom
Department of Physics, North Carolina State University,
Raleigh, NC 27695, USA*

Quantum systems with multiple degenerate classical harmonic minima exhibit new non-perturbative phenomena which are not present for the double-well and periodic potentials. The simplest characteristic example of this family is the triple-well potential. Despite the fact that instantons are exact semiclassical solutions with finite and minimal action, they do not contribute to the energy spectrum at leading order in the semiclassical analysis. This is because the instanton fluctuation prefactor vanishes, which can be interpreted as the action becoming infinite quantum mechanically. Instead, the non-perturbative physics is governed by different types of *bion* configurations. A generalization to supersymmetric and quasi-exactly soluble models is also discussed. An interesting pattern of interference between topological and neutral bions, depending on the hidden topological angle, the discrete θ angle and the perturbative level number, leads to an intricate pattern of divergent/convergent expansions for low-lying states, and provides criteria for the exact solvability of some of the states. We confirm these semiclassical bion predictions using the BenderWu Mathematica package to study the structure of the associated perturbative expansions. It also turns out that all the systems we study have a curious exact one-to-one relationship between the perturbative coefficients of the three wells, which we check using the BenderWu package.

Dedicated to Roman Jackiw on the occasion of his 80th birthday

1. Introduction

Much of our physical intuition about instantons is derived from quantum mechanical instanton examples such as the symmetric double-well or periodic cosine potential, for which the instanton analysis is standard textbook material.[1–10] The subject of instantons is one in which Roman Jackiw has made many important contributions, both original discoveries,[11–13] and also crystal clear pedagogical expositions that get to the heart of the physics.[14–16] We dedicate this paper to Roman, in recognition of his profound influence on the many facets of non-perturbative physics. In this paper we describe several new and intriguing aspects of instantons, critical points at infinity and non-trivial configurations that live on their thimbles, known as bion configurations, that arise in multi-well potentials with degenerate minima. The simplest example that captures the new non-trivial aspects of such systems is the triple-well potential.

It is first useful to recall the salient features of the symmetric double-well potential, both the bosonic system and its supersymmetric (SUSY) extension, to exhibit the sharp contrast with triple-well and multi-well potentials. Interestingly, many things that we learn in textbooks concerning the degenerate double-well system are not general rules, but exceptions that arise in the limiting case of identical neighboring wells, i.e. neighboring wells which are related by a symmetry.

The symmetric double-well potential can be expressed as

$$V = \frac{\omega^2}{2}x^2(x-1)^2 \tag{1}$$

Familiar non-perturbative features of this symmetric double-well system include:

(1) In the semiclassical limit of deep wells, the low-lying states are split into doublets with exponentially small energy splitting, $\Delta E \sim e^{-S_I}$, associated non-perturbatively with the existence of instanton solutions that tunnel between the two minima. These instanton configurations have action S_I.

(2) Perturbation theory in each well is asymptotic, with expansion coefficients that are factorially divergent and non-alternating in sign: $c_n \sim \frac{n!}{(2S_I)^n}$. This behavior is associated with the existence of non-perturbative neutral bion configurations, correlated instanton/anti-instanton configurations, which live on the thimble of critical points at infinity.[17]

(3) In the SUSY extension of this model, the ground state energy vanishes to all orders perturbatively, but SUSY is broken non-perturbatively: $E_0^{NP} \sim -e^{-2S_I+i\pi}$. This behavior is associated with a non-perturbative neutral bion solution to the second-order equations of motion associated with the classical plus quantum potential, $\frac{1}{2}(W')^2 \pm \frac{\hbar}{2}W''$, having action $2S_I$ and hidden topological angle $\theta_{\mathrm{HTA}} = \pi$. In the original instanton language, this can be viewed as a configuration on the Lefschetz thimble of an instanton/anti-instanton critical point at infinity. The quasi-zero-mode thimble integral lives in the complex plane, which is responsible for the $\theta_{\mathrm{HTA}} = \pi$, and which is the semiclassical origin of the positivity of the non-perturbative energy shift.

Many of the characteristic features of multiple-well systems appear already for the triple-well potential. For comparison purposes we concentrate on the symmetric triple-well potential (see Fig. 1):

$$V = \frac{\omega^2}{2}x^2(x^2-1)^2 \tag{2}$$

The most interesting non-perturbative features of this symmetric triple-well system are:

(1) There are exact instanton solutions tunneling between neighboring minima, with finite action S_I, but they play no role for the leading non-perturbative effects! This is due to the vanishing of the prefactor coming from the instanton fluctuation determinant.

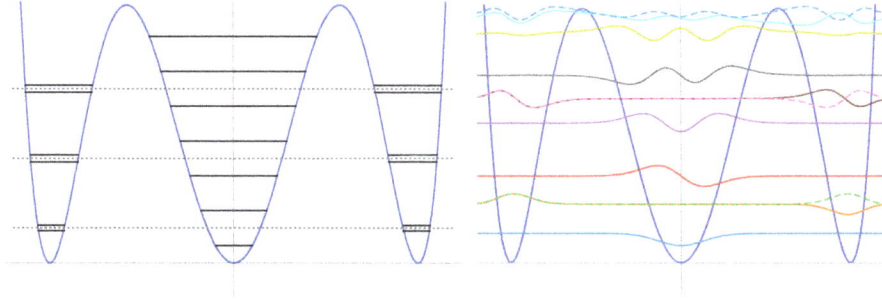

Fig. 1. Energy levels in the symmetric triple-well potential (2). The outer wells have classical frequency $\omega_{\text{outer}} = 2\omega$, while the inner well has frequency $\omega_{\text{inner}} = \omega$. The perturbative energy levels for states localized in the outer wells are split into doublets by tunneling, while those for states localized in the inner well are not split into doublets. Note that the unperturbed harmonic levels for states localized in the inner and outer wells are interlaced in a systematic way. The right hand side shows the numerical solutions of the low-energy wavefunctions. Note that for nearly degenerate wavefunctions we have represented the one with a higher energy with a dashed line. Notice also that for the lowest doublet the even, rather than the odd, eigenfunction has higher energy.

(2) The low-lying energy levels for states localized in the outer wells are exponentially split by a *two-instanton* effect rather than a one-instanton effect, $\Delta E_{\text{outer}} \sim e^{-2S_I}$, while the energy levels for states localized in the inner well are not split at all, $\Delta E_{\text{inner}} = 0$. Rather, they are shifted up or down by the neutral bion contribution, $E_{\text{inner}}^{\text{shift}} \sim e^{-2S_I}$. In the first doublet, the lower state has an *antisymmetric* wavefunction, in contrast to the situation in the double-well potential where the lower state wavefunction is symmetric. In fact, the pattern of wavefunction symmetries is quite different, since the pattern must respect the oscillation theorem (increase of the number of nodes with energy) and the parity symmetry properties of the wavefunctions. See Fig. 1.

(3) There are two identical barriers, but two different types of wells: inner and outer. See Fig. 1. Nevertheless, the perturbative expansions in the inner and outer wells are explicitly related: see Eq. (20) below.

(4) Perturbation theory in each kind of well is asymptotic, with expansion coefficients that are factorially divergent and non-alternating in sign: $c_n \sim \frac{n!}{(2S_I)^n}$. This behavior is associated with the existence of non-perturbative neutral bion configurations with zero topological charge but action equal to twice the in-

stanton action. These can be interpreted as configurations on the thimble of an instanton/anti-instanton critical point at infinity.

(5) In the SUSY extension of this triple-well model, the ground state energy vanishes to all orders perturbatively for both $(-1)^F$ even/odd sectors $H^\pm$. In the H^- sector, the ground state energy is $E_0^{NP} = 0$, due to the fact that there is a normalizable zero mode of H^-. Semiclassically this vanishing of ΔE_0^{NP} is due to cancellation between topological and neutral bion contributions; while in the H^+ sector, $E_0^{NP} \sim -e^{-2S_I + i\pi}$, due to an unpaired neutral bion. The hidden topological angle is crucial for both these results.

(6) In the SUSY extension there is again a simple explicit relation between the perturbative expansion coefficients in the inner and outer wells. See Eq. (41).

(7) There is a ζ-deformed generalization[4,34,47,48] of the SUSY triple-well system, discussed in Sec. 4, in which the fermion number parameter is deformed from $\zeta = 1$. When $\zeta = \frac{2m+1}{3}, m = 1, 2, \ldots$ curious cancellations arise in the semiclassical analysis, and result in convergent perturbation theory for part of the spectrum. These special ζ values correspond to the quasi-exactly solvable potentials.[18–20] Namely the corresponding $H^{-,\zeta}$ system has $2m$ lowest states for which perturbation theory is convergent, while for higher states one finds the generic divergent perturbative expansion. Of these $2m$ states, m are exactly solvable by methods of Refs. 18–20, with a convergent perturbative expansion which sums to the correct result. In this case the two types of bion contribution cancel against each other. The other m states are not exactly solvable, and there is a non-vanishing combination of topological and neutral bion contributions. For $H^{+,\zeta}$, there is an alternating pattern for the lowest m states. $\lceil \frac{m}{2} \rceil$ of these states have convergent perturbative expansions and non-perturbative contributions from neutral and topological bions, and $m - \lceil \frac{m}{2} \rceil$ have asymptotic divergent expansions. None of these states are exactly solvable non-perturbatively. The interplay between the topological and neutral bions provides a path integral semiclassical explanation of certain puzzles about non-perturbative effects in quasi-exactly solvable systems, for which a finite number of energy levels can be found algebraically.

These physical features of the triple-well potential (2) are explained semiclassically in the following sections. They illustrate that the double-well potential (1) is in fact quite special, in the sense that neighboring wells have the same frequency, and this is what makes instanton solutions physically relevant. In the case for which neighboring wells do not have the same frequency, such as the triple-well potential (2), despite the fact that the instanton action is finite, the instanton amplitude is zero due to the vanishing of the fluctuation prefactor.

2. Instantons and bions in the triple-well system

2.1. *Vanishing of instanton amplitudes for inequivalent degenerate wells*

In this section we show that given two consecutive degenerate harmonic wells with frequencies $\omega_1 \neq \omega_2$, despite the fact that a classical finite action instanton solution exists for the Euclidean BPS equation, the amplitude for such an instanton is zero due to quantum mechanical effects. Therefore, the instantons in general do not contribute to the energy spectrum at leading order in semi-classics, even though they are exact solutions with minimal action. Below, we describe this effect for the symmetric triple-well potential, which captures the essence of the general case.

The BPS equations for the symmetric triple-well potential (2), $\dot{x} = \mp \omega\, x(x^2 - 1)$, have instanton and anti-instanton solutions:

$$\text{instantons}:\quad x_I^{(\pm)}(t) = \pm \frac{1}{\sqrt{1 + e^{-2\omega(t-t_0)}}} \tag{3}$$

$$\text{anti-instantons}:\quad x_{\bar{I}}^{(\pm)}(t) = \pm \frac{1}{\sqrt{1 + e^{2\omega(t-t_0)}}} \tag{4}$$

Here, the $(\pm)$ notation denotes the minimum (at $x = \pm 1$) to or from which the

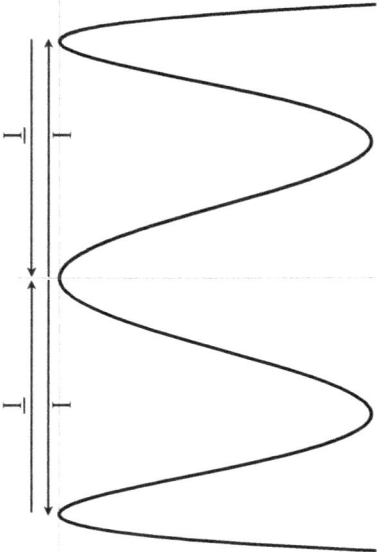

Fig. 2.   Instantons and anti-instantons in the symmetric triple well (2). Instantons tunnel from the inner minimum to the outer minima $(\pm)$, and anti-instantons tunnel from the outer minima $(\pm)$ to the inner minimum.

solution tunnels, and t_0 denotes the zero-mode degree of freedom. The instanton solutions, $x_I^{(\pm)}(t)$ tunnel from the inner vacuum at $x = 0$ to the outer vacua at $x = \pm 1$, while the anti-instanton solutions $x_{\bar{I}}^{(\pm)}(t)$ tunnel from the outer vacua at $x = \pm 1$ to the inner vacuum at $x = 0$. See Fig. 2. The instanton and anti-instanton action is

$$S_I = S_{\bar{I}} = \frac{\omega}{4} \tag{5}$$

The quadratic fluctuation operator in the background of an instanton is $F = (-\partial_t^2 + V_{\text{fluc}}(t))$ where

$$V_{\text{fluc}}(t) = [V''(x)]_{x=x_I^{(\pm)}(t)}$$
$$= \frac{\omega^2 \left(-10\, e^{2\omega(t-t_0)} + 4\, e^{4\omega(t-t_0)} + 1\right)}{\left(e^{2\omega(t-t_0)} + 1\right)^2}. \qquad (6)$$

See Fig. 3. (For an anti-instanton: $t \to -t$.) The non-perturbative amplitude asso-

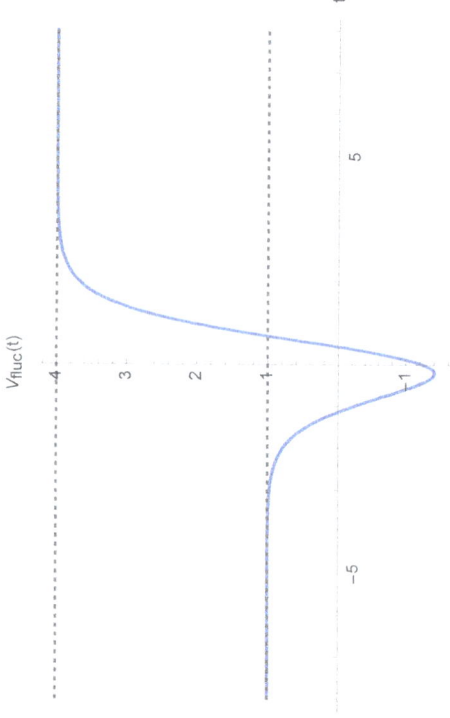

Fig. 3.   The instanton fluctuation operator potential (6) for the symmetric triple well. The mismatch between the $t \to \pm\infty$ limits leads to a vanishing fluctuation determinant: see Eq. (8).

ciated with an instanton or anti-instanton is a standard textbook computation:[1-8]

$$[I] \sim J_{\tau_0} \left[\frac{\det' F}{\det F_0}\right]^{-1/2} e^{-S_I} \qquad (7)$$

Here, prime indicates that the zero mode is omitted from the determinant, as it must be integrated over exactly. The Jacobian factor is given by $J_{t_0} = \sqrt{\frac{S_I}{2\pi}}$, and $\det F_0$ is the normalization by the free fluctuation operator, which if one is interested in the ground state should have the frequency of the middle well. The determinant factor of this instanton amplitude can be expressed in terms of asymptotic values of the instanton and anti-instanton solutions:[1,7,22]

$$\left[\frac{\det' F}{\det F_0}\right]^{-1/2} \sim \lim_{\beta \to \infty} e^{-\frac{\beta}{2} \times \frac{1}{2}(\omega_1 - \omega_0)} = 0 \qquad (8)$$

Here $\omega_0 = \omega$ is the frequency of the outer wells, and $\omega_1 = 2\omega$ is the frequency of the inner well, and β is the regulated length of the Euclidean time direction. The determinant factor in (8) vanishes, due to the mismatch of the frequencies in the inner and outer wells. Therefore

$$[I] = 0 = [\bar{I}] \qquad (9)$$

This means that the instanton solutions, despite being the non-perturbative objects with the lowest action, do not contribute to the energy spectrum of the ground state and low-lying states. This is also clear from looking at the numerical wavefunctions shown in Fig. 1, where the states are either localized in the outer wells or the inner wells, but not in both. Generalizing this argument, we see that this is actually true whenever classical degeneracy is not accidental, i.e. not related by a symmetry.

2.2. *Perturbation theory in the triple-well system*

There are two different wells in the triple-well potential (2), so we might expect that perturbation theory is different in the inner and outer wells. Indeed, the perturbative expansions for low-lying levels *look* different. We can study the various perturbative expansions using the BenderWu Mathematica package.[21] Comparing the perturbative expansion coefficients for the first few levels in the inner and outer wells suggests no particular relation between them beyond the obvious relation for the first term, the unperturbed energy:

inner-well perturbative expansion coefficients:

$$\nu_{inner} = 0 : \left\{ \frac{1}{2}, -\frac{3}{4}, -\frac{27}{16}, -\frac{153}{16}, -\frac{20385}{256}, -\frac{27027}{32}, \dots \right\} \tag{10}$$

$$\nu_{inner} = 1 : \left\{ \frac{3}{2}, -\frac{15}{4}, -\frac{225}{16}, -\frac{2025}{16}, -\frac{411075}{256}, -\frac{799875}{32}, \dots \right\} \tag{11}$$

$$\nu_{inner} = 2 : \left\{ \frac{5}{2}, -\frac{39}{4}, -\frac{855}{16}, -\frac{10809}{16}, -\frac{3009285}{256}, -\frac{7884891}{32}, \dots \right\} \tag{12}$$

outer-well perturbative expansion coefficients:

$$\nu_{outer} = 0 : \left\{ 1, -\frac{15}{8}, -\frac{45}{8}, -\frac{5265}{128}, -\frac{6885}{16}, -\frac{5735205}{1024}, \dots \right\} \tag{13}$$

$$\nu_{outer} = 1 : \left\{ 3, -\frac{111}{8}, -\frac{711}{8}, -\frac{165969}{128}, -\frac{412695}{16}, -\frac{628455429}{1024}, \dots \right\} \tag{14}$$

$$\nu_{outer} = 2 : \left\{ 5, -\frac{303}{8}, -\frac{3105}{8}, -\frac{1132497}{128}, -\frac{4310145}{16}, -\frac{9871632837}{1024}, \dots \right\} \tag{15}$$

To recognize the relation between the perturbative expansions in the inner and outer wells, we write the expansion coefficients in terms of the perturbative level number ν for the respective well. The first five orders of the perturbative expansions for the inner and outer wells, as a function of the perturbative level number ν in the respective well, are:

$$\text{inner} : \left\{ \nu + \frac{1}{2}, -\frac{3\nu^2}{2} - \frac{3\nu}{2} - \frac{3}{4}, -3\nu^3 - \frac{9\nu^2}{2} - \frac{39\nu}{8} - \frac{27}{16}, \right.$$
$$-\frac{105\nu^4}{8} - \frac{105\nu^3}{4} - \frac{363\nu^2}{8} - \frac{129\nu}{4} - \frac{153}{16},$$
$$\left. -\frac{603\nu^5}{8} - \frac{3015\nu^4}{16} - \frac{3645\nu^3}{8} - 495\nu^2 - \frac{39897\nu}{128} - \frac{20385}{256}, \dots \right\} \tag{16}$$

$$\text{outer}: \left\{ 2\nu + 1, -6\nu^2 - 6\nu - \frac{15}{8}, -24\nu^3 - 36\nu^2 - \frac{93\nu}{4} - \frac{45}{8}, \right.$$

$$-210\nu^4 - 420\nu^3 - \frac{1671\nu^2}{4} - \frac{831\nu}{4} - \frac{5265}{128},$$

$$\left. -2412\nu^5 - 6030\nu^4 - \frac{16335\nu^3}{2} - \frac{24885\nu^2}{4} - \frac{20259\nu}{8} - \frac{6885}{16}, \dots \right\} \quad (17)$$

These inner-well and outer-well expansion coefficients do not immediately look like they are related, but when expressed as functions of the action parameter, $B \equiv \nu + \frac{1}{2}$, we find:

$$\text{inner}: \left\{ B, -\frac{3B^2}{2} - \frac{3}{8}, -3B^3 - \frac{21B}{8}, -\frac{105B^4}{8} - \frac{411B^2}{16} - \frac{297}{128}, \right.$$

$$\left. -\frac{603B^5}{8} - \frac{4275B^3}{16} - \frac{351B}{4}, \dots \right\} \quad (18)$$

$$\text{outer}: \left\{ 2B, -6B^2 - \frac{3}{8}, -24B^3 - \frac{21B}{4}, -210B^4 - \frac{411B^2}{4} - \frac{297}{128}, \right.$$

$$\left. -2412B^5 - \frac{4275B^3}{2} - \frac{351B}{2}, \dots \right\} \quad (19)$$

We recognize that the perturbative expansions are related by a simple explicit map:

$$E_{\text{pert}}^{\text{outer}}(B, \hbar) = E_{\text{pert}}^{\text{inner}}(2B, \hbar) \quad (20)$$

We have verified this result to very high orders using the BenderWu package[21] for computing perturbative expansions.

The origin of relation (20) is not immediately obvious using direct Rayleigh–Schrödinger perturbation theory, but it can be understood straightforwardly using the relation to exact WKB. In this approach, ordinary perturbation theory can be generated by the following procedure.[23,24] First, compute the formal series for the all-orders WKB action

$$a(E, \hbar) = \sum_{n=0}^{\infty} \hbar^{2n} a_{2n}(E) \quad (21)$$

where $a_{2n}(E)$ are the WKB actions:[25,26]

$$a_0(E) = \sqrt{2} \oint_{\text{tp}} dx \sqrt{E - V} \quad ; \quad a_2(E) = -\frac{\sqrt{2}}{2^6} \oint_{\text{tp}} dx \frac{(V')^2}{(E - V)^{5/2}} \quad ; \quad \dots \quad (22)$$

In the triple-well system, all the $a_{2n}(E)$ are expressed in terms of simple hypergeometric functions.[24] Next, impose the all-orders Bohr–Sommerfeld quantization condition

$$a(E, \hbar) = 2\pi\hbar \left(\nu + \frac{1}{2} \right) \quad , \quad \nu = 0, 1, 2, \dots \quad (23)$$

Finally, expand each $a_{2n}(E)$ at small E and then invert (23) to express the energy as a function of ν and $\hbar$. This inversion produces an expression of the form $E = E(\nu, \hbar)$, rather than (23) which is of the form $\nu = \nu(E, \hbar)$. Furthermore, the expression $E = E(\nu, \hbar)$ coincides with standard Rayleigh–Schrödinger perturbation theory expanded about the νth harmonic unperturbed state. This perturbative procedure can be implemented for actions with turning points for the inner well or outer wells, and the coefficients of the $\hbar$ expansion are polynomials in the respective ν label, as listed in Eqs. (16) and (17).

The triple-well potential in (2) has the interesting geometric property[24] that at each order of the WKB expansion, the actions are *equal* in the inner and outer wells, up to a simple factor of 2:

$$a_{2n}^{\text{inner}}(E) = \frac{1}{2} a_{2n}^{\text{outer}}(E) \quad , \quad n = 0, 1, 2, \ldots \tag{24}$$

For example, at the classical level it is clear that the frequencies and actions in the inner and outer wells differ only by factors of 2. Remarkably, this behavior persists to all orders for the triple-well system.[24] This means that in this all-orders Bohr–Sommerfeld approach to perturbation theory, the symmetry under $B \to 2B$ in (20) follows immediately from (23), which is equivalent to perturbation theory.

It is well known that important connections between perturbative and non-perturbative physics are encoded in the large-order behavior of perturbation theory.[27–30] For the triple-well potential (2) the high orders of the perturbative expansions can be studied efficiently using the BenderWu Mathematica package.[21] It is simple to generate many hundreds of terms in these expansions, which permits high-precision analysis of the large-order growth. The leading growth for the perturbative expansion coefficients in (16) and (17) is as follows:

$$\text{inner:} \quad c_n^{(\text{level } \nu), \, \text{inner}} \sim \beta_\nu \, \frac{2^{n + \frac{3\nu}{2} - \frac{1}{4}} \Gamma\left(n + \frac{3\nu}{2} - \frac{1}{4}\right)}{\pi \, \Gamma\left(\frac{3\nu}{2} + \frac{3}{4}\right)} \quad , \quad n \to \infty \tag{25}$$

$$\text{outer:} \quad c_n^{(\text{level } \nu), \, \text{outer}} \sim \gamma_\nu \, \frac{2^{n + 3\nu + \frac{1}{2}} \Gamma\left(n + 3\nu + \frac{1}{2}\right)}{\pi \, \Gamma\left(3\nu + \frac{3}{2}\right)} \quad , \quad n \to \infty \tag{26}$$

up to some n-independent rational normalization factors. Note that these asymptotic behaviors are consistent with the exact relation (20), and the expression (25) is consistent with an expression in Ref. 30.

For our purposes here, the most important facts about large-order perturbation theory for the triple-well system are:

- the coefficients grow factorially in magnitude;
- the coefficients do not alternate in sign;
- the factor 2^n in (25) and (26) corresponds to $1/(2S_I)^n$ in our normalization, the nth power of the inverse of *twice* the instanton action (5).

These facts imply that naive Borel summation of these formal perturbative series produces ambiguous imaginary non-perturbative terms with exponential factor e^{-2S_I}. These are cancelled by contributions from instanton/anti-instanton interaction effects,[17,31–35] which we refer to as quantum bions, and which are analyzed in the next subsection. This cancellation mechanism is one of the clearest examples of the application of resurgent trans-series in physics, where a trans-series combines both perturbative and non-perturbative contributions in such a way that naive imaginary terms are cancelled in the full trans-series, producing real and unambiguous physical results.[36,37]

2.3. *Critical points at infinity, Lefschetz thimbles and bions in the triple-well system*

In the two-instanton sector of a generalized instanton gas analysis, we must carefully consider the critical points at infinity, and their Lefschetz thimble contribution.[17] The main results of Ref. 17 for the double-well potential and its $\hbar$-tilting are the following:

- An instanton and anti-instanton is a critical point at infinite separation. Starting with the theory compactified on a circle with size β, the classical + quantum interaction between the two is of the form $V_{\text{eff}}(\tau) = \left(-A(e^{-\omega\tau} + e^{-\omega(\beta-\tau)}) + \hbar\omega\tau\right)$. The critical point is determined by the classical action, and located at $\omega\tau^* = \frac{\beta}{2}$. As $\beta \to \infty$, the classical interaction between the two instantons dies off and the configuration becomes a genuine saddle point.

- The thimble associated with the critical $\omega\tau^*$ is given by (for $\hbar \to e^{i\theta}\hbar$)

$$\Gamma_{\text{QZM}}^{\theta=0^+} = \gamma_1^+ + \gamma_2^+ + \gamma_3^+ \tag{27}$$

  where the segments are $\gamma_1^+ = (-\infty + i\pi, \frac{\beta}{2} + i\pi)$, $\gamma_2^+ = [\frac{\beta}{2} + i\pi, \frac{\beta}{2} - i\pi]$, $\gamma_3^+ = (\frac{\beta}{2} - i\pi, \infty - i\pi)$. In the $\beta \to \infty$ limit, the integration is equal to the contribution of the γ_1^+ segment.

- These critical points at infinity are non-Gaussian. Therefore, to reproduce the correct NP-contributions at second order in semi-classics, the quasi-zero-mode (QZM) integrals need to be done exactly, not in the Gaussian approximation.

- The configurations that dominate the integration over $\Gamma_{\text{QZM}}^{\theta=0^+}$ are the neutral and topological bion configurations.

In a semiclassical treatment of the quantum mechanical path integral, we must take into account the physical effect of all saddles and their thimbles, including the critical points at infinity, and their thimbles: see e.g. Refs. 17, 33–35, 38–42. In what follows we analyze the triple-well system using the notions of topological bions and neutral bions.

There are two different kinds of correlated instanton configurations: topological[a] bions and neutral bions. We begin by discussing neutral bions of the classical triple-well system. These can be thought of as the solutions of the inverted potential which

---

[a]In some of our previous work these were called the "real bions". We have renamed them "topological bions" here because they need not be real saddles of the classical equations of motion. Their most important distinguishing feature is that they have a non-zero topological charge, in contrast to neutral bions whose topological charge vanishes.

roll down from one of the peak of the inverted potential, to an adjacent peak of the inverted potential, and then roll back. It is clear that we can find a solution which performs this motion any number of times in time β. We will refer to an object which performs this motion once (down the hill, up the hill and back) as a neutral bion. The term neutral indicates that there is no topological charge associated with this object.

The topological bion is more interesting. Intuitively such an object corresponds to a tunneling event between the far left well and the far right well. Indeed looking at Fig. 1, such an object must exist on physical grounds in order to explain the energy splitting between the states localized on the left and on the right. The name "topological" emphasizes the topological stability of the configuration which has its two ends in two classical minima related by a symmetry. However the classical potential clearly does not have real finite action solutions of this kind.[b] Candidate solutions of the topological bion can be found in various ways as limits approaching this functional separatrix. One possibility is to make a small shift of the minimum of the middle well, in which case a solution appears. Another possibility is to take a limit of the complex solutions at energy just above the top of the inverted potential. Another approach, which we adopt here, is to employ minimization of the quantum action, which yields physically consistent results.[33–35] These quantum saddles correspond to the tails of the Lefshitz thimbles of the appropriate saddles minimizing the classical action, in the spirit of Ref. 17. The full details of the connection between the aforementioned limiting procedures is interesting and deserves further future study. However, for the purposes of this work we consider both the neutral bion and the topological bion to be solutions of the equations of motion with the quantum corrected potential $\frac{1}{2}(W')^2 + \frac{\hbar}{2}W''$, as in Refs. 33–35. The neutral bions are also well-defined objects on the thimble as in Ref. 17, and we expect that the same holds for the topological bions. We show that this combined approach is self-consistent.

The characteristic size of bions in the Euclidean time direction is given by

$$t_* \sim \ln \frac{1}{\hbar} \tag{28}$$

The topological and neutral bions have finite action, and their fugacities are given by

$$[\mathcal{TB}] \propto e^{-2S_I} \quad , \quad [\mathcal{NB}] \propto e^{-2S_I \pm i\frac{\pi}{4}} \tag{29}$$

for the topological bion $[\mathcal{TB}]$ and the neutral bion $[\mathcal{NB}]$ respectively. The complex phase, the hidden topological angle θ_{HTA}, comes from the fact that these configurations are in fact complex configurations which saturate the thimble integral. The physical effects of the topological and neutral bions are very different.

---

[b]Thinking of the inverted potential of the triple well, the only classical solutions which go from one outer maximum to another would be flying off to infinity, and would hence have infinite action.

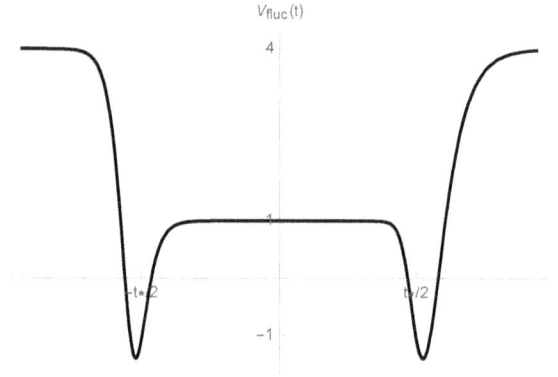

Fig. 4.   Sketch of the structure of the fluctuation potential for the bion configurations.

2.3.1. Physics of the topological bion

The non-perturbative effect of the topological bion is analogous to that of a single instanton in the symmetric double-well potential in the sense that it is responsible for the non-perturbative splitting, $\Delta E \sim e^{-2S_I}$, of the energy levels for low-lying states localized in the outer wells. The prefactor is computed from the determinant factors as in expression (7), but now with action $2S_I$, and with the fluctuation potential of the form in Fig. 4. Since the asymptotic values of the fluctuation potential coincide, the prefactor is finite and non-zero, unlike for the single instanton where it vanishes (8):

$$\text{topological bion amplitude:}\quad \left[\bar{I}_{\pm} I_{\mp}\right] \sim e^{-2S_I} \tag{30}$$

2.3.2. Physics of the neutral bion

The neutral bions play a very different role. Their non-perturbative character is similar to that of a correlated instanton/anti-instanton molecule in the symmetric double-well system.[17,33–35] The associated amplitude can be computed in an instanton gas picture as follows. Consider a widely separated (separation much greater than the instanton size scale) instanton/anti-instanton molecule, interacting via the effective potential

$$V_{\text{eff}}(\tau) = -\omega_1 c_1^2 e^{-\omega_1 \tau} + \frac{\hbar}{2}(\omega_1 - \omega_0)\tau \tag{31}$$

where we write $\omega_0 = \omega$ for the frequency of the inner well, and $\omega_1 = 2\omega$ for that of the outer wells. The first term in $V_{\text{eff}}(\tau)$ is the classical interaction, arising from the overlap of the tails of two consecutive instantons, as is familiar from the usual symmetric double-well potential.[17,31–35] The second term is the action accumulated during the intermediate Euclidean time regime spent in the other minimum. The difference in energy between the true and false vacua is $\Delta E = \frac{1}{2}(\omega_1 - \omega_0)$, and the time spent is τ. Hence the action cost of this intermediate regime is $\Delta S = \Delta E \tau = \frac{\tau}{2}(\omega_1 - \omega_0)$. So for the symmetric triple-well potential in (2), $\Delta S = \frac{\tau}{2}\omega$.

Unlike instantons, whose amplitude vanishes at the one-loop level (since the time spent in the false vacuum is infinite), the amplitude for a neutral bion is non-zero. To see this, consider a correlated instanton/anti-instanton pair, starting in the central well at the true vacuum with frequency ω, and taking a journey to the neighboring degenerate minimum with frequency $\omega_1 > \omega_0$, spending Euclidean time τ (which is the separation quasi-zero mode that we will integrate over), and returning back to the original vacuum. To account for this correlated event, we evaluate the $[I_+ \bar{I}_+]$ amplitude as follows

$$[I_+ \bar{I}_+] \sim e^{-2S_I} \int_{\Gamma_{\text{QZM}}} d(\omega_1 \tau) e^{-\frac{1}{\hbar}\left(-\omega_1 c_1^2 e^{-\omega_1 \tau} + \frac{\hbar}{2}(\omega_1 - \omega_0)\tau\right)} \tag{32}$$

where the integral is to be taken along an appropriate thimble,[17] which, for $\beta \to \infty$, passes along the Im $\tau = \pi$ line. This integral encodes the effect of the neutral bion fluctuation determinant.

The critical point of the effective potential $V_{\text{eff}}(\tau)$ determines the characteristic size of the neutral bion:

$$V'_{\text{eff}}(\tau) = 0 \Rightarrow \omega_1 \tau^* = \log\left[\frac{\omega_1 c_1^2}{\hbar \mathcal{D}}\right] \mp i\pi \tag{33}$$

We have defined the "deficit parameter"

$$\mathcal{D} = \frac{1}{2}\left(1 - \frac{\omega_0}{\omega_1}\right) \tag{34}$$

The integration Γ_{QZM} over the quasi-zero-mode degree of freedom yields

$$[I_+ \bar{I}_+]_{\pm} \sim e^{-2S_I} e^{\pm i\pi\mathcal{D}}\left(\frac{\hbar}{\omega_1 c_1^2}\right)^{\mathcal{D}} \Gamma(\mathcal{D}) \tag{35}$$

This result identifies the hidden topological angle (HTA)[43,44] as

$$\theta_{\text{HTA}} = \frac{\pi}{2}\left(\frac{\omega_1 - \omega_0}{\omega_1}\right) \equiv \pi\mathcal{D} \tag{36}$$

In general, there would be a different HTA for a neutral bion connecting with another neighboring minimum with a different characteristic frequency. But in the symmetric triple-well potential (2) the frequencies of the two outer wells are equal, so the amplitude for the different types of neutral bions are the same. Since $\omega_0 = \omega$ and $\omega_1 - 2\omega$, we obtain a hidden topological angle given by:

$$\theta_{\text{HTA}} = \frac{\pi}{4} \tag{37}$$

Hence the quantum neutral bion amplitude is

$$[I_+ \bar{I}_+]_{\pm} \sim e^{-2S_I} e^{\pm i\frac{\pi}{4}}\left(\frac{\hbar}{\omega c_1^2}\right)^{1/4} \Gamma\left(\frac{1}{4}\right) \tag{38}$$

A similar discussion applies to all neutral bion amplitudes: $[I_\pm \bar{I}_\pm]$ and $[\bar{I}_\pm I_\pm]$. Note that these amplitudes have an imaginary two-fold ambiguity. The ambiguity cancels the ambiguity associated with the Borel resummation of perturbation theory,[17,31-35] and the real part provides a non-perturbative shift to the ground state energy.

In Sec. 4 we consider a deformation of the triple-well system in which the hidden topological angles differ by special fractionally quantized amounts. This has the effect that perturbation theory is convergent for a finite set of states, but divergent for all other states.

3. Supersymmetric extension of the triple-well system

3.1. *SUSY symmetric triple well*

Integrating out the fermions in SUSY quantum mechanics produces a pair of bosonic partner Hamiltonians:[4,17,33,34,45]

$$H^\pm = -\frac{\hbar^2}{2}\frac{d^2}{dx^2} + \frac{\omega^2}{2}x^2(x^2-1)^2 \pm \frac{\hbar\,\omega}{2}(3\,x^2-1) \tag{39}$$

The potential has a classical term and a quantum induced $O(\hbar)$ term coming from integrating out the fermions. These partner Hamiltonians can be factored as

$$H^\pm = \frac{1}{2}Q^\pm\,Q^\mp \quad , \quad Q^\pm \equiv \pm\hbar\,\frac{d}{dx} + W' \tag{40}$$

where the superpotential term is $W'(x) = \omega\,x(x^2-1)$. The associated partner potentials, $V_\pm(x) = \frac{1}{2}(W')^2 \pm \frac{\hbar}{2}W''$, are shown in Fig. 5. Note that it is important that the magnitude of the fermionic contribution, $\hbar\,W''(x)$ is parametrically small, proportional to $\hbar$, but still much greater than $e^{-2/\hbar}$, at which point the non-perturbative effects of the non-SUSY system dominate the physics.

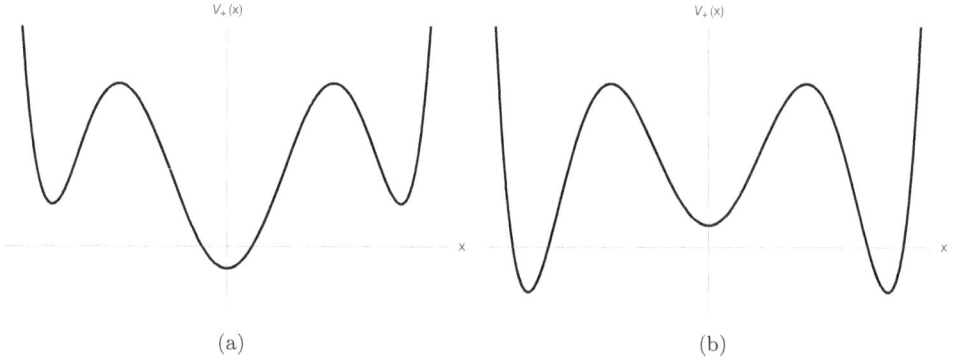

$V_+(x)$ $V_+(x)$

         (a)   (b)

Fig. 5.   The form of the SUSY partner potentials $V_\pm(x)$ for the SUSY triple-well system.

As expected, the ground state of each potential $V_\pm$ has zero energy to all orders in perturbation theory. All other states have divergent perturbative expansions. (Nevertheless, there are still resurgent relations between perturbative and non-perturbative sectors.[46]) The perturbative structure of the SUSY partner Hamiltonians $H^\pm$ can be studied with the BenderWu Mathematica package.[21] This enables one to verify explicitly the generalization of the relation (20) between the perturbative expansions in the inner and outer wells extended to the SUSY potentials. We find that (recall $B \equiv \nu + \frac{1}{2}$, where ν is the perturbative level number for states localized in that well) the perturbative expansions for states localized in the inner and outer wells are related as:

$$E_{\text{pert}}^{\text{outer},\pm}(B,\hbar) = E_{\text{pert}}^{\text{inner},\pm}\left(2B \pm \frac{3}{2},\hbar\right) \tag{41}$$

This is easy to check using the BenderWu package.[c]

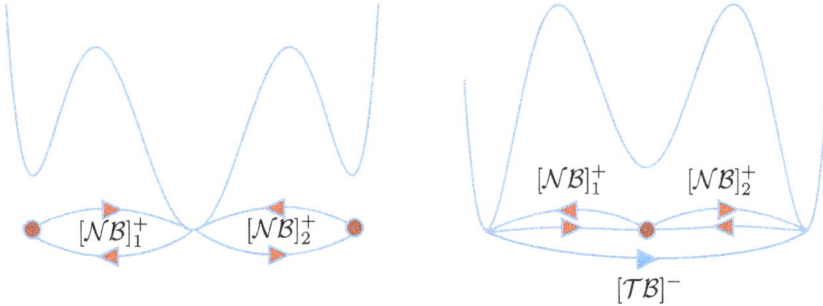

Fig. 6.   Sketch of the neutral and topological bions in the H^+ (left) and H^- (right) sectors.

The semiclassical analysis of the SUSY triple-well system can be described as follows. Recall that near the classical minima $x = 0, \pm 1$, the bosonic potential behaves as

$$V_{\text{bosonic}} \equiv \frac{1}{2} W'(x)^2 \approx \begin{cases} \frac{4\omega^2}{2} x^2 & , \quad x \approx \pm 1 \\ \frac{\omega^2}{2} x^2 & , \quad x \approx 0 \end{cases} \tag{42}$$

The classical harmonic oscillator frequencies are $\omega_0 = \omega$, and $\omega_1 = 2\omega$. Therefore a classical path fixed near $x = 0, \pm 1$ will contribute $\beta \omega_{0,1}/2$ to the action density, to one loop in perturbation theory in $\hbar$. However, such paths also get a contribution from the fermionic potential, $W''(x)$:

$$\pm \frac{\hbar}{2} W''(x) \approx \begin{cases} \pm \frac{2\hbar\omega}{2} & , \quad x \approx \pm 1 \\ \mp \frac{\hbar\omega}{2} & , \quad x \approx 0 \end{cases} \tag{43}$$

Thus, for H^+ the action contributions for paths near $x \approx 0$ cancel between the one-loop bosonic fluctuation and the fermionic contribution, but they add near $x \approx \pm 1$. On the other hand, for H^- the action contributions for paths near $x \approx \pm 1$ cancel, while they add near $x \approx 0$. In fact one can show that this cancellation is exact to any order, and so the energies in the corresponding local minima are all zero to any order in perturbation theory.

Thus, for the ground state a non-perturbative path must start and end at $x = 0$ for the Hamiltonian H^+, but it must start and end at $x = \pm 1$ for H^-. See Fig. 6. Therefore, H^+ allows only an instanton $I_\pm$ followed by an anti-instanton $\bar{I}_+$, i.e. a correlated $[I_\pm \bar{I}_\pm]$ configuration, interpolating from $x = 0$ to $x = \pm 1$ and then back to $x = 0$. We refer to these configurations as *neutral bions*, because their path is topologically trivial in the sense that they start and end at the same perturbative vacuum. The H^- case also allows for a neutral bion: i.e. an instanton $I_\pm$ followed by an anti-instanton $\bar{I}_\pm$. However since H^- has two degenerate minima to all orders in perturbation theory, we also have a topologically nontrivial configuration:

---

[c]To check this one can use the `BenderWuLevelPolynomial` function to compute the level-number dependence. See the documentation of the BenderWu package.[21]

an anti-instanton of type $\bar{I}_\pm$, followed by an instanton of type $I_\mp$, thereby taking $x : \pm 1 \to 0 \to \mp 1$. We refer to such a configuration as a *topological bion*.

The correlated semiclassical configurations which contribute to the partition function are therefore:

$$\text{neutral bion:} \quad [\mathcal{NB}]^+ \equiv [I_\pm \bar{I}_\pm] \quad \text{for } H^+ \tag{44}$$

$$\text{neutral bion:} \quad [\mathcal{NB}]^- \equiv [I_\pm \bar{I}_\pm] \quad \text{for } H^- \tag{45}$$

$$\text{topological bion:} \quad [\mathcal{TB}]^- \equiv [\bar{I}_\pm I_\mp] \quad \text{for } H^- \tag{46}$$

These configurations are parametrized by the distance τ between the constituent instanton and anti-instanton. To compute their effect, one integrates over the separation τ:

$$[I_\pm \bar{I}_\pm] = C_+ e^{-2S_I} \int d\tau e^{-S_{nb}^+(\tau) - \omega_1 \tau} \quad \text{, for } H^+ \tag{47}$$

$$[I_\pm \bar{I}_\pm] = C_- e^{-2S_I} \int d\tau e^{-S_{nb}^-(\tau) - \omega_0 \tau} \quad \text{, for } H^- \tag{48}$$

$$[\bar{I}_\pm I_\mp] = C_- e^{-2S_I} \int d\tau e^{-S_{tb}^-(\tau) - \omega_0 \tau} \quad \text{, for } H^- \tag{49}$$

The constants $C_\pm$ factorize to a pair of one-loop determinants around single (anti-) instantons, provided that the divergent part due to the frequency mismatch is subtracted, as it is explicitly taken into account above. The difference between C_+ and C_- is determined entirely by the difference of the terms $\pm \int W''(x)$ in the action, evaluated in the background of the relevant configurations. Furthermore $S_{nb}(\tau)$ and $S_{tb}(\tau)$ are the "classical" (i.e. leading order in $\hbar$) interactions, and the subscript refers to the "neutral bion" and "topological bion", respectively. As usual, these interaction terms can be found from the asymptotic behavior of the bion solutions.[17,33,34] A straightforward computation yields:

$$S_{nb}^+(\tau) = -2\omega_1 c_1^2 e^{-\omega_1 \tau} , \tag{50}$$

$$S_{nb}^-(\tau) = -2\omega_0 c_0 e^{-\omega_0 \tau} , \tag{51}$$

$$S_{tb}^-(\tau) = 2\omega_0 c_0 e^{-\omega_0 \tau} = -S_{nb}^-(\tau) \tag{52}$$

where the coefficients, $c_0 = 1/\sqrt{3}$ and $c_1 = -3/2$, come from the details of the asymptotics of the instanton solution. Note that, as expected, for the neutral bion the classical interaction between constituents is attractive, while for the topological bion it is repulsive. Furthermore, also as expected, in the H^- system the interaction between the topological bion constituents is minus that of the neutral bion constituents.

As explained in Sec. 2.3, these bions are the dominant configurations of a critical point at infinity.[17] The saddle point at infinity is a non-Gaussian critical point with vanishing contribution, but its thimble contributes non-trivially. The evaluation over the thimble amounts to integrating over a contour $\omega_1 \tau \in \mathbb{R} + i\pi$, $\omega_0 \tau \in \mathbb{R} + i\pi$ in the first and second cases (neutral bions), and $\omega_0 \tau \in \mathbb{R}$ in the third case (topological bions). Performing the τ integrals *exactly* yields the amplitude and also the phase associated with each bion configuration:

$$H^+ : \quad [\mathcal{N}\mathcal{B}]^+ \equiv [I_\pm \bar{I}_\pm] = \frac{\hbar}{\omega_0 \omega_1 c_0 c_1} e^{-2S_I + i\pi} \tag{53}$$

$$H^- : \quad [\mathcal{N}\mathcal{B}]^- \equiv [\bar{I}_\pm I_\pm] = \frac{\hbar}{\omega_0 \omega_1 c_0 c_1} e^{-2S_I + i\pi} \tag{54}$$

$$H^- : \quad [\mathcal{T}\mathcal{B}]^- \equiv [\bar{I}_\pm I_\mp] = \frac{\hbar}{\omega_0 \omega_1 c_0 c_1} e^{-2S_I} \tag{55}$$

The phase is given by $\theta_{\mathrm{HTA}} = \pi$ both for the neutral bion $[\mathcal{N}\mathcal{B}]^+$ in the H^+ sector, and for the neutral bion $[\mathcal{N}\mathcal{B}]^-$ in the H^- sector. In contrast, the phase associated with the topological bion is zero. The quantization of the HTA in units of π for the neutral bion implies that there is no ambiguity in the neutral bion amplitude, and hence there should not be any ambiguity in perturbation theory for the ground state for H^-. Indeed, perturbation theory for the ground state is convergent in either sector (in fact, it vanishes to all perturbative orders due to supersymmetry).

3.2. *Non-perturbative cancellations and the Witten index*

We can now combine the contributions of the neutral and topological bions. Consider first the H^+ sector. In the leading order semiclassical approximation, we sum over the neutral bion configurations. There are two distinguishable configurations of equal magnitude which contribute to the partition function (see Fig. 6). The partition function can be viewed as a dilute gas of neutral bions of these two types. Summing over all such configurations, we find

$$Z^+ \approx \sum_{n=0}^{\infty} \frac{1}{n!} (\beta[\mathcal{N}\mathcal{B}])^n \sum_{m=0}^{\infty} \frac{1}{m!} (\beta[\mathcal{N}\mathcal{B}])^m$$
$$= e^{2\beta[\mathcal{N}\mathcal{B}]} = e^{-2\beta K e^{-2S_I}} \tag{56}$$

with K a positive constant. The non-perturbative ground state ($N = 0$) energy in the H^+ sector is

$$E^{\mathrm{np},+}(N = 0) - -2[\mathcal{N}\mathcal{B}] = -2K e^{-2S_I + i\pi} > 0 \tag{57}$$

Here the label N refers to all states in the spectrum, not just the perturbative levels in one of the potential wells. Note that the positive semi-definiteness of the spectrum [guaranteed by the SUSY factorization (40)] arises semiclassically in (57) as a result of the hidden topological angle: $\theta_{\mathrm{HTA}} = \pi$. The HTA arises here because the integration cycle is in the complex domain, hence, the neutral bion configuration (or any other configuration on this thimble that contributes to semi-classics at this order) is manifestly complex. Since the contribution of real saddles or real configurations to path integrals are manifestly negative, the existence of $\theta_{\mathrm{HTA}} = \pi$ is strictly necessary in this case for compatibility with the supersymmetry algebra, which requires $E^{\mathrm{np},+}(N = 0) \geq 0$.

For the H^- sector we sum over both the neutral bions and also the topological bion. However the topological bion starts in one vacuum and ends in the other, so only even powers of these configurations can contribute to the thermal partition function [which demands periodicity in Euclidean time]. In addition there is a neutral bion, the sum over which is not constrained. Therefore,

$$Z^- \approx 2 \sum_{n=0}^{\infty} \frac{1}{n!} (\beta[\mathcal{N}\mathcal{B}])^n \sum_{m=0}^{\infty} \frac{1}{(2m)!} (\beta[\mathcal{T}\mathcal{B}])^{2m}$$
$$= 2 \cosh([\mathcal{T}\mathcal{B}]\beta) e^{[\mathcal{N}\mathcal{B}]\beta} . \tag{58}$$

The overall factor of two in front is due to the fact that there is a sum over the two perturbative supersymmetric vacua, i.e. the outer-well harmonic vacua. However, notice that since $[\mathcal{N}\mathcal{B}]^+ = [\mathcal{N}\mathcal{B}]^- = -[\mathcal{T}\mathcal{B}]^- = -Ke^{-2S_I}$, where K is a (positive) constant, we have that

$$Z^- \approx 1 + e^{-2\beta K e^{-2S_I}} . \tag{59}$$

This has the consequence that the non-perturbative energies of the two lowest levels in the H^- sector are

$$E^{\mathrm{np},-}(N=0) = -([\mathcal{T}\mathcal{B}] + [\mathcal{N}\mathcal{B}]) = 0$$
$$E^{\mathrm{np},-}(N=1) = -(-[\mathcal{T}\mathcal{B}] + [\mathcal{N}\mathcal{B}]) = 2Ke^{-2S_I} \tag{60}$$

Observe that $E^{\mathrm{np},-}(N=1)$ is degenerate with $E^{\mathrm{np},+}(N=0)$, as we know from supersymmetry.[45] The supersymmetric Witten index is therefore given semiclassically by

$$I_W = Z^- - Z^+ = 1 \tag{61}$$

which agrees with the well known result from SUSY.[45] Semiclassically, the cancellation of the ground state energy is due to the opposite sign contributions of $[\mathcal{T}\mathcal{B}]$ and $[\mathcal{N}\mathcal{B}]$, which arises from the hidden topological angle $\theta_{\mathrm{HTA}} = \pi$.[33]

4. Zeta-deformed theories and quasi-exactly soluble models

The fermionic contribution to the partner Hamiltonians (39) can be extended to multiple flavors of fermions, where we will also take the "fermion number" parameter $\varsigma > 0$ to be non-integer, constructing $H^{\pm,\varsigma}$ pairs of partner Hamiltonians.[4,34,47,48] In this section, we examine novel properties of these paired Hamiltonians:

$$H^{\pm,\varsigma} = -\frac{\hbar^2}{2} \frac{d^2}{dx^2} + \frac{\omega^2}{2} x^2 (x^2 - 1)^2 \pm \varsigma \frac{\omega}{2} \hbar (3x^2 - 1) \tag{62}$$

Notice that the form mimics the supersymmetric form of the potential

$$V^{\pm,\varsigma}(x) = \frac{1}{2} W'(x)^2 \pm \frac{\varsigma}{2} \hbar W''(x) \tag{63}$$

with $W'(x)$ being analogous to the "superpotential"

$$W'(x) = \omega x (x^2 - 1) \tag{64}$$

4.1. *Perturbative expansions for the zeta-deformed theories*

We first observe that the perturbative expansion around the inner and outer wells are once again related. We employ the BenderWu mathematica package to facilitate these expansions, and discover that the energy expansion coefficients around the inner and outer wells, as a function of the harmonic oscillator level number, given in terms of $B = \nu + \frac{1}{2}$ for the respective well, are given by

$$E^{\text{inner},\pm,\zeta} : \left\{ B \mp \frac{\zeta}{2}, -\frac{3B^2}{2} \pm \frac{3B\zeta}{2} - \frac{3}{8}, \right.$$

$$\left. -3B^3 \pm \frac{9B^2\zeta}{2} - \frac{9B\zeta^2}{8} - \frac{21B}{8} \pm \frac{9\zeta}{8}, \ldots \right\} \tag{65}$$

$$E^{\text{outer},\pm,\zeta} : \left\{ 2B \pm \zeta, -6B^2 \mp 6B\zeta - \frac{9\zeta^2}{8} - \frac{3}{8}, \right.$$

$$\left. -24B^3 \mp 36B^2\zeta - \frac{63B\zeta^2}{4} - \frac{21B}{4} \mp \frac{27\zeta^3}{16} \mp \frac{45\zeta}{16}, \ldots \right\} \tag{66}$$

These expansion coefficients imply that the perturbative energies in the inner and outer wells are related as follows:

$$E^{\text{outer},\pm,\zeta}_{\text{pert}}(B,\hbar) = E^{\text{inner},\pm,\zeta}_{\text{pert}}\left(2B \pm \frac{3\zeta}{2}, \hbar\right). \tag{67}$$

These relations generalize the inner–outer perturbative relations for the non-SUSY case in (20), and for the SUSY case in (41).

There is further interesting structure in the perturbative expansions for the partner Hamiltonians $H^{\pm,\zeta}$, which we now use to give a semiclassical understanding of the rich algebraic structure of quasi-exactly-solvable (QES) Hamiltonians.[6,18–20,50] After suitable rescaling, the sextic potential analyzed in Ref. 20 matches the form in (62). In our notation, the QES systems arise for special rational values of ζ:

$$\zeta_{\text{QES}} = \frac{2m+1}{3}, \quad m = 1,2,3,\ldots \tag{68}$$

Using the BenderWu package, we have studied the perturbative expansions for the low-lying states of the $H^{\pm,\zeta}$ partner Hamiltonians. Note that for $H^{+,\zeta}$ the middle well is lowered, while for $H^{-,\zeta}$ the outer wells are symmetrically lowered, analogous to Fig. 5 for the SUSY models.

4.1.1. *Perturbative structure for $H^{+,\zeta_{\text{QES}}}$*

For $H^{+,\zeta_{\text{QES}}}$ the middle well is lowered, so the low-lying states are localized in the middle well for $0 \leq \nu_{\text{inner}} \leq m - 1$.

- $m = 1$: The lowest QES case has $m = 1$, which means $\zeta = 1$, which is the SUSY case. The ground state has $E_{\text{pert}}(\nu_{\text{inner}} = 0, \hbar) = 0$ to all orders of perturbation theory. But it receives a non-perturbative shift, which is positive, as discussed in Sec. 3.2. For $\nu_{\text{inner}} \geq 1$, the perturbative expansions are all divergent asymptotic expansions, with non-alternating expansion coefficients that grow factorially in magnitude.

$$\nu_{\text{inner}} = 0 : \text{convergent expansion} \quad E_{\text{pert}} = 0 \tag{69a}$$

$$\nu_{\text{inner}} \geq 1 : \text{divergent non-alternating expansion} \tag{69b}$$

- $\underline{m = 2}$: In this case the deformation parameter has a non-integer rational value, $\zeta = \frac{5}{3}$, and we find that the ground state ($\nu_{\text{inner}} = 0$), has a divergent and non-alternating perturbative expansion, while the series for the first excited state ($\nu_{\text{inner}} = 1$), truncates and is thus convergent:

$$\nu_{\text{inner}} = 0 : \text{divergent non-alternating expansion} \tag{70a}$$

$$\nu_{\text{inner}} = 1 : \text{convergent expansion} \quad E_{\text{pert}} = \frac{2}{3} \tag{70b}$$

$$\nu_{\text{inner}} \geq 2 : \text{divergent non-alternating expansion} \tag{70c}$$

- $\underline{m = 3}$: In this case the deformation parameter has a non-integer rational value, $\zeta = \frac{7}{3}$, and we find the following pattern:

$$\nu_{\text{inner}} = 0 : \text{convergent expansion} \quad E_{\text{pert}} = \frac{1}{3} - \sqrt{1 - 2\hbar} \tag{71a}$$

$$\nu_{\text{inner}} = 1 : \text{divergent non-alternating expansion} \tag{71b}$$

$$\nu_{\text{inner}} = 2 : \text{convergent expansion} \quad E_{\text{pert}} = \frac{1}{3} + \sqrt{1 - 2\hbar} \tag{71c}$$

$$\nu_{\text{inner}} \geq 3 : \text{divergent non-alternating expansion} \tag{71d}$$

- $\underline{m = 4}$: In this case the deformation parameter has an integer value, $\zeta = 3$, and we find the following pattern:

$$\nu_{\text{inner}} = 0 : \text{divergent non-alternating expansion} \tag{72a}$$

$$\nu_{\text{inner}} = 1 : \text{convergent expansion} \quad E_{\text{pert}} = 1 - \sqrt{1 - 6\hbar} \tag{72b}$$

$$\nu_{\text{inner}} = 2 : \text{divergent non-alternating expansion} \tag{72c}$$

$$\nu_{\text{inner}} = 3 : \text{convergent expansion} \quad E_{\text{pert}} = 1 + \sqrt{1 - 6\hbar} \tag{72d}$$

$$\nu_{\text{inner}} \geq 4 : \text{divergent non-alternating expansion} \tag{72e}$$

This pattern continues for higher m: the states with $0 \leq \nu_{\text{inner}} \leq m - 1$ alternate between convergent and divergent perturbative expansions, and all states receive non-perturbative corrections of the form: $\Delta E^{\text{NP}} \sim e^{-2S_I}$.

4.1.2. Perturbative structure for $H^{-,\zeta_{\text{QES}}}$

For $H^{-,\zeta_{\text{QES}}}$ the outer wells are lowered symmetrically, so the low-lying states are localized in the outer wells for $0 \leq \nu_{\text{outer}} \leq m - 1$. These states have an additional parity structure because of the parity symmetry between the two outer wells.

- $\underline{m = 1}$: The lowest QES case has $m = 1$, which means $\zeta = 1$, which is the SUSY case. The lowest perturbative state in each outer well has $E_{\text{pert}}(\nu_{\text{outer}} = 0, \hbar) =$

0 to all orders of perturbation theory. This perturbative level is split by non-perturbative effects. The parity-symmetric ground state remains zero, while the parity-antisymmetric first excited state receives a positive non-perturbative shift, $\Delta E^{\mathrm{NP}} \sim e^{-2S_I}$, as discussed in Sec. 3.2. For $\nu_{\mathrm{outer}} \geq 1$, the perturbative expansions are all divergent asymptotic expansions, with non-alternating expansion coefficients that grow factorially in magnitude.

$$\nu_{\mathrm{outer}} = 0 : \text{convergent expansion} \quad E_{\mathrm{pert}} = 0 \tag{73a}$$

$$\nu_{\mathrm{outer}} \geq 1 : \text{divergent non-alternating expansion} \tag{73b}$$

- $\underline{m = 2}$: In this case the deformation parameter has a non-integer rational value, $\zeta = \frac{5}{3}$, and we find that the lowest perturbative level ($\nu_{\mathrm{outer}} = 0$) has a convergent (indeed, truncating) expansion,

$$E_{\mathrm{pert}}^{-,\zeta=\frac{5}{3}}(\nu_{\mathrm{outer}} = 0, \hbar) = -\frac{2}{3} \tag{74}$$

This lowest perturbative level is split into a doublet by non-perturbative effects, with the lower (ground) doublet state receiving a negative shift, $\Delta E^{\mathrm{NP}} \sim -e^{-2S_I}$, while the higher (first excited) doublet state receives no non-perturbative shift. The general perturbative pattern for $m = 2$ is:

$$\nu_{\mathrm{outer}} = 0 : \text{convergent perturbative expansion} \tag{75a}$$

$$\nu_{\mathrm{outer}} \geq 1 : \text{divergent non-alternating expansion} \tag{75b}$$

- $\underline{m = 3}$: In this case the deformation parameter has a non-integer rational value, $\zeta = \frac{7}{3}$, and we find that the first two perturbative levels for states localized in the outer wells have non-truncating but *convergent* (indeed, exactly summable) expansions:

$$E_{\mathrm{pert}}^{-,\zeta=\frac{7}{3}}(\nu_{\mathrm{outer}} = 0, \hbar) = -\frac{1}{3} - \sqrt{1 + 2\hbar} \tag{76a}$$

$$E_{\mathrm{pert}}^{-,\zeta=\frac{7}{3}}(\nu_{\mathrm{outer}} = 1, \hbar) = -\frac{1}{3} + \sqrt{1 + 2\hbar} \tag{76b}$$

$$E_{\mathrm{pert}}^{-,\zeta=\frac{7}{3}}(\nu_{\mathrm{outer}} \geq 2, \hbar) : \text{divergent non-alternating expansion} \tag{76c}$$

These should be contrasted with the $\nu_{\mathrm{inner}} = 0, 2$ perturbative energies for the $II^{+,\zeta}$ sector with $\zeta = \frac{7}{3}$ in the previous subsection. The parity-symmetric forms of the $\nu_{\mathrm{outer}} = 0, 1$ perturbative states in (76a) and (76b) are exactly solvable, and receive no non-perturbative corrections. On the other hand, the parity-antisymmetric forms of the $\nu_{\mathrm{outer}} = 0, 1$ states in (76a) and (76b) receive non-perturbative corrections with a positive shift: $\Delta E^{\mathrm{NP}} \sim e^{-2S_I} > 0$. For $\nu_{\mathrm{outer}} \geq 2$, the perturbative energy expansions are divergent with non-alternating coefficients growing factorially fast in magnitude.

- $\underline{m = 4}$: In this case the deformation parameter has an integer value, $\zeta = 3$, and we find that the first two perturbative levels for states localized in the outer wells have non-truncating but *convergent* (indeed, exactly summable) expansions:

$$E_{\mathrm{pert}}^{-,\zeta=3}(\nu_{\mathrm{outer}} = 0, \hbar) = -1 - \sqrt{1 + 6\hbar} \tag{77a}$$

$$E_{\mathrm{pert}}^{-,\zeta=3}(\nu_{\mathrm{outer}} = 1, \hbar) = -1 + \sqrt{1 + 6\hbar} \tag{77b}$$

$$E_{\mathrm{pert}}^{-,\zeta=3}(\nu_{\mathrm{outer}} \geq 2, \hbar) : \text{divergent non-alternating expansion} \tag{77c}$$

These should be contrasted with the $\nu_{\text{inner}} = 1, 3$ perturbative energies for the $H^{+,\varsigma}$ sector with $\varsigma = 3$ in the previous subsection. The parity-symmetric forms of the $\nu_{\text{outer}} = 0, 1$ perturbative states in (77a) and (77b) are exactly solvable, and receive no non-perturbative corrections. On the other hand, the parity-antisymmetric forms of the $\nu_{\text{outer}} = 0, 1$ states in (77a) and (77b) receive non-perturbative corrections with a positive shift: $\Delta E^{\text{NP}} \sim e^{-2S_I} > 0$. For $\nu_{\text{outer}} \geq 2$, the perturbative energy expansions are divergent with non-alternating coefficients growing factorially fast in magnitude.

4.2. Semiclassical bion analysis of QES spectra: The hidden topological angle and the discrete θ angle

In this section we present a bion explanation of the intricate spectral patterns found for the $H^{\pm,\varsigma}$ QES systems in the previous section.

4.2.1. $H^{+,\varsigma}$: Physics of the hidden topological angle

The partner Hamiltonian $H^{+,\varsigma}$ is the ς-deformation of the partner Hamiltonian in SUSY QM which does not have a normalizable zero-energy state, with potential shown in the left hand side of Fig. 5. For $H^{+,\varsigma}$, the inner well is lowered and the outer wells are lifted. The harmonic states in the inner well have energies $\omega_0 \left(\frac{1}{2} - \frac{\varsigma}{2} + \nu_{\text{inner}} \right)$, where $\nu_{\text{inner}} = 0, 1, 2, \ldots$, and the lowest harmonic state in each of the outer wells has energy $\omega_1 \left(\frac{1}{2} + \frac{\varsigma}{2} \right)$. Recall that for our sextic potential, $\omega_0 = \omega$, and $\omega_1 = 2\omega$. We wish to explain semiclassically the alternating pattern structure of the low-lying states in the inner well found in Sec. 4.1.1, where for half of the states perturbation theory is convergent, while for the other half it is an asymptotic divergent series, with non-alternating coefficients. This arises due to an interesting structure of the HTA. However, for the convergent states, the result will not converge to the physical non-perturbative answer, as there are only neutral bion contributions, and no topological bions to cancel them. Hence the non-perturbative effects are present, unlike for the symmetric states in the $H^{-,\varsigma}$ system which we discuss in the next subsection.

The semiclassical amplitude for the neutral bion configuration can be computed via a generalization of the method in Sec. 3.1, extended to include the dependence on the perturbative level number ν. This analysis builds on earlier work,[23,34,48,49] and full details of the modern bion approach will be presented elsewhere.[51] The main result is that the thimble integration leads to the following expression for the neutral bion amplitude:

$$[I_+\bar{I}_+]_\pm \sim e^{-2S_I} e^{\pm i\pi \mathcal{D}^{+,\varsigma}} \left(\frac{\hbar}{\omega_1 c_1^2} \right)^{\mathcal{D}^{+,\varsigma}} \Gamma(\mathcal{D}^{+,\varsigma}) \tag{78a}$$

Here the "deficit angle" $\mathcal{D}^{+,\varsigma}$ is (compare with Eqs. (34) and (84)):

$$\mathcal{D}^{+,\varsigma} = \frac{1}{2} \left[(1+\varsigma) - \frac{\omega_0}{\omega_1} (1 - \varsigma + 2\nu_{\text{inner}}) \right] \tag{79}$$

The neutral bion amplitude has an imaginary ambiguous part

$$\text{Im}([I_+\bar{I}_+]_\pm) \sim \pm ie^{-2S_I}\frac{\pi}{\Gamma(1-\mathcal{D}^{+,\varsigma})}\left(\frac{\hbar}{\omega_0 c_0^2}\right)^{\mathcal{D}^{+,\varsigma}} \tag{80}$$

For the symmetric triple well, for which $\omega_1 = 2\omega_0 = 2\omega$, we deduce the hidden topological angle:

$$\theta_{\text{HTA}}^{+,\varsigma} = \pi\mathcal{D}^{+,\varsigma} = \pi\left(\frac{1}{4} + \frac{3\varsigma}{4} - \frac{\nu_{\text{inner}}}{2}\right) = \frac{\pi}{2}(m+1-\nu_{\text{inner}}) \tag{81}$$

Thus, for the $H^{+,\varsigma}$ sector, the HTA is quantized in units of $\pi/2$, rather than in units of π, as arises for the $H^{-,\varsigma}$ sector: compare with Eq. (86) for $\theta_{\text{HTA}}^{-,\varsigma}$.

The result (81) for $\theta_{\text{HTA}}^{+,\varsigma}$ has several important implications for the structure of the low-lying levels for which $0 \le \nu_{\text{inner}} \le m-1$. There are two distinct cases, depending on whether $m+1-\nu_{\text{inner}}$ is even or odd.

- *m* **odd**: The imaginary ambiguous part vanishes for $\nu_{\text{inner}} \in \mathcal{S}_1 = \{0, 2, \ldots, m-1\}$, and it does not vanish for $\nu_{\text{inner}} \in \mathcal{S}_2 = \{1, 3, \ldots, m-2\}$. This comes about because $\theta_{\text{HTA}}^{+,\varsigma}$ is quantized in units of π for $\mathcal{S}_1$, but $\theta_{\text{HTA}}^{+,\varsigma}$ is quantized in odd multiples of $\pi/2$ for $\mathcal{S}_2$. Therefore, perturbation theory must be convergent for $\mathcal{S}_1$, and it must be divergent for $\mathcal{S}_2$. Furthermore, for the states in $\mathcal{S}_1$, since $\Delta E^{\text{NP}} = -[\mathcal{NB}]$, the non-perturbative energy shift is positive if the HTA is an odd integer multiple of π, and is negative if the HTA is an even integer multiple of π.

- *m* **even**: The imaginary ambiguous part vanishes for $\nu_{\text{inner}} \in \mathcal{S}_1 = \{1, 3, \ldots, m-1\}$, and it does not vanish for $\nu_{\text{inner}} \in \mathcal{S}_2 = \{0, 2, 4, \ldots, m-2\}$. This comes about because $\theta_{\text{HTA}}^{+,\varsigma}$ is quantized in units of π for $\mathcal{S}_1$, but $\theta_{\text{HTA}}^{+,\varsigma}$ is quantized in odd multiples of $\pi/2$ for $\mathcal{S}_2$. Therefore, perturbation theory must be convergent for $\mathcal{S}_1$, and it must be divergent for $\mathcal{S}_2$. Note that this has the interesting implication that for the ground state, perturbation theory is asymptotic, while for the first excited state it is convergent. Furthermore, for the states in $\mathcal{S}_1$,

$$\Delta E^{\text{NP}} = -[\mathcal{NB}] = -e^{i\pi\left(\frac{m+1}{2} - \frac{\nu_{\text{inner}}}{2}\right)}2Ke^{-2S_I} \tag{82}$$

Therefore, the non-perturbative energy shift is positive if the HTA is an odd integer multiple of π, and it is negative if the HTA is an even integer multiple of π. For level number $\nu_{\text{inner}} \ge m$, all the states have divergent asymptotic expansion.

These predictions of the neutral bion analysis explain the patterns found in Sec. 4.1.1 using the BenderWu Mathematica package.[21]

4.2.2. $H^{-,\varsigma}$: Physics of the hidden topological angle and the discrete θ angle

When ς is close to unity, the deformation (62) serves as a soft SUSY breaking deformation which reveals how the hidden resurgent structure of SUSY quantum mechanics is present, disappearing at the non-generic SUSY value of $\zeta = 1$.[46] Consider the role of the neutral bions in this ζ-deformed theory. The neutral bion configuration starts at an outer well, interpolates to the inner well, and then interpolates back again. At the harmonic level, the lowest state localized in the inner well has energy $\omega_0 \left(\frac{1}{2} + \frac{\zeta}{2} \right)$, and the ground state and low-lying states localized in the outer wells have energies $\omega_1 \left(\frac{1}{2} - \frac{\zeta}{2} + \nu_{\text{outer}} \right)$, where $\nu_{\text{outer}} = 0, 1, 2, \ldots$ is the harmonic energy level label. For a certain number of low-lying states localized in the outer wells, we will show that the perturbation theory is convergent. For half of these states, perturbation theory converges to the exact physical result, while for the other half there is a non-perturbative bion contribution on top of the convergent perturbative sum, generalizing the result of the supersymmetric model.

As in the $H^{+,\zeta}$ sector discussed in the previous section, the semiclassical amplitude for the neutral bion configuration can be computed via a generalization[51] of the method in Sec. 3.1, extended to include the dependence on the perturbative level number ν. The main result is that the thimble integration leads to the following expression for the neutral bion amplitude:

$$[I_+\bar{I}_+]_\pm \sim e^{-2S_I} e^{\pm i\pi \mathcal{D}^{-,\zeta}} \left(\frac{\hbar}{\omega_0 c_0^2} \right)^{\mathcal{D}^{-,\zeta}} \Gamma(\mathcal{D}^{-,\zeta}) \tag{83}$$

Here $\mathcal{D}^{-,\zeta}$ is the zeta-deformed "deficit angle" (compare with Eqs. (34) and (79)):

$$\mathcal{D}^{-,\zeta} \equiv \frac{1}{2} \left[1 - \frac{\omega_1}{\omega_0} (1 + 2\nu_{\text{outer}}) + \zeta \left(1 + \frac{\omega_1}{\omega_0} \right) \right] \tag{84}$$

The imaginary part of the neutral bion amplitude can be written as

$$\text{Im}([I_+\bar{I}_+]_\pm) \sim \pm i e^{-2S_I} \frac{\pi}{\Gamma(1 - \mathcal{D}^{-,\zeta})} \left(\frac{\hbar}{\omega_0 c_0^2} \right)^{\mathcal{D}^{-,\zeta}} \tag{85}$$

Thus, for the symmetric triple-well, for which $\omega_1 = 2\omega_0 = 2\omega$, we deduce the hidden topological angle associated with the neutral bion configuration:

$$\theta_{\text{HTA}}^{-,\zeta} = \pi \mathcal{D}^{-,\zeta} = \pi \left(-\frac{1}{2} + \frac{3\zeta}{2} - 2\nu_{\text{outer}} \right) = \pi \left(m - 2\nu_{\text{outer}} \right) \tag{86}$$

Note that $\theta_{\text{HTA}}^{-,\zeta}$ is quantized in integer units of π, rather than in integer units of $\frac{\pi}{2}$, as for $\theta_{\text{HTA}}^{+,\zeta}$ in (81). Furthermore, the phase $\theta_{\text{HTA}}^{-,\zeta}$ does not have a level number dependence, since the angles are identified by 2π shifts. This is in contrast to the $H^{+,\zeta}$ sector, where there is a non-trivial level number dependence in (81), leading to an alternating convergent/divergent perturbative pattern for the low-lying states of the $H^{+,\zeta}$ sector. For the states for which $0 \leq \nu_{\text{outer}} \leq \left\lfloor \frac{3(\zeta-1)}{4} \right\rfloor = \left\lfloor \frac{m-1}{2} \right\rfloor$, the amplitude of the topological bions and neutral bions are related by

$$[\mathcal{NB}] = e^{i\pi m} [\mathcal{TB}] \tag{87}$$

for the QES ζ values in (68).

Each value of ν_{outer} gives rise to two eigenstates of the $H^{-,\varsigma}$ Hamiltonian, one symmetric combination ($N = 2\nu_{\text{outer}}$) and one antisymmetric ($N = 2\nu_{\text{outer}} + 1$), where N is the fundamental quantum number associated with $H^{-,\varsigma}$. Here the level label N refers not to a perturbative level in a given well, but to the states of the whole potential. The states for which there is no non-perturbative contribution, i.e, the non-perturbative contribution cancels, are determined by an interference pattern sourced by the hidden topological angle θ_{HTA} and by a discrete theta angle $\theta_{\text{disc.}}$.

The appearance of the discrete theta angle can be seen as follows. The triple-well potential has a parity symmetry, P. Therefore, one can consider two types of partition functions: $\text{tr}\left(e^{-\beta H^{-,\varsigma}}\right)$ and $\text{tr}\left(Pe^{-\beta H^{-,\varsigma}}\right)$ (see also[23]). One can now gauge parity. By this one usually means the following

$$\text{tr}\left[\left(\frac{1+P}{2}\right)e^{-\beta H^{-,\varsigma}}\right] = \sum_{N\in\mathcal{H}_{P\,\text{even}}} e^{-\beta E_N} \tag{88}$$

where in the state sum we sum over only parity even states, i.e. gauging the parity is equivalent to this projection. However when gauging one has a choice to project to a state sum over only parity odd states, and we can identify this with turning on the discrete theta angle:[d]

$$\theta_{\text{disc.}} = \pi \tag{89}$$

Then

$$\text{tr}\left[\left(\frac{1-P}{2}\right)e^{-\beta H^{-,\varsigma}}\right] = \sum_{N\in\mathcal{H}_{P\,\text{odd}}} e^{-\beta E_N} \tag{90}$$

On the other hand,

$$Z^- = \text{tr}\left(e^{-\beta H^{-,\varsigma}}\right) \approx e^{-\beta E_0} 2\cosh([\mathcal{TB}]\beta)\, e^{[\mathcal{NB}]\beta}$$
$$= e^{-\beta E_0}\left(e^{\beta([\mathcal{TB}]+[\mathcal{NB}])} + e^{\beta(-[\mathcal{TB}]+[\mathcal{NB}])}\right)$$

$$Z_P^- = \text{tr}\left(Pe^{-\beta H^{-,\varsigma}}\right) \approx e^{-\beta E_0} 2\sinh([\mathcal{TB}]\beta)\, e^{[\mathcal{NB}]\beta}$$
$$= e^{-\beta E_0}\left(e^{\beta([\mathcal{TB}]+[\mathcal{NB}])}\quad e^{\beta(-[\mathcal{TB}]+[\mathcal{NB}])}\right) \tag{91}$$

Therefore, the $[\mathcal{TB}]$ contribution to parity even and parity odd states has an overall sign difference. As a result, we can write the leading non-perturbative contribution at energy level N as

$$E_N^{\text{NP}} = -(e^{i\pi N}[\mathcal{TB}] + [\mathcal{NB}])$$
$$= -(e^{i\pi N} + e^{i\pi m})[\mathcal{TB}]$$
$$= -(e^{i\pi N} + e^{i\pi m})K_{N,m}e^{-2S_I}, \tag{92}$$

[d]Here the terminology[52] is analogous to the one in periodic potentials. The discrete translation symmetry of a periodic potential can be gauged by projecting to a particular charge θ which is angle valued for a $\mathbb{Z}$ symmetry. This is the usual θ angle, or Bloch angle, for a particle on a circle.

where $K_{N,m} > 0$. Therefore, the contribution of bion configurations to an energy level can be positive, negative or zero, depending on the even/odd parity of N and m. This result has the following implications:

- $m = $ odd, $N = 2\nu_{\text{outer}}=$ even (S): the NP contribution vanishes for such a symmetric state. These states are exactly solvable.
- $m = $ odd, $N = 2\nu_{\text{outer}} + 1=$ odd (AS): the NP shift is $E_N^{\text{NP}} = 2K_N e^{-2S_I} > 0$. The AS partner is lifted up non-perturbatively.
- $m = $ even, $N = 2\nu_{\text{outer}}=$ even (S): the NP shift is $E_N^{\text{NP}} = -2K_N e^{-2S_I} < 0$. The symmetric state is pushed down. These states are not exactly solvable.
- $m = $ even, $N = 2\nu_{\text{outer}} + 1=$ odd (AS): the NP contribution is $E_N^{\text{NP}} = 0$. These states are exactly solvable.

This bion analysis explains the spectral structure in Sec. 4.1.2, which was found using the BenderWu package. Our construction provides a semiclassical path integral explanation of the remarkable phenomena observed in quasi-exactly-solvable (QES)[18,50] systems. In QES systems, for the $m = $ odd (even) case, a number $\lfloor \frac{m+1}{2} \rfloor$ of symmetric (antisymmetric) states are exactly solvable and free of non-perturbative contributions. It was a puzzle why these states do not receive non-perturbative contributions despite the existence of obvious non-perturbative solutions. The above bion analysis, culminating in the expression (92), resolves this puzzle. First, apart from the obvious configurations (the topological bions), there are also neutral bions. Their net effect involves a subtle interference effect between the hidden topological angle θ_{HTA} and the discrete theta angle $\theta_{\text{disc.}}$. When this expression vanishes in the path integral, the corresponding state is exactly solvable in the Hamiltonian formulation.

5. Conclusions

We have shown that several new non-perturbative effects arise in the triple-well system that have no analogue in the familiar symmetric double-well system. The main results are the following.

- In potentials with harmonic, classically degenerate minima not related by a symmetry, despite the fact that instantons are exact solutions, they do not contribute to the energy spectrum at leading order in semi-classics. The fluctuation prefactor of the instanton amplitude vanishes if the frequencies in two consecutive wells are not equal.
- Quite generally the leading order semiclassical configurations contributing to the spectrum of such systems are bion configurations. These are dominant configurations that live on the Lefschetz thimble of critical points at infinity.
- Even though the inner and outer wells have different shapes and different curvatures, the perturbative expansions for states localized in the inner and outer wells are related by an exact mapping; in the bosonic system (20) as well as in

the supersymmetric (41) and quasi-exactly-solvable systems (67). This fact has a simple explanation in terms of the all-orders WKB approach to perturbation theory.

- In the SUSY and QES systems, there is an intriguing pattern of interference between the neutral bions and topological bions, which arises from the interplay of the hidden topological angle and the discrete theta angle. Whenever the non-perturbative effects cancel precisely, the corresponding state in the Hilbert space is exactly solvable.

- The bion analysis resolves an old puzzle concerning quasi-exactly-solvable systems. Despite the presence of obvious non-perturbative configurations which would contribute to the spectrum, the spectrum turns out to be algebraic, and does not include non-perturbative factors. This is the result of interference between different bions, and the analysis shows that there also exist complex configurations, and there are exact non-perturbative cancellations among them. This is a clear demonstration of resurgent structure in the SUSY and QES systems.

- Semiclassical analysis based on the hidden topological angle θ_{HTA} predicts that the character of the perturbative expansion has an alternating pattern of convergent/divergent states, and then changes from convergent to divergent after a certain energy level. These predictions have been confirmed by a large-order analysis of the associated perturbative expansions using the BenderWu Mathematica package.[21]

Our bion construction shows that the semiclassical analysis for general quantum potentials is far more intricate than the paradigmatic textbook examples of the double-well potential and the periodic potential. Perhaps, the most interesting lessons concern the important roles played by complex configurations and the remarkable interference patterns induced by the hidden topological angle and the discrete theta angle. We believe that there are many other further phenomena waiting to be explored. In particular, it would be interesting to investigate the appearance of the hidden topological angle, and the interference between saddles, in the exact WKB formulation.

Acknowledgments

This material is based upon work supported by the U.S. Department of Energy, Office of Science, Division of High Energy Physics under Award Number DE-SC0010339 (GD), and by the U.S. Department of Energy, Office of Science, Division of Nuclear Physics under Award DE-SC0013036 (MU). We thank Yuya Tanizaki for useful discussions and comments. TS is funded by the Royal Society University Research Fellowship.

References

1. S. R. Coleman, "The uses of instantons," in *Aspects of Symmetry* (Cambridge University Press, 1985).
2. L. S. Schulman, *Techniques and Applications of Path Integration* (Wiley, New York, 1981).
3. A. I. Vainshtein, V. I. Zakharov, V. A. Novikov and M. A. Shifman, "ABC's of instantons," *Sov. Phys. Usp.* **25**, 195 (1982) [*Usp. Fiz. Nauk* **136**, 553 (1982)].
4. I. I. Balitsky and A. V. Yung, "Instanton molecular vacuum in $N = 1$ supersymmetric quantum mechanics," *Nucl. Phys. B* **274**, 475 (1986).
5. M. E. Peskin and D. V. Schroeder, *An Introduction to Quantum Field Theory* (Addison-Wesley, Reading, MA, 1995).
6. M. A. Shifman, *ITEP Lectures on Particle Physics and Field Theory*, Vols. 1 and 2, World Scientific Lecture Notes in Physics (World Scientific, Singapore, 1999).
7. J. Zinn-Justin, *Quantum Field Theory and Critical Phenomena* (Oxford University Press, 2002).
8. T. Schäfer and E. V. Shuryak, "Instantons in QCD," *Rev. Mod. Phys.* **70**, 323 (1998), arXiv:hep-ph/9610451.
9. M. Mariño, *Instantons and Large N : An Introduction to Non-Perturbative Methods in Quantum Field Theory* (Cambridge University Press, 2015).
10. N. Nekrasov, "Tying up instantons with anti-instantons," in *Ludwig Faddeev Memorial Volume*, M.-L. Ge, A. J. Niemi, K. K. Phua and L. A. Takhtajan (eds.) (World Scientific, Singapore, 2018), arXiv:1802.04202.
11. R. Jackiw and C. Rebbi, "Vacuum periodicity in a Yang-Mills quantum theory," *Phys. Rev. Lett.* **37**, 172 (1976).
12. R. Jackiw, C. Nohl and C. Rebbi, "Conformal properties of pseudoparticle configurations," *Phys. Rev. D* **15**, 1642 (1977).
13. R. Jackiw and C. Rebbi, "Spinor analysis of Yang-Mills theory," *Phys. Rev. D* **16**, 1052 (1977).
14. R. Jackiw, "Quantum meaning of classical field theory," *Rev. Mod. Phys.* **49**, 681 (1977).
15. R. Jackiw, "Introduction to the Yang-Mills quantum theory," *Rev. Mod. Phys.* **52**, 661 (1980).
16. R. Jackiw, "Topological investigations of quantized gauge theories," in *Current Algebra and Anomalies*, S. B. Treiman *et al.* (eds.) (Princeton University Press, 1985).
17. A. Behtash, G. V. Dunne, T. Schäfer, T. Sulejmanpasic and M. Ünsal, "Critical points at infinity, non-Gaussian saddles, and bions," *JHEP* **1806**, 068 (2018), arXiv:1803.11533.
18. A. V. Turbiner, "Quasiexactly solvable problems and SL(2) group," *Commun. Math. Phys.* **118**, 467 (1988).
19. A. V. Turbiner and A. G. Ushveridze, "Spectral singularities and quasiexactly solvable quantal problem," *Phys. Lett. A* **126**, 181 (1987).
20. A. Turbiner, "Quasiexactly solvable differential equations," arXiv:hep-th/9409068.
21. T. Sulejmanpasic and M. Ünsal, "Aspects of perturbation theory in quantum mechanics: The BenderWu Mathematica package," *Comput. Phys. Commun.* **228**, 273 (2018), arXiv:1608.08256.
22. G. V. Dunne, "Functional determinants in quantum field theory," *J. Phys. A* **41**, 304006 (2008), arXiv:0711.1178.
23. J. Zinn-Justin and U. D. Jentschura, "Multi-instantons and exact results I: Conjectures, WKB expansions, and instanton interactions," *Annals Phys.* **313**, 197 (2004),

arXiv:quant-ph/0501136. "Multi-instantons and exact results II: Specific cases, higher-order effects, and numerical calculations," *Annals Phys.* **313**, 269 (2004), arXiv:quant-ph/0501137.

24. G. Başar, G. V. Dunne and M. Ünsal, "Quantum geometry of resurgent perturbative/nonperturbative relations," *JHEP* **1705**, 087 (2017) arXiv:1701.06572.
25. J. L. Dunham, "The Wentzel-Brillouin-Kramers method of solving the wave equation", Phys. Rev. **41**, 713 (1932).
26. C. M. Bender and S. A. Orszag, *Advanced Mathematical Methods for Scientists and Engineers* (Springer, 1999).
27. C. M. Bender and T. T. Wu, "Anharmonic oscillator," *Phys. Rev.* **184**, 1231 (1969).
28. C. M. Bender and T. T. Wu, "Large order behavior of perturbation theory," *Phys. Rev. Lett.* **27**, 461 (1971).
29. J. C. Le Guillou and J. Zinn-Justin, *Large Order Behavior of Perturbation Theory* (North-Holland, Amsterdam, 1990).
30. E. Brezin, G. Parisi and J. Zinn-Justin, "Perturbation theory at large orders for potential with degenerate minima," *Phys. Rev. D* **16**, 408 (1977).
31. E. B. Bogomolny, "Calculation of instanton–anti-instanton contributions in quantum mechanics," *Phys. Lett.* **91B**, 431 (1980).
32. J. Zinn-Justin, "Multi-instanton contributions in quantum mechanics," *Nucl. Phys. B* **192**, 125 (1981).
33. A. Behtash, G. V. Dunne, T. Schäfer, T. Sulejmanpasic and M. Ünsal, "Complexified path integrals, exact saddles and supersymmetry," *Phys. Rev. Lett.* **116**, 011601 (2016), arXiv:1510.00978.
34. A. Behtash, G. V. Dunne, T. Schäfer, T. Sulejmanpasic and M. Ünsal, "Toward Picard-Lefschetz theory of path integrals, complex saddles and resurgence," *Ann. Math. Sci. Appl.* **02**, 95 (2017), arXiv:1510.03435.
35. T. Fujimori, S. Kamata, T. Misumi, M. Nitta and N. Sakai, "Exact resurgent transseries and multibion contributions to all orders," *Phys. Rev. D* **95**, 105001 (2017), arXiv:1702.00589.
36. M. Mariño, "Lectures on non-perturbative effects in large N gauge theories, matrix models and strings," *Fortsch. Phys.* **62**, 455 (2014), arXiv:1206.6272.
37. I. Aniceto and R. Schiappa, "Nonperturbative ambiguities and the reality of resurgent transseries," *Commun. Math. Phys.* **335**, 183 (2015), arXiv:1308.1115.
38. J. L. Richard and A. Rouet, "Complex saddle points versus dilute gas approximation in the double well anharmonic oscillator," *Nucl. Phys. B* **185**, 47 (1981).
39. A. Lapedes and E. Mottola, "Complex path integrals and finite temperature," *Nucl. Phys. B* **203**, 58 (1982).
40. G. V. Dunne and M. Ünsal, "New nonperturbative methods in quantum field theory: From large-N orbifold equivalence to bions and resurgence," *Ann. Rev. Nucl. Part. Sci.* **66**, 245 (2016), arXiv:1601.03414.
41. M. Serone, G. Spada and G. Villadoro, "Instantons from perturbation theory," *Phys. Rev. D* **96**, 021701 (2017), arXiv:1612.04376.
42. M. Serone, G. Spada and G. Villadoro, "The power of perturbation theory," *JHEP* **05**, 056 (2017), arXiv: 1702.04148.
43. M. Ünsal, "Theta dependence, sign problems and topological interference," *Phys. Rev. D* **86**, 105012 (2012), arXiv:1201.6426.
44. A. Behtash, T. Sulejmanpasic, T. Schäfer and M. Ünsal, "Hidden topological angles and Lefschetz thimbles," *Phys. Rev. Lett.* **115**, 041601 (2015), arXiv:1502.06624.
45. E. Witten, "Dynamical breaking of supersymmetry," *Nucl. Phys. B* **188**, 513 (1981).
46. G. V. Dunne and M. Ünsal, "Deconstructing zero: Resurgence, supersymmetry and

complex saddles," *JHEP* **1612**, 002 (2016), arXiv:1609.05770.

47. J. J. M. Verbaarschot, P. C. West and T. T. Wu, "Large order behavior of the super-symmetric anharmonic oscillator," *Phys. Rev. D* **42**, 1276 (1990).

48. J. J. M. Verbaarschot and P. C. West, "Instantons and Borel resummability for the perturbed supersymmetric anharmonic oscillator," *Phys. Rev. D* **43**, 2718 (1991).

49. A. Behtash, "More on homological supersymmetric quantum mechanics," *Phys. Rev. D* **97**, 065002 (2018), arXiv:1703.00511.

50. C. Kozçaz, T. Sulejmanpasic, Y. Tanizaki and M. Ünsal, "Cheshire cat resurgence, self-resurgence and quasi-exact solvable systems," *Commun. Math. Phys.* **364**, 835 (2018), arXiv:1609.06198.

51. G. V. Dunne, T. Sulejmanpasic and M. Ünsal, in preparation.

52. O. Aharony, N. Seiberg and Y. Tachikawa, "Reading between the lines of four-dimensional gauge theories," *JHEP* **1308**, 115 (2013) arXiv:1305.0318.

Chapter 16

On the Sudakov Form Factor, and a Factor of Two*

Stefano Forte

*Tif Lab, Dipartimento di Fisica, Università di Milano and
INFN, Sezione di Milano,
Via Celoria 16, I-20133 Milano, Italy*

I answer a question that Roman asked me, and I draw some lessons from the answer. The question is: why is the Sudakov form factor larger by a factor of two, if computed for off-shell fermions in comparison to the on-shell case? The answer sheds some light on the interplay between infrared and collinear singularities — and the importance of factors of two.

1. Master of scientific style

Much can be said about working under Roman's supervision in the mid-eighties: it was an absorbing, intense, at times exhilarating, at times stressful experience. A formidable array of ideas to take in, concepts to grasp, and good practices to learn, often delivered as a side remark, accompanied by a grin.[a] Ranging from the way to write displayed equations in a paper ("it is called an equation because it has an equal sign"), to the importance of choosing the symbols when performing a calculation ("one should not pick letters at random from the alphabet"). Some perks, too, such as going for dinner at the Harvard Faculty Club with Stephen Hawking and Sidney Coleman — including the task of steadying the former's wheelchair in the minivan that took us there.

Overall, it amounted to a lesson of scientific method, and of scientific style: delivered mostly by example. One thing I understood — the painful way, as I am prone to algebraic mistakes — is the importance of details when performing a computation. Roman used to retell the story of someone who published a perturbative computation in which he had guessed the value of a high-order term without actually calculating it — only to be belied by the explicit result once Roman got round to determine it.

Not so long ago, I came to think about it again. It was the summer of 2017, I was spending some time at the Aspen Center for Physics, and Roman, coincidentally also there, took me out for lunch. The conversation at some point revolved on

---

*Contribution to the volume *Roman Jackiw: 80th Birthday Festschrift*.
[a]See the contribution by Michael Bos in this volume.

some then-recent work of mine[1] on QCD resummation. Roman mentioned that he had worked on related topics around the time of his PhD thesis:[2] he had computed perturbatively the high-momentum transfer limit of the QED vertex function, which is double logarithmic, and correctly guessed the exponentiation of the double logs. As our lunch was going on, Roman then abruptly asked me whether I knew that the coefficient of the double log is by a factor of two larger off-shell in comparison to the on-shell result, and whether I knew a simple physical reason for that.

I didn't know.

Continuing the discussion with Roman via email, after I got back home, I realized that I couldn't immediately come up with an answer. I also subsequently realized that this point is typically not discussed, or even mentioned in textbooks. In fact, what is commonly known as "the Sudakov form factor" is the on-shell result, yet the original[3] Sudakov calculation applies to the off-shell case, and the factor two difference usually goes unnoticed. Indeed, as Roman pointed out to me, in a recent paper from the Russian school[4] it is incorrectly stated[b] that the same result applies in the on-shell and off-shell cases, only with a different choice of infrared regulator. I asked various experts on QCD, where the Sudakov exponentiation plays an important role, and none was aware of this.

Answering Roman's question is the purpose of this note.

2. The vertex function and the Sudakov form factor

The computation performed by Roman[2] determines to all orders the high-energy behavior of the vertex function in quantum electrodynamics (QED) (see Fig. 1). The genesis of this paper has been recounted by Roman:[5] his advisor, Ken Wilson, suggested him as a thesis project to derive this high-energy behavior, which had been previously obtained by Sudakov[3] in the off-shell case, using renormalization-group (RG) methods: both as a way of validating the then-novel RG techniques, and also, of obtaining the on-shell result.

The way to attack and solve this problem using RG techniques was only found several years later[6,7] (see also Sec. 4.3 below), but Roman did manage to tackle it by direct computation using an eikonal approximation, which was known to Wilson, and systematically developed by Weinberg[8] into what is now known as light-cone field theory.[c]

The result found by Roman[2] is that in the high-energy limit the vertex function is equal to

$$\Gamma^{\mu}(p_1, p_2) = \gamma^{\mu} \Gamma(p_1^2, p_2^2, q^2); \tag{1}$$

where to one loop (Fig. 1)

$$\Gamma^{(1),\,\text{off}}(p_1^2, p_2^2, q^2) = -\frac{\alpha}{2\pi} \ln\left|\frac{q^2}{p_1^2}\right| \ln\left|\frac{q^2}{p_2^2}\right| \tag{2}$$

---

[b]See in particular the discussion after Eq. (34) of Ioffe's paper.[4]
[c]A textbook discussion of the eikonal approximation is e.g. given by Sterman,[9] while a presentation of light-cone field theory can be found in recent summer school proceedings.[10]

for off-shell fermions with virtualities p_i^2, while for on-shell fermions

$$\Gamma^{(1),\,\text{on}}(m^2, m^2, q^2) = -\frac{\alpha}{4\pi}\ln^2\frac{|q|^2}{\mu^2} \tag{3}$$

where μ is an infrared regulator (i.e. a photon mass).

To all perturbative orders the one-loop result exponentiates:

$$\Gamma^{\text{off}}(p_1^2, p_2^2, q^2) = \exp\left(-\frac{\alpha}{2\pi}\ln\left|\frac{q^2}{p_1^2}\right|\ln\left|\frac{q^2}{p_2^2}\right|\right), \tag{4}$$

$$\Gamma^{\text{on}}(m^2, m^2, q^2) = \exp\left(-\frac{\alpha}{4\pi}\ln^2\frac{|q|^2}{\mu^2}\right). \tag{5}$$

All these results hold to double-logarithmic accuracy, i.e. up to terms with a lower power of $\ln|q|^2$.

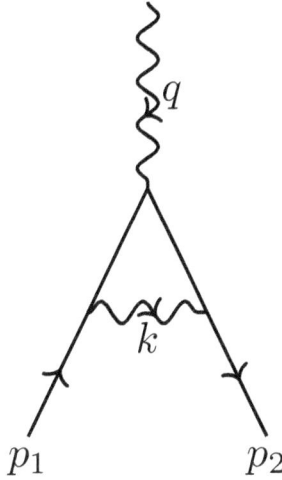

Fig. 1. The one-loop vertex function.

The result Eq. (4) is in agreement with the previous result of Sudakov,[3] which had been subsequently reproduced by others,[11] who attempted to determine the on-shell result but did not obtain the correct answer and failed to prove exponentiation. The results Eqs. (4) and (5) are also given in the volume devoted to QED of "Landau's" theoretical physics course,[12] first published in 1974 (after Landau's death): for the off-shell result Sudakov[3] is cited, while the on-shell result is written in the form

$$\bar{\Gamma}^{(1),\,\text{on}}(m^2, m^2, q^2) = -\frac{\alpha}{4\pi}\left(\ln^2\left|\frac{q^2}{m^2}\right| + 4\ln\left|\frac{q^2}{m^2}\right|\ln\left|\frac{m}{\mu}\right|\right), \tag{6}$$

which of course coincides with Eq. (3) up to terms which are not logarithmic in q^2:

$$\bar{\Gamma}^{(1),\,\text{on}}(m^2, m^2, q^2) = \Gamma^{(1),\,\text{on}}(m^2, m^2, q^2) + \frac{\alpha}{4\pi}\ln^2\frac{m^2}{\mu^2}. \tag{7}$$

It is clear that, contrary to what one might naively think (and contrary to what sometimes stated explicitly[4]) the on-shell result is not simply obtained by setting $p_1^2 = p_2^2 = \mu^2$ in the off-shell one — rather, it is twice as large. Why?

Answering this question requires a computation of the vertex function, starting at one loop. Rather than the vertex function itself, however, it is more instructive to look at its real-emission counterpart. Indeed, it is the imaginary part of the photon propagator in the diagram of Fig. 1 which leads to the double-logarithmic behavior Eqs. (2)–(5).[2,12] This imaginary part can be extracted using the standard cutting rule

$$\text{Disc}\frac{1}{k^2 + i\epsilon} = -2\pi i\delta(k^2)\Theta(k^0).\tag{8}$$

This transforms the vertex function into the interference of two real-emission diagrams (see Fig. 2). Of course, it is this one-to-one correspondence of virtual and real-emission contributions which guarantees the cancellation of infrared singularities for sufficiently inclusive physical observables.[d]

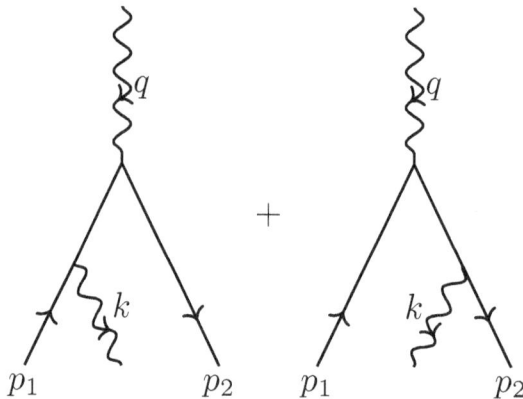

Fig. 2.   The real-emission diagrams corresponding to the vertex function of Fig. 1.

I will first, present in Sec. 3 a direct computation of these real-emission contributions, viewed as contributions to the decay amplitude of a virtual photon into a fermion–antifermion pair, in the rest frame of the virtual photon. This computation reproduces the result of Eqs. (2) and (3), and it gives a first hint on its origin. For a complete clarification, however, it is useful to look at the problem in a different frame: namely, by viewing the diagrams of Fig. 2 as contributions to Drell–Yan-like production of a virtual photon in a fermion–antifermion collision, in the center-of-mass frame of the colliding fermions. In Sec. 4 I will show that the origin of the difference between on-shell and off-shell can be traced to a different interplay of soft and collinear singularities in either case. Exponentiation then ensues from

[d]See Chapter 13 of Weinberg's treatise[13] for a modern discussion.

the factorized structure of phase space, which can be proved using an argument[1] developed in order to combine soft and collinear resummation.

3. Computing the vertex function[e]

Consider the decay of an off-shell photon with momentum q. Of course, this means that the incoming fermion of Fig. 2 now is an outgoing antifermion, and the result in the kinematics of Fig. 2 can be recovered by crossing. However, because the double log behavior only depends on the modulus of the momentum transfer, the computation can be indifferently performed in either kinematics. For simplicity we consider the case of massless fermions, though it can be checked explicitly[14] that results are unchanged with a finite fermion mass.

Double logs Eqs. (2) and (3) arise due to the soft region of integration over the momentum k of the emitted photon. The amplitude for emission of a soft photon is obtained by multiplying the amplitude for the process without soft photon by an eikonal factor,[f] so that in this limit the amplitude for the process of Fig. 2 is given by

$$M = M_0 e \left(\frac{2p_1^\mu}{(p_1 + k)^2} - \frac{2p_2^\mu}{(p_2 + k)^2} \right), \tag{9}$$

where M_0 is the amplitude without the extra photon. The square amplitude is then

$$|M|^2 = -|M_0|^2 e^2 \frac{8p_1 \cdot p_2}{(p_1 + k)^2 (p_2 + k)^2}, \tag{10}$$

and the desired real-emission amplitude is found integrating this over the phase space of the emitted photon:

$$d\Phi_k = \frac{k^2 dk d\cos\theta d\phi}{2E(2\pi)^3} = \frac{kdEd\cos\theta}{8\pi^2}, \tag{11}$$

where E and k are respectively the energy and modulus of the three-momentum of the emitted photon, and in the last step we have used $\frac{dE}{k} = \frac{dk}{E}$, and integrated over the azimuth ϕ.

The calculation is performed in the on-shell case by assuming a small photon mass μ as a regulator, so $(p_i+k)^2 = 2p_i \cdot k + \mu^2$, and in the off-shell case by assuming $p_i^2 > 0$ so $(p_i + k)^2 = 2p_i \cdot k + p_i^2$.

3.1. *On-shell fermions*

In the rest frame of the decaying photon the square amplitude is

$$|M|^2 = -|M_0|^2 e^2 \frac{4s}{[E\sqrt{s}(1 - \beta\cos\theta) + \mu^2][E\sqrt{s}(1 + \beta\cos\theta) + \mu^2]}$$
$$= -|M_0|^2 e^2 \frac{4}{E^2\left[(1 - \beta\cos\theta) + \frac{\mu^2}{E\sqrt{s}}\right]\left[(1 + \beta\cos\theta) + \frac{\mu^2}{E\sqrt{s}}\right]}, \tag{12}$$

---

[e]The computation presented in this section is based on unpublished notes by Paolo Nason.[14]
[f]See Chapter 13 of Weinberg's treatise.[13]

where $E = k^0$ is the emitted photon's energy, $\sqrt{s} = \sqrt{(p_1 + p_2)^2} = 2p_i^0 = \sqrt{|q^2|}$ is the energy of the fermion–antifermion system, and

$$\beta = \sqrt{1 - \frac{\mu^2}{E^2}} = 1 - \frac{\mu^2}{2E^2}\left(1 + O(\mu^2/E^2)\right). \tag{13}$$

Integrating over the emitted photon's phase space Eq. (11) we get

$$\int d\Phi_k |M|^2$$

$$= -|M_0|^2 \frac{e^2}{2\pi^2} \int \frac{E dE d\cos\theta}{E^2\left[(1 - \cos\theta + \cos\theta\frac{\mu^2}{2E^2} + \frac{\mu^2}{E\sqrt{s}})\right]\left[(1 + \cos\theta - \cos\theta\frac{\mu^2}{2E^2} + \frac{\mu^2}{E\sqrt{s}})\right]}. \tag{14}$$

This leads to logarithmic behavior either when $\theta \to 1$ or $\theta \to -1$, corresponding to the region in which the emitted photon is respectively collinear to p_1 or p_2. These two collinear and anticollinear contributions are the same, and we get

$$\int d\Phi_k |M|^2 = -2|M_0|^2 \frac{e^2}{4\pi^2} \int \frac{dE d\cos\theta}{E(1 - \cos\theta)} + \text{non log}, \tag{15}$$

where we have for definiteness written the collinear contribution, while introducing a factor of two in order to account for the anticollinear one, and we have retained the leading term in an expansion in $\cos\theta$ about $\cos\theta = 1$, as well as in an expansion of μ^2 about $m^2 = 0$, so the second square bracket in the denominator of Eq. (14) just reduces to a factor of two.

Performing the angular integral we immediately get

$$\int d\Phi_k |M|^2 = -2|M_0|^2 \frac{\alpha}{\pi} \int \frac{1}{2}\frac{dE^2}{E^2} \ln\left[\left(\frac{\mu^2}{2E^2} + \frac{\mu^2}{E\sqrt{s}}\right)^{-1}\right], \tag{16}$$

where we have introduced the fine-structure constant $\alpha = \frac{e^2}{4\pi}$. The double integral comes from the infrared region of integration over the energy E of the emitted fermion, hence we can neglect the second term in the argument of the log, and we get, keeping only double logarithmic terms,

$$\int d\Phi_k |M|^2 = -|M_0|^2 \frac{\alpha}{2\pi} \ln^2 \frac{|q|^2}{\mu^2}, \tag{17}$$

where the upper limit of integration over energy is of course just $\sqrt{s}/2$; indeed, the argument of the log is fixed by dimensional analysis.

The real-emission contribution should be compared to the square of the virtual one, which leads to an extra factor of two in the real emission case, so this result exactly matches Roman's result[2] for the on-shell vertex function Eq. (3).

3.2. Off-shell fermions

If the fermions are off-shell, it is now the fermion virtuality which regulates the collinear singularity, so that no massive photon regulator is needed. The square amplitude is then

$$|M|^2 = -|M_0|^2 e^2 \frac{4s}{[E\sqrt{s}(1-\beta\cos\theta)+p_1^2][E\sqrt{s}(1+\beta\cos\theta)+p_2^2]} \left[1+O(p_i^2/s)\right] \tag{18}$$

where now the equalities $2p_i^0 \approx \sqrt{|q^2|} \approx \sqrt{s}$ all hold up to terms of order p_i^2/s, and

$$\beta_i = \sqrt{1 - \frac{p_i^2}{(p_i^0)^2}} = 1 - \frac{2p_i^2}{s}\left[1+O(p_i^2/s)\right]. \tag{19}$$

Integrating over the photon's phase space we have again a pair of collinear and anticollinear singularities:

$$\int d\Phi_k |M|^2$$

$$= -|M_0|^2 \frac{e^2}{2\pi^2} \int \frac{EdEd\cos\theta}{\left[E(1-\cos\theta)+\cos\theta E\frac{2p_1^2}{s}+\frac{p_1^2}{\sqrt{s}}\right]\left[E(1+\cos\theta)-\cos\theta E\frac{2p_2^2}{s}+\frac{p_2^2}{\sqrt{s}}\right]}. \tag{20}$$

However, there are two differences: the form of the collinear cutoff, which now depends on the virtuality, p_i^2 rather than the photon mass μ^2 and also, the form of the energy denominator — the second square bracket in the denominator of Eq. (20) — which is now also cut off. Indeed, focusing as before on the collinear contribution (with a factor of two accounting for the anticollinear one) we get

$$\int d\Phi_k |M|^2 = -2|M_0|^2 \frac{e^2}{4\pi^2} \int \frac{EdEd\cos\theta}{E\left[(1-\cos\theta)+\frac{2p_1^2}{s}+\frac{p_1^2}{E\sqrt{s}}\right]\left[E+\frac{2p_2^2}{2\sqrt{s}}\right]} + \text{non log}, \tag{21}$$

where again we have kept the leading terms as $\cos\theta \to 1$.

Performing the angular integral we now get

$$\int d\Phi_k |M|^2 = -2|M_0|^2 \frac{\alpha}{\pi} \int \frac{dE}{E+\frac{p_2^2}{\sqrt{s}}} \ln\left[\left(\frac{2p_1^2}{s}+\frac{p_1^2}{E\sqrt{s}}\right)^{-1}\right], \tag{22}$$

so it is apparent that the logarithmic integration over E is cut off by $\frac{p_2^2}{\sqrt{s}}$. It is now the first term in the argument of the log which is subleading, and performing the integral over the energy gives

$$\int d\Phi_k |M|^2 = -|M_0|^2 \frac{\alpha}{\pi} \ln\frac{s}{p_1^2} \ln\frac{s}{p_2^2} = -|M_0|^2 \frac{\alpha}{\pi} \ln\frac{|q|^2}{p_1^2} \ln\frac{|q|^2}{p_2^2} \tag{23}$$

up to single logarithmic terms. This is indeed twice as big as the on-shell result Eq. (17), and thus it exactly matches Roman's result.[2]

4. Infrared and collinear singularities

Having reproduced the result of Eqs. (2) and (3), and in particular the factor two difference between on- and off-shell, we would now like to understand the origin of this difference. Comparing Eqs. (14)–(21), it is clear that in both cases the double log stems from a collinear and an infrared singularity, respectively coming from the integral over the angle and the energy of the emitted photon. The difference resides in the way the singularities are regulated by the photon mass, or by the virtuality: however, the factor two appears somewhat haphazard, as it looks like the reason why Eqs. (3) is twice as large is that the integration variable is the energy E, rather than E^2, with the remaining $\sqrt{s}$ dependence contained in the cutoff.

However, a more transparent physical interpretation appears if we consider the same computation, but in a different frame. Namely, we view the amplitude of Fig. 2 as the production of an off-shell photon in the annihilation of a fermion–antifermion pair, in the center-of-mass reference frame of the incoming fermions. The physics is then similar to the familiar one of Drell–Yan production in QCD (in which the fermions are quarks).

4.1. The Sudakov parametrization

It is then convenient to introduce a Sudakov-like parametrization of the momentum k of the emitted photon (which, in the QCD analogy, would be an emitted gluon):

$$k = (1 - x)\frac{p_1 + p_2}{2} + y\frac{p_1 - p_2}{2} + k_\mathrm{T}$$

$$= x_1 p_1 + x_2 p_2 + k_\mathrm{T} \tag{24}$$

where $k_\mathrm{T} \cdot p_1 = k_\mathrm{T} \cdot p_2 = 0$ is a space-like transverse momentum vector, such that $k_T^2 = -|k_T|^2$, and of course

$$x_1 = \frac{1}{2}\left[(1 - x) + y\right], \tag{25}$$

$$x_2 = \frac{1}{2}\left[(1 - x) - y\right], \tag{26}$$

so that either (x_1, x_2) or (x, y) can be used according to convenience.

In the center-of-mass frame of the incoming fermion–antifermion pair the energy of the emitted photon is

$$E = (1 - x)\frac{\sqrt{s}}{2} \tag{27}$$

while its longitudinal momentum component

$$k_z \equiv y\frac{p_1 - p_2}{2} \tag{28}$$

is entirely fixed by the on-shell condition

$$|k_z| = \sqrt{E^2 - |k_\mathrm{T}|^2} = y\frac{\sqrt{|p_1 - p_2|^2}}{2} \tag{29}$$

where in the general off-shell case $|p_1 - p_2|^2 = s - 2(p_1^2 + p_2^2)$. Of course in the off-shell case k_z is the longitudinal momentum only up to terms proportional to the difference of the two virtualities. Solving for y we get

$$y = \pm\sqrt{(1-x)^2 - \frac{4|k_T|^2}{s}} \left(1 + O(p_i^2/s)\right). \tag{30}$$

The advantage of this choice of parametrization is seen by writing the phase space of the emitted photon, which now takes the form

$$d\Phi_k = \frac{|k_T|d|k_T|d\phi dk_z}{2E(2\pi)^3} = \frac{d|k_T|^2 dE}{4|k_z|(4\pi^2)}, \tag{31}$$

instead of the previous Eq. (11). Using Eqs. (27) and (29) we get

$$d\Phi_k = \frac{1}{4(4\pi^2)} \frac{dx d|k_T|^2}{\sqrt{(1-x)^2 - \frac{4|k_T|^2}{s}}} \left(1 + O(p_i^2/s)\right). \tag{32}$$

This last form exposes the phase space origin[1,7] of the soft and collinear singularity: if the squared amplitude behaves as $|M|^2 \underset{|k_T| \to 0}{\sim} \frac{1}{|k_T|}$, the k_T integration is logarithmic; but then as $|k_T| \to 0$ the square root factor in the denominator reduces to $1 - x$ and the x integration also becomes logarithmic, in the $x \to 1$ limit in which the energy of the emitted photon Eq. (27) vanishes.

This can be exposed by rewriting, in the limit as $|k_T|^2 \to 0$

$$d\Phi_k = \frac{1}{4(4\pi^2)} dx d|k_T|^2 \frac{1}{\sqrt{(1-x)^2 - \frac{4|k_T|^2}{s}}}$$
$$= \frac{1}{4(4\pi^2)} dx d|k_T|^2 \left[\frac{1}{(1-x)_+} - \frac{1}{2}\delta(1-x) \ln \frac{4|k_T|^2}{s} \right] + O(|k_T|^2), \tag{33}$$

where we have introduced the standard plus distribution, implicitly defined by the distributional identity

$$\int_0^1 dx \frac{1}{(1-x)_+} f(x) = \int_0^1 dx \frac{f(x) - f(1)}{1-x}. \tag{34}$$

Note that because $|k_T|^2 \le s/4$ the sign of the log in Eq. (33) is such that the contribution proportional to the delta is always positive. If, as mentioned, the squared amplitude behaves as $|M|^2 \sim \frac{1}{|k_T|^2}$, when integrating over the phase space Eq. (33), the k_T integration leads to a double log, which is now clearly seen to arise when both $|k_T|^2$ but also $x \to 1$, because of the delta.

We now show this explicitly. The amplitude has the form of Eq. (10), but with $k \to -k$ because the fermions are in the final state, so with the Sudakov parametrization Eq. (24)

$$|M|^2 = -|M_0|^2 e^2 \frac{8p_1 \cdot p_2}{(p_1^2 - 2k \cdot p_1)(p_2^2 - 2k \cdot p_2)}, \tag{35}$$

with

$$k \cdot p_1 = \frac{1}{2} \left[(1-x) - y \right] p_1 \cdot p_2 + \frac{1}{2} \left[(1-x) + y \right] p_1^2 = x_2 p_1 \cdot p_2 + x_1 p_1^2$$

$$k \cdot p_2 = \frac{1}{2} \left[(1-x) + y \right] \left(p_1 \cdot p_2 + p_2^2 \right) = x_1 p_1 \cdot p_2 + x_2 p_1^2. \tag{36}$$

In the $|k_T| \to 0$ limit, using Eq. (30) we get

$$x_1 = \frac{1}{2} \left[(1-x) + y \right] = (1-x) + O(|k_T|^2/s) \tag{37}$$

$$x_2 = \frac{1}{2} \left[(1-x) - y \right] = \frac{|k_T|^2}{(1-x)s} (1 + O(|k_T|^2/s)), \tag{38}$$

where we have assumed for definiteness $y > 0$, and the opposite sign would simply amount to interchanging x_1 and x_2 (i.e. the collinear and anticollinear limits). Equations (37-38) show that $x_2 \to 0$ corresponds to the collinear limit, while $x_1 \to 0$ to the soft limit, with the two limits interchanged in the anticollinear case in which the negative y solution is chosen.

4.2. On-shell and off-shell

For on-shell fermions, $2p_1 \cdot p_2 = s$ and $p_i^2 = 0$, so that the matrix element is then given by

$$|M|^2 = -2|M_0|^2 e^2 \frac{16}{s[(1-x)^2 - y^2]}$$

$$= -2|M_0|^2 e^2 \frac{4}{|k_T|^2}, \tag{39}$$

where we have used Eq. (30) and we have provided a factor of 2 in order to account for the two solutions for y, which correspond respectively to the collinear or anticollinear regions when $|k_T| \to 0$.

Hence, integrating over x with the phase space Eq. (33) we get

$$\int d\Phi_k |M|^2 = -|M_0|^2 \frac{e^2}{4(4\pi^2)} \int dx \, d|k_T|^2 \left[\frac{1}{(1-x)_+} - \frac{1}{2}\delta(1-x) \ln \frac{4|k_T|^2}{s} \right] \frac{8}{|k_T|^2}$$

$$= -|M_0|^2 \frac{\alpha}{2\pi} \ln^2 \frac{s}{\mu^2}, \tag{40}$$

where the first equality holds up to non-logarithmic terms, and the second equality, which holds to double-logarithmic accuracy, is found cutting off the logarithmic integration over $|k_T|^2$ with an infrared regulator (photon mass) μ^2; note that the sign follows from the fact that it is the lower limit of integration which provides the μ^2 dependence.

We thus get the same double-log result as Eq. (17). The advantage of this choice of frame is that origin of the double log can be traced to the behavior of the phase space Eq. (33) in the simultaneous infrared $x \to 1$ and collinear $|k_T| \to 0$ limit.

Let us now turn to the off-shell case. The denominator of the amplitude is now given by

$$D = (p_1^2 - 2k \cdot p_1)(p_2^2 - 2k \cdot p_2)$$

$$= s[(1-x)^2 - y^2] \left[\left(\frac{s}{2} + p_1^2 \right) - \frac{p_1^2}{x_2} \right] \left[\left(\frac{s}{2} + p_2^2 \right) - \frac{p_2^2}{x_1} \right] (1 + O(p_i^2/s)). \quad (41)$$

This immediately implies that when integrating with the phase space Eq. (33) the term proportional to the delta does not contribute: x_1 vanishes in the $x \to 1$ limit, so $\lim_{x \to 1} D = \infty$ because of the second factor in square brackets in Eq. (41). Indeed, there is no longer an infrared singularity when $x \to 1$, because the off-shellness regulates it.

It is then convenient to write the denominator as

$$D = \left(x_2 \frac{s}{2} - p_1^2 \right) \left(x_1 \frac{s}{2} - p_2^2 \right) (1 + O(p_i^2/s))$$

$$= \frac{1}{1-x} \left[|k_T|^2 - p_1^2(1-x) \right] s \left[(1-x) - \frac{p_2^2}{s} \right] (1 + O(p_i^2/s) + O(|k_T|^2/s)), \quad (42)$$

where in the second step we have used Eqs. (37) and (38), in the small $|k_T|$ limit.

The integrated square amplitude is thus given by

$$\int d\Phi_k |M|^2 = -|M_0|^2 \frac{e^2}{4(4\pi^2)} \int dx d|k_T|^2 \frac{1}{(1-x)_+} \frac{4s}{D}$$

$$= -2|M_0|^2 \frac{\alpha}{\pi} \int dx d|k_T|^2 \frac{1}{\left[|k_T|^2 - p_1^2(1-x) \right] \left[(1-x) - \frac{p_2^2}{s} \right]} + \text{non log}, \quad (43)$$

where again we have provided a factor of 2 in order to account for the two (collinear and anticollinear) solutions for y. Note that the plus prescription in the first line of Eq. (43) has no effect because the integrand vanishes at $x = 1$, as it is clear from Eq. (42).

The integral over $|k_T|^2$ in Eq. (43) is logarithmic about $|k_T|^2 \sim p_1^2(1-x)$, where the first factor in square brackets in the denominator D (Eq. (42)) vanishes, thus leading to

$$\int d\Phi_k |M|^2 = -2|M_0|^2 \frac{\alpha}{\pi} \int dx \ln \left(\frac{s}{2p_1^2(1-x)} \right) \frac{1}{(1-x) - \frac{p_2^2}{s}}. \quad (44)$$

The integral over x has an infrared singularity regulated by p_2^2 when $(1-x) \sim \frac{p_2^2}{s}$. The integral over x is thus again double-logarithmic, leading to

$$\int d\Phi_k |M|^2 = -|M_0|^2 \frac{\alpha}{\pi} \ln \frac{s}{p_1^2} \ln \frac{s}{p_2^2} = -|M_0|^2 \frac{\alpha}{\pi} \ln \frac{|q|^2}{p_1^2} \ln \frac{|q|^2}{p_2^2} \quad (45)$$

as in Eq. (23).

It is now clear that the factor two difference between the on-shell result Eq. (40) and the off-shell result Eq. (45) reveals a different underlying physics. In the on-shell case, the double log stems from the soft-collinear region, corresponding to the last term in the expression (33) of the phase space. This is a genuine double log, in that it is due to the square-root factor in the phase space being singular when both the transverse momentum $|k_T| \to 0$ and the energy of the emitted photon $E \to 0$ [i.e. $x \to 1$, recalling Eq. (27)]. In the off-shell case instead the double log is really coming from the interference of two logarithmic integration regions when the two propagators go on shell, with the phase space now playing no role. These two integration regions correspond to an integral over energy (or x) Eqs. (31) and (32) and transverse momentum $|k_T|$, but they are now decoupled.

4.3. *Exponentiation*

The argument presented in this section so far concerns only the one-loop or single-emission contributions of Figs. 1 and 2, so one may wonder whether they apply to all orders, and if so why. Clearly, multiple eikonal emission does exponentiate, as textbook arguments show, but the nontrivial question is what happens to the phase space structure. However, it was recently[1] shown that the phase space for n-gluon (and thus also photon) emission in the small $|k_T|$ has a factorized form which reproduces iteratively the structure of Eq. (33).

Specifically, the momenta of the emitted photons can be parametrized as

$$k_i = \alpha_i \frac{p_1 + p_2}{2} + y_i \frac{p_1 - p_2}{2} + k_T^i, \tag{46}$$

so that of course

$$y_i = \pm \sqrt{\alpha_i^2 - \frac{4|k_T^i|^2}{\hat{s}}}. \tag{47}$$

Introducing new variables z_i through

$$\alpha_1 = 1 - z_1; \qquad \alpha_i = z_1 \ldots z_{i-1}(1 - z_i), \ i \geq 2 \tag{48}$$

it can then be shown[1] that

$$\prod_{i=1}^{n} \frac{d\alpha_i}{\sqrt{\alpha_i^2 - 4\frac{|k_T|^2}{s}}} = \prod_{i=1}^{n} \frac{dz_i}{\sqrt{(1 - z_i)^2 - \frac{4|k_T|^2}{z_1^2 \ldots z_{i-1}^2 s}}}, \tag{49}$$

and that the phase space can be written as

$$d\Phi_{n+1}(p_1, p_2; q, k_1, \ldots, k_n) = \frac{8\pi^3}{[4(2\pi)^2]^{n+1}} \frac{dq_t^2}{s} \int db^2 \, J_0\left(b|q_t|\right)$$

$$\times J_0\left(b|k_T^1|\right) d|k_T^1|^2 dz_1 \left[\frac{1}{(1 - z_1)_+} - \delta(1 - z_1)\frac{1}{2}\ln\frac{|k_T^1|^2}{s}\right] \cdots$$

$$\times J_0\left(b|k_T^n|\right) d|k_T^n|^2 dz_n \left[\frac{1}{(1 - z_n)_+} - \delta(1 - z_n)\frac{1}{2}\ln\frac{|k_T^n|^2}{s}\right]$$

$$\times \delta(\tau - z_1 \ldots z_n) + O\left(\frac{1}{b}\right), \tag{50}$$

where $\tau = \frac{|q^2|}{s}$. The Fourier transform with respect to transverse momentum is necessary in order to factorize the delta function which ensures transverse momentum conservation, but it is clear that the structure of Sec. 4 is then preserved and simply iterated, thereby leading to exponentiation through arguments that are now textbook[15] matter.

Clearly, once the problem is viewed in this way, the exponentiation is seen to have the same origin, both on-shell and off-shell: even though the different origins of the double log are manifested by the factor two difference that we discussed.

As for the RG argument that Wilson asked Roman to construct, it was eventually presented thirty years later,[6] as a consequence of the factorization of the integrated amplitude in terms of a factor which contains the soft emissions, and the rest. The underlying physical reason is[7] that in the soft limit the amplitude depends on the variables $|k_T|^2$, x and s only through the combination $\frac{4|k_T|^2}{s(1-x)^2}$, essentially because $|k_T^2|^{\max} = \frac{s(1-x)^2}{4}$ is the upper limit of the logarithmic transverse momentum integration. The fact that the square amplitude only depends on one variable then allows for RG improvement with respect to it.

5. Conclusion

In summary, Roman's computation amounted to what in modern language would be called the determination of the high-energy behavior of the Drell–Yan process using reverse unitarity:[16] the inclusive production of a gauge boson in fermion–antifermion annihilation, computed using the cutting rule Eq. (8) to express real radiation phase-space integrals in terms of loops.

What I have done here instead is to compute the real-emission contributions directly. With this, I have answered the question that Roman had asked me: the factor two difference between the coefficients of the double logarithm in on-shell and off-shell vertex functions (which then to all orders exponentiates) does have a simple physical interpretation. Namely, in the on-shell case the double log is a soft-collinear log, coming from the behavior of phase space in the region in which the energy and momentum of the emitted photon simultaneously go to zero; while in the off-shell case the double log is the interference of two independent logarithmic integrations, over the emitted photon energy and transverse momentum, coming from the region where two propagator denominators vanish. It is only in the on-shell case that there exists a genuinely infrared and collinear region.

Note added

After publication on the arXiv of the first version of this paper, John Collins pointed out to me that the factor of two disscussed in the present note is also discussed in his QCD treatise,[17] and it was surely known in the early days of perturbative QCD: indeed, it is mentioned (in passing) in the introduction of the seminal paper by Mueller[18] in which the Sudakov form factor is computed in QED on-shell, for massive fermions (and photons), beyond the double logarithmic approximation, to all logarithmic orders. This paper was at the origin of subsequent generalizations to QCD by Collins himself,[19] which are at the basis of the celebrated Collins–Soper–Sterman early results on QCD factorization.[20,21] Interestingly, Mueller's paper does already introduce and exploit renormalization group methods.

The discussion in the QCD book[17] presents a computation of the loop diagram, similar to that performed by Roman[2] and presented in "Landau";[12] the origin of the factor of two is explained in terms of regions which contribute to the loop integral in the on-shell vs. off-shell case (see in particular Sec. 10.5.3 of the book[17]). In this sense the discussion presented here provides a complementary, possibly more "physical" interpretation, to the extent that real emission is physically more intuitive than virtual corrections.

Also, it was recently found that a curious difference of a factor two in Sudakov form factors appears when comparing initial and final state radiation.[22]

Interestingly, Mueller[18] does not cite Roman's result. We will leave it to the reader to draw a lesson from this sequence of oblivions and re-discoveries — including my own.

Note added in proof:

A few months after submission of this contribution, I was looking into an old review paper that I should know quite well, by my late mentor Guido Altarelli.[23] I realized that it contains a section called "The Sudakov form factor of partons", something that I had completely forgotten. I also realized that in this section Eqs. (4) and (5) are both to be found, accompanied by the following sentence: "It is well known that for an off shell quark the exponent differs by a factor of 2." No reference is given, presumably because this fact is so well known.

Acknowledgments

I am grateful to Giancarlo Ferrera and Paolo Nason for several discussions on the content of this paper, and also for a critical reading of the manuscript. In particular, Giancarlo pointed out to me the discussion of the Sudakov form factor in Landau's treatise,[12] while the computation presented in Sec. 2 is due to Paolo, whom I also thank for providing me with detailed notes. I am very grateful to John Collins for correspondence on the subject of this note and for calling my attention to Refs. 18,19 and especially to the discussion in his book.[17] I also thank Giampiero Passarino for interesting comments and for spotting some typos, and Jeffrey Forshaw for pointing out his recent work.[22] I acknowledge financial support from the European Research Council under the European Union's Horizon 2020 research and innovation Programme (grant agreement n. 740006).

References

1. C. Muselli, S. Forte, and G. Ridolfi, Combined threshold and transverse momentum resummation for inclusive observables, *JHEP* **03** 106, (2017). doi: 10.1007/JHEP03(2017)106.
2. R. Jackiw, Dynamics at high momentum and the vertex function of spinor electrodynamics, *Annals Phys.* **48**, 292–321 (1968). doi: 10.1016/0003-4916(68)90087-0.
3. V. V. Sudakov, Vertex parts at very high-energies in quantum electrodynamics, *Sov. Phys. JETP* **3**, 65–71 (1956). [Zh. Eksp. Teor. Fiz. 30, 87 (1956)].
4. B. L. Ioffe, The first dozen years of the history of ITEP Theoretical Physics Laboratory, *Eur. Phys. J.* **H38**, 83–135 (2013). doi: 10.1140/epjh/e2012-30008-3.
5. R. Jackiw, Ken Wilson — The early years, *Int. J. Mod. Phys.* **A29**, 1430008 (2014). doi: 10.1142/S0217751X14300087.
6. H. Contopanagos, E. Laenen, and G. F. Sterman, Sudakov factorization and resummation, *Nucl. Phys.* **B484**, 303–330 (1997). doi: 10.1016/S0550-3213(96)00567-6.
7. S. Forte and G. Ridolfi, Renormalization group approach to soft gluon resummation, *Nucl. Phys.* **B650**, 229–270 (2003). doi: 10.1016/S0550-3213(02)01034-9.
8. S. Weinberg, Dynamics at infinite momentum, *Phys. Rev.* **150**, 1313–1318 (1966). doi: 10.1103/PhysRev.150.1313.
9. G. F. Sterman, *An Introduction to Quantum Field Theory* (Cambridge University Press, 1993).
10. R. Venugopalan, Introduction to light cone field theory and high-energy scattering, *Lect. Notes Phys.* **516**, 89 (1999). doi: 10.1007/BFb0107312.
11. M. Cassandro and M. Cini, Asymptotic limit of vertex functions in perturbation theory, *Nuovo Cim.* **34**, 1719 (1964). doi: 10.1007/BF02750567.
12. V. B. Berestetskii, E. M. Lifshitz, and L. P. Pitaevskii, *Quantum Electrodynamics. Vol. 4, Course of Theoretical Physics* (Pergamon Press, Oxford, 1982).
13. S. Weinberg, *The Quantum Theory of Fields. Vol. 1: Foundations* (Cambridge University Press, 2005).
14. P. Nason, Sudakov wih photon mass or off-shell fermion cutoff (2019).
15. M. E. Peskin and D. V. Schroeder, *An Introduction to Quantum Field Theory* (Addison-Wesley, Reading, USA, 1995).
16. C. Anastasiou, K. Melnikov, and F. Petriello, A new method for real radiation at NNLO, *Phys. Rev.* **D69**, 076010 (2004). doi: 10.1103/PhysRevD.69.076010.
17. J. Collins, Foundations of perturbative QCD, *Camb. Monogr. Part. Phys. Nucl. Phys. Cosmol.* **32**, 1–624 (2011).
18. A. H. Mueller, On the asymptotic behavior of the Sudakov form-factor, *Phys. Rev.* **D20**, 2037 (1979). doi: 10.1103/PhysRevD.20.2037.
19. J. C. Collins, Algorithm to compute corrections to the Sudakov form-factor, *Phys. Rev.* **D22**, 1478 (1980). doi: 10.1103/PhysRevD.22.1478.
20. J. C. Collins and D. E. Soper, Back-to-back jets in QCD, *Nucl. Phys.* **B193**, 381 (1981). doi: 10.1016/0550-3213(81)90339-4. [Erratum: Nucl. Phys. B213, 545 (1983)].
21. J. C. Collins, D. E. Soper, and G. F. Sterman, Does the Drell–Yan cross-section factorize?, *Phys. Lett.* **109B**, 388–392 (1982). doi: 10.1016/0370-2693(82)91097-8.
22. J. R. Forshaw, J. Holguin, and S. Plätzer, Parton branching at amplitude level, *JHEP* **08**, 145 (2019). doi: 10.1007/JHEP08(2019)145.
23. G. Altarelli, Partons in quantum chromodynamics, *Phys. Rept.* **81**, 1 (1982). doi: 10.1016/0370-1573(82)90127-2.

Chapter 17

Emerging Majorana Modes in Junctions of One-Dimensional Spin Systems

Domenico Giuliano[(1,2)], Andrea Trombettoni[(3,4)], and Pasquale Sodano[(5)]

[(1)] *Dipartimento di Fisica, Università della Calabria, Arcavacata di Rende I-87036, Cosenza, Italy*

[(2)] *I.N.F.N., Gruppo collegato di Cosenza, Arcavacata di Rende I-87036, Cosenza, Italy*

[(3)] *Department of Physics, University of Trieste, Strada Costiera 11, I-34151 Trieste, Italy*

[(4)] *CNR-IOM DEMOCRITOS Simulation Center and SISSA, Via Bonomea 265, I-34136 Trieste, Italy*

[(5)] *I.N.F.N., Sezione di Perugia, I-64100 Perugia, Italy*

The non-local effects induced by Majorana fermions in field theories for condensed matter systems are deeply related to the fermion charge fractionalization discovered by Roman Jackiw in relativistic field theories. We show how the presence of Majorana fermions may be mimicked in pertinent networks of spin chains inducing a spin analogue of the multi-channel Kondo effect. The relevance of this spin analogue of the Kondo effect for networks of Josephson arrays and Tonks–Girardeau gases is highlighted.

1. Introduction

Low energy neutral fermionic excitations (Majorana modes), deeply related to the charged fermion zero modes discovered by Roman Jackiw in pioneering topological investigations of relativistic field theories,[1,2] have been claimed to be relevant in a variety of strongly correlated condensed matter systems, providing new insights also for the investigation of non-Fermi liquid states.

Majorana fermions were first proposed in 1937 by Ettore Majorana[3] who considered a modification to the relativistic Dirac equation for conventional spin-1/2 particles (Dirac fermions) giving purely real (as opposed to complex) solutions. These Majorana fermions are particles coinciding with their own antiparticles since their creation operator is equal to their annihilation operator. In spite of the beautiful simplicity of this idea, Majorana fermions are not easy to come by in nature. One could, for example, decompose a relativistic electron, whose wave equation does have charge conjugation symmetry, into its real and imaginary parts. However, the interaction of the electron with photons is not diagonal in this decomposition.

The real and the imaginary components will be readily remixed by the electromagnetic interaction: they cannot be stationary states of the full Hamiltonian of Quantum Electrodynamics. In condensed matter systems, however, it may be easier to look for emergent Majorana fermions, since usually electromagnetic interactions are screened.

Our particular interest in the following is in situations where the fermion spectrum has mid-gap, or zero energy, states. Already for complex electrons, mid-gap states give rise to fractional quantum numbers[1,4–6] relevant for the studies on polyacetylene.[7] With Majorana fermions, these states lead to peculiar representations of the anticommutator algebra, which can violate basic symmetries, such as fermion parity symmetry.[8–12]

The huge interest in Majorana fermions goes beyond fundamental curiosity since there is an enormous potential for applications to quantum technology and to devices based on the manipulation of Majorana fermions. For example, this could allow an electron to be splitted into a pair of widely separated Majorana bound states,[13] which could be less sensitive to the effect of localized sources of decoherence. Current research attempts to develop integrated devices suitable for detecting, storing and manipulating Majorana fermions.[14]

Junctions of one-dimensional (1D) wires seem to possess several distinct advantages when it comes to fabrication and subsequent detection of Majorana zero modes. In these 1D devices, zero energy Majorana modes are confined either at the wire edges or at a domain wall between topological and non-topological regions of the wire and, due to the 1D confinement, there are very few modes to "disturb" their observable signatures. From the experimental side, there have been recent advances in fabrication and manipulation of clean quantum wires allowing for an unprecedented level of control and analysis of these devices in a wide range of settings.

Low energy Majorana modes have been recently the object of many theoretical[14–17] and experimental[18] investigations. Located at the edges of 1D devices, they are responsible for the emergence of stretched nonlocal electron states[13,19] allowing for distance independent tunnelling,[20] crossed Andreev reflection,[21] teleportation-like coherent transfer of a fermion[22] and fractional Josephson effects.[23,24] Their effects emerge in a variety of platforms: quantum wires immersed in a *p*-wave superconductor,[13,19,20] cold atomic systems,[25] topological insulator–superconductor magnetic structures,[22,26,27] semiconductor heterostructures,[28–31] superconducting wires,[32] Josephson arrays[33] and spin systems.[34] In addition, they may be relevant excitations also in conventional *s*-wave superconductors.[35]

It is worth to stress here that, when a superconductor is coupled to a conducting wire in an SN-junction, the Majorana mode hybridizes with the conducting electrons at the normal side of the junction. Such a feature eventually yields to a "Majorana hybridization" with the electrons in the normal wire,[36] a phenomenon strikingly similar with the emergence of the "electronic Kondo cloud" at the Kondo fixed point.[37]

Majorana low energy modes are expected to be relevant for applications to topological quantum computing[38] and quantum interferometry[39,40] and their manipulation is under current investigation.[38,41–44] In junctions of quantum wires, Majorana edge modes induce, for various network topologies, remarkable even–odd effects on the tunneling conductance.[45] In addition, Coulomb charging effects cause conductance oscillations and resonances connected to teleportation as well as peculiar finite-bias peaks[46,47] and may trigger the flow towards exotic Kondo fixed points when the center island has a finite charging energy.[48]

Majorana fermions in condensed matter are not fundamental particles. Rather, they are effective degrees of freedom, emerging in the presence of a degeneracy in the ground state which is topologically protected. One paradigmatic example is the Kitaev chain,[19] a tight-binding model for the effectively spinless fermions in 1D p-wave superconductor. In this model, a topologically protected phase (i.e., a phase which cannot be changed by any local operation, and thus robust against interaction with the environment) can be described in terms of a pair of Majorana modes localized at the ends of the chain. Together, they form a non-local fermionic degree of freedom, which can be used to encode a qubit. As a solid-state realization of this model, a set of nanowires with strong Rashba coupling (InAs, InSb), laid on a conventional BCS superconductor (Al, Nb) and subject to a suitably tuned magnetic field can develop Majorana ending modes,[30,31] for which experimental evidence has been provided (see e.g. Ref. 18).

The setup described above is used in the so-called Majorana–Coulomb box, or topological Kondo model (TKM).[49] It is obtained connecting a set of M effectively 1D wires to a set of nanowires supporting Majorana modes at their ends, hosted on a mesoscopic superconducting substrate with a large charging energy, and subject to an applied voltage potential (see Fig. 1). At low temperatures, the model is integrable,[50] despite not being 1D. It is possible, via conformal field theory or Bethe ansatz, to compute thermodynamic quantities such as free energy, specific heat and entropy contribution from the central region.[51]

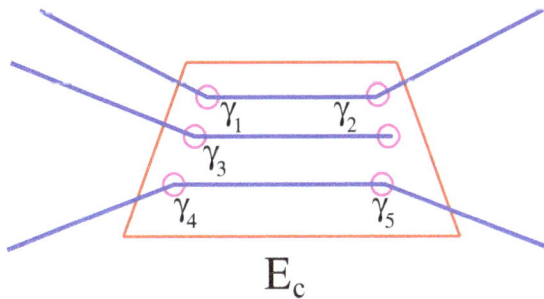

Fig. 1.   Schematic representation of the topological Kondo model: Majorana edge modes of quantum wires are hosted on an s-wave superconductor and are capacitively coupled with the ground. These modes are proximity-coupled with the external leads.

It should be noticed that conventional multi-channel Kondo models are difficult to attain,[52] due to the need for a perfect symmetry between the couplings of the spin density from the channels to the spin of the impurity. The TKM provides a shortcut to circumvent this problem,[49] relating the symmetry between the channels to the topological degeneracy characterizing a 1D quantum wire into a topological phase, and associated with the emergence of localized Majorana modes at the edges of the wire.

Neglecting the overlap between Majorana modes lying at the edges of the wire, the authors of Ref. 49 showed that the ground state of the wire exhibits a topological degeneracy and that, on coupling M topological quantum wires to a central superconducting island with a finite charging energy, for $M \geq 3$, one may realize over screened multi-channel Kondo physics, with the spin of the isolated impurity given by a pertinent combination of the real fermionic operators of the Majorana modes localized at the inner edges. The emerging Kondo Hamiltonian is topological as the central spin is a nonlocal combination of emerging Majorana modes. This allows for a robust realization of multi-channel Kondo physics where the effects of the potentially dangerous anisotropies in the couplings between the spin impurity and the various channels are suppressed due to topology, being irrelevant in the renormalization group flow towards the Kondo regime.[49,53]

Finally, one should observe that the competition between superconductivity and Kondo physics leads to unusual Josephson current-phase relations and to a remarkable transfer of fractionalized charges.[54] Also interesting is the study of the concurrence of entanglement between two quantum dots in contact with Majorana bound states on a floating superconducting island. One finds that the distance between the Majorana states, the charging energy of the island and the average island charge are decisive parameters for the efficiency of the entanglement generation. This leads to the possibility of long range entanglement with distance independent concurrence over wide parameter regions.[55]

A different platform for the multi-channel Kondo model has been recently proposed in Ref. 56 for a Y-junction of three 1D XX models and by Tsvelik in a Y-junction of three 1D quantum Ising models, in both cases with the 1D chains joined at their inner edges.[53] This proposal, when the relevant parameters are pertinently tuned, is particularly attractive since the behavior of the uncoupled spin models is known and it provides reliable and effective descriptions of strongly correlated phenomena in condensed matter systems. As a result one may envision to probe the multi-channel Kondo effect in a variety of controllable, and yet "topologically robust", experimental settings such as the ones provided by degenerate Bose gases confined in an optical lattice[57,58] or quantum Josephson junction networks (see Refs. 59–61 and references therein). It should be noticed that spin realizations of multi-impurity[62] and multi-channel[63] Kondo models revealed very useful not only for quantifying the entanglement,[64] but also for characterizing quantum phase transitions.[65]

Remarkably, the TKM behavior is that of a non-Fermi liquid. Configurations based on plaquette schemes are probably the simplest way of realizing a topologically protected memory,[66] and, within a surface code, all operations required for quantum computation. These configuration can be realized using as basic unit the Majorana–Coulomb box.[67] Remarkably, the TKM can be realized by a junction of spin chains[57] and it could be obtained in laboratory with the use of holographic optical traps for cold atomic gases, which paves the way for a whole new family of holographic devices.[57]

The plan of the paper is the following. Using a pertinently generalized Jordan–Wigner transformation, Sec. 2 shows how a Y-junction of XY spin chains may be transformed in a fermionic model with a non-trivial boundary term containing Klein factors (KF), i.e. real fermion modes, introduced to ensure the appropriate commutation relations on the Y-graph. Section 3 is meant to clarify how the emergence of real fermion KF opens the possibility to engineer a spin-chain version of the TKM. In addition, one sees that a Y-junction of quantum Ising chains leads to a two-channel version of the topological Kondo effect, while a Y-junction of XX spin chains leads to a four-channel Kondo model. Section 4 shows how one can map networks of quantum spin chains — in a pertinent range of parameters — in networks of Josephson junction arrays. In Sec. 5 one sees how a Y-junction of XX spin chains may be used to describe cold atomic systems in holographic traps. Section 6 summarizes our findings.

To help following the various abbreviations, we list in Table 1 the meaning of the ones we use most commonly throughout the paper.

Table 1. Glossary of most commonly used abbreviations.

1D	One-dimensional
TKM	Topological Kondo model
KF	Klein factors
QIC	Quantum Ising chain
JW	Jordan–Wigner
YSC	Y-junction of spin chains
JJC	Josephson junction chains
TG	Tonks–Girardeau

2. Emerging Majorana modes at junctions of quantum spin chains

In this section we address networks of spin chains with nearest-neighbour magnetic exchange terms in the x- and y-directions in spin space (not necessarily equal to each other); in addition, we alledge for a nonzero uniform magnetic field applied in the z-direction. Specifically, each arm of the network is described by an XY spin-1/2 Hamiltonian given by

$$H_{XY} = -J \sum_{j=1}^{\ell-1}(S_j^x S_{j+1}^x + \gamma S_j^y S_{j+1}^y) - H \sum_{j=1}^{\ell} S_j^z \quad . \tag{1}$$

In Eq. (1) $\vec{S}_j$ is a quantum spin-1/2 operator acting at site-j of an ℓ-site chain (with ℓ kept finite and eventually sent to infinite at the end of the calculations), obeying the algebra

$$[S_j^a, S_{j'}^b] = i\delta_{j,j'} \, \epsilon^{abc} S_j^c \quad . \tag{2}$$

J is the magnetic exchange strength between spins, H is the applied, uniform magnetic field in the z-direction and γ is the anisotropy parameter of the spin exchange interaction. In the following γ is chosen so that $0 \le \gamma \le 1$. By tuning γ between the extreme values $\gamma = 0$ and $\gamma = 1$, the H_{XY} in (1) continuously interpolates between the quantum Ising chain (QIC) and the XX chain in transverse magnetic field.

To describe a junction of several XY chains, we introduce the boundary Hamiltonian H_Δ, given by

$$H_\Delta = -J_\Delta \sum_{\alpha \neq \beta}(S_{1,\alpha}^x S_{1,\beta}^x + \gamma S_{1,\alpha}^y S_{1,\beta}^y) \quad . \tag{3}$$

In Eq. (3) the j-th spin of the α-th chain (with $\alpha = 1, \cdots, M$) is denoted by $\vec{S}_{j,\alpha} = (S_{j,\alpha}^x, S_{j,\alpha}^y, S_{j,\alpha}^z)$. In H_Δ, each spin in position $j = 1$ of a chain α is coupled with all the others spins in the position $j = 1$ of all other chains $\beta \neq \alpha$. The advantage of the form (3) is that one can safely take the continuous limit in each chain, and the wires are coupled by a tunneling term.

The standard approach to the single chain described by the Hamiltonian (1) consists in mapping X_{XY} onto a quadratic, spinless fermion Hamiltonian via the Jordan–Wigner (JW) transformations.[68,69] The latter allows for rewriting the quantum spin operators S_j^z and $S_j^\pm$, where $S_j^\pm = S_j^x \pm iS_j^y$, in terms of spinless lattice operators $\{a_j, a_j^\dagger\}$ as

$$S_j^+ = a_j^\dagger e^{i\pi \sum_{r=1}^{j-1} a_r^\dagger a_r}$$
$$S_j^- = a_j e^{i\pi \sum_{r=1}^{j-1} a_r^\dagger a_r}$$
$$S_j^z = a_j^\dagger a_j - \frac{1}{2} \quad . \tag{4}$$

In terms of the JW fermions, one obtains

$$H_{XY} = -\frac{J(1+\gamma)}{2}\sum_{j=1}^{\ell-1}\{a_j^\dagger a_{j+1} + a_{j+1}^\dagger a_j\} + \frac{J(1-\gamma)}{2}\sum_{j=1}^{\ell-1}\{a_j a_{j+1} + a_{j+1}^\dagger a_j^\dagger\} + H \sum_{j=1}^{\ell} a_j^\dagger a_j \quad . \tag{5}$$

Once written in terms of JW fermions, from Eq. (5) one has that, for $\gamma = 1$, H_{XY} reduces to the Hamiltonian for free fermions on a 1D lattice with a chemical potential $-H$ in the XX-limit, and to the Hamiltonian for a 1D Kitaev model for a p-wave superconductor when $\gamma = 0$.[19]

Remarkably, while the above JW transformations (4) are sufficient to rewrite the generic XY-Hamiltonian for a single chain in fermionic coordinates, a fundamental problem arises when several chains are connected to each other to form a junction of spin chains. Of course, one should have $M \geq 3$, since a junction of two chains can always be mapped onto one chain. Thus, for the sake of simplicity, in the following we shall take $M = 3$. The network with $M = 3$ is called a Y-junction of spin chains.

In general, models of Y-junctions have been studied at the crossing of three, or more, Luttinger liquids,[59–61,70–76] in Bose gases in star geometries,[77–79] and in superconducting Josephson junctions.[61,72,80] When introducing more than one chain, the Hamiltonian of the uncoupled chains (henceforth referred to as the "leads"), $H^{(0)}$, will be given by the sum of many Hamiltonians like the one in Eq. (5). In the specific case of the Y-junction of spin chains (YSC), one has

$$H = H^{(0)} + H_\Delta \quad , \tag{6}$$

where H_0 is the bulk Hamiltonian of the uncoupled chains and H_Δ is the junction Hamiltonian. The bulk Hamiltonian is given by

$$H^{(0)} = \sum_{\alpha=1,2,3} \left\{ -J \sum_{j=1}^{\ell-1} (S_{j,\alpha}^x S_{j+1,\alpha}^x + \gamma S_{j,\alpha}^y S_{j+1,\alpha}^y) - H \sum_{j=1}^{\ell} S_{j,\alpha}^z \right\} \equiv \sum_{\alpha=1,2,3} H_{XY;\alpha}, \tag{7}$$

with α being the chain index, while the junction Hamiltonian H_Δ is given by (3). Furthermore, in order to obtain the Kondo effect in spin chains, one should assume $|J_\Delta|/|J| \leq 1$.

The YSC-Hamiltonian $H \equiv H^{(0)} + H_\Delta$ provides the natural generalization of the Hamiltonian for a Y-junction of QICs introduced in Ref. 53 and later on generalized in Refs. 67 and 81, to which it reduces for $\gamma = 0$, as well as of the Hamiltonian for a Y-junction of three quantum XX-spin chains discussed in Ref. 56, to which it reduces for $\gamma = 1$.

In view of the successfull application of the JW transformation to solving a single XY-chain, one may attempt to look for an appropriate generalization to the YSC of the mapping onto a spinless fermionic model, but a simple-minded generalized JW representation of the spin coordinates such as

$$S_{j,\alpha}^+ = a_{j,\alpha}^\dagger e^{i\pi \sum_{r=1}^{j-1} a_{r,\alpha}^\dagger a_{r,\alpha}}$$

$$S_{j,\alpha}^- = a_{j,\alpha} e^{i\pi \sum_{r=1}^{j-1} a_{r,\alpha}^\dagger a_{r,\alpha}}$$

$$S_{j,\alpha}^z = a_{j,\alpha}^\dagger a_{j,\alpha} - \frac{1}{2} \quad , \tag{8}$$

would eventually lead to operators that properly commute with each other, if they belong to the same chain, but such that $\{S_{j,\alpha}^{\pm}, S_{j',\alpha'}^{\pm}\} = 0$, if $\alpha \neq \alpha'$, instead than $[S_{j,\alpha}^{\pm}, S_{j',\alpha'}^{\pm}] = 0$.

The solution to this problem was first put forward in Ref. 56, in analogy to what is done when bosonizing spinful fermionic operators in interacting one-dimensional electronic systems.[82] Basically, one introduces the so-called KF, that is, real fermion modes, one for each chain. In the specific case of the YSC, one has to introduce three KF, η^1, η^2, η^3, obeying the appropriate anticommutation relations with each other and anticommuting with all the other JW fermion operators used to represent the spin operators in the leads. Specifically, one sets

$$\{\eta^{\alpha}, \eta^{\alpha'}\} = 2\delta_{\alpha,\alpha'}$$
$$\{\eta^{\alpha}, a_{j,\alpha'}\} = \{\eta^{\alpha}, a_{j,\alpha'}^{\dagger}\} = 0 \quad , \tag{9}$$

with $\alpha, \alpha' = 1, 2, 3$.

Using the KF, one modifies Eqs. (8) so as to take into account the commutation relations between the spin operators; that is, one sets[53,56,83]

$$S_{j,\alpha}^{+} = i a_{j,\alpha}^{\dagger} e^{i\pi \sum_{r=1}^{j-1} a_{r,\alpha}^{\dagger} a_{r,\alpha}} \eta^{\alpha}$$
$$S_{j,\alpha}^{-} = i a_{j,\alpha} e^{i\pi \sum_{r=1}^{j-1} a_{r,\alpha}^{\dagger} a_{r,\alpha}} \eta^{\alpha}$$
$$S_{j,\alpha}^{z} = a_{j,\alpha}^{\dagger} a_{j,\alpha} - \frac{1}{2} \quad . \tag{10}$$

Apparently, the KF are introduced as a mathematical tool to ensure the correct commutation relations between operators acting on sites belonging to different chains. Equivalently, one may just state that their introduction corresponds to an "artificial" enlargement of the Hilbert space, which is yet necessary, in order to recover the correct operator algebra.[56]

When joining together the leads into the YSC, the KF conspire to actually interact with the dynamical degrees of freedom of the leads, thus being legitimately promoted to "physical" degrees of freedom of the system, whose presence, or absence, can potentially strongly affect the behavior of the junction. Thus, in this respect, they can definitely be regarded as "emerging real-fermion degrees of freedom", that is, as emerging Majorana modes, by no means different from the "actual" Majorana modes emerging in e.g. a 1D topological superconductor in its topological phase.[19] Notice that a number of exotic phases have been predicted to emerge in pertinently designed junctions of 1D leads, due to the effect of coupling the "bulk" degrees of freedom of the system to the KF,[60] to the so-called Klein–Majorana hybridization between the two kinds of real-fermion degrees of freedom entering the junction Hamiltonian,[84] or to both phenomena taking place at the same time.[85]

Once the generalized JW transformations, Eqs. (10), are inserted into $H^{(0)}$ in Eq. (7), one obtains the fermionic version of the disconnected lead Hamiltonian as

$$H^{(0)} = \sum_{\alpha=1,2,3} \left\{ -\frac{J(1+\gamma)}{2} \sum_{j=1}^{\ell-1} \{a_{j,\alpha}^\dagger a_{j+1,\alpha} + a_{j+1,\alpha}^\dagger a_{j,\alpha}\} \right.$$

$$\left. + \frac{J(1-\gamma)}{2} \sum_{j=1}^{\ell-1} \{a_{j,\alpha} a_{j+1,\alpha} + a_{j+1,\alpha}^\dagger a_{j,\alpha}^\dagger\} + H \sum_{j=1}^{\ell} a_{j,\alpha}^\dagger a_{j,\alpha} \right\} . \quad (11)$$

We see that the KF do not appear in the "bulk" lead Hamiltonian in Eq. (11). At variance, they do appear in the junction Hamiltonian H_Δ in Eq. (3) which, once rewritten in fermionic coordinates, takes the form

$$H_\Delta = 2J_\Delta \left(\vec{\Sigma}_1 + \gamma \vec{\Upsilon}_1\right) \cdot \vec{\mathcal{R}} . \quad (12)$$

In particular, the KF determine the "topological" spin-1/2 operator $\vec{\mathcal{R}}$,[53,56] which is given by

$$\vec{\mathcal{R}} = -\frac{i}{2} \begin{pmatrix} \sigma^2 \sigma^3 \\ \sigma^3 \sigma^1 \\ \sigma^1 \sigma^2 \end{pmatrix} . \quad (13)$$

It is, indeed, a straightforward check to verify that the components of $\vec{\mathcal{R}}$ satisfy the appropriate $su(2)$-commutation relation for a spin-1/2 angular-momentum operator. Following the notation of Ref. 83, in Eq. (13) we have also introduced the lattice spin- and isospin-density operators, $\vec{\Sigma}_j, \vec{\Upsilon}_j$, defined as

$$\vec{\Sigma}_j = -\frac{i}{2} \begin{pmatrix} (a_{j,2} + a_{j,2}^\dagger)(a_{j,3} + a_{j,3}^\dagger) \\ (a_{j,3} + a_{j,3}^\dagger)(a_{j,1} + a_{j,1}^\dagger) \\ (a_{j,1} + a_{j,1}^\dagger)(a_{j,2} + a_{j,2}^\dagger) \end{pmatrix} , \quad \vec{\Upsilon}_j = \frac{i}{2} \begin{pmatrix} (a_{j,2} - a_{j,2}^\dagger)(a_{j,3} - a_{j,3}^\dagger) \\ (a_{j,3} - a_{j,3}^\dagger)(a_{j,1} - a_{j,1}^\dagger) \\ (a_{j,1} - a_{j,1}^\dagger)(a_{j,2} - a_{j,2}^\dagger) \end{pmatrix} . \quad (14)$$

Eqs. (11) and (12) are the key result of the generalized JW transformations applied to a YSC. They evidence the emergence of actual degrees of freedom at the junction (the topological spin $\vec{\mathcal{R}}$) which, despite being determined by operators originally introduced as a mathematical mean to assure the consistency of the JW transformations, eventually become true "physical" degrees of freedom.

3. Kondo effect in spin systems and the topological Kondo model

The most striking physical consequences of the emergence of real-fermion KF at a junction of bosonic quantum spin chains is the possibility of using them to engineer spin-chain version of the so-called topological Kondo effect.

Historically, the Kondo effect was discovered as a low-temperature upturn in the resistivity of conducting metals doped with magnetic impurities.[86–88] It is determined by the relevance of nonperturbative spin-flip processes due to the magnetic dipole interaction between the spin of a magnetic impurity and of the itinerant conduction electrons in the metal, which, as the temperature goes down, determines the formation of a strongly correlated Kondo state between the impurity and the conduction electrons (the "Kondo cloud").[86,87,89]

In the last decades, the development in quantum device fabrication techniques has triggered a novel interest in the Kondo effect, as it has been possible to realize it in quantum dots with metallic,[90–93] as well as superconducting leads.[94–96]

Of particular relevance is the proposal of realizing the Kondo impurity spin in terms of Majorana fermion modes emerging at the endpoints of one-dimensional quantum wires connected to topological superconductors. Such a peculiar realization of Kondo effect is characterized by the "topological" nature of the impurity spin, each component of which is determined by Majorana modes belonging to different quantum wires (leads). Thus, we see that in such a topological version of Kondo effect (the TKM), the "emerging" magnetic impurity is nonlocal in the lead index. Therefore, disconnecting even a single lead from the junction fully destroys the effect, which is the main signature of the topological nature of the phenomenon.[50,51,97,98]

In Refs. 48 and 49, it was shown that networks of quantum wires supporting edge Majorana modes[45] provide a possible experimentally attainable realization of the TKM. Subsequently, the spin dynamics,[50] as well as the exact solution for finite number of electrons,[99] were investigated.

Typically, a TKM is realized at a junction of quantum wires connected to a superconducting island with a finite charging energy E_c. In such a system, Coulomb interactions play an essential role for the host superconducting region, since they determine E_c.[50,97] Focusing on the temperature regime $T \ll E_c$, one obtains that in all the allowed physical processes at the junction the number of electrons on the island is conserved. Under these conditions, the effective Hamiltonian describing the TKM at low energy scales can be written in the form[48,97,100,101]

$$H = -iv_F \sum_{\alpha=1}^{M} \int dx \psi_\alpha^\dagger(x) \partial_x \psi_\alpha(x) + \sum_{\alpha \neq \beta} \lambda_{\alpha\beta} \gamma_\alpha \gamma_\beta \psi_\alpha^\dagger(0) \psi_\beta(0) + i \sum_{\alpha \neq \beta} h_{\alpha\beta} \gamma_\alpha \gamma_\beta .$$

(15)

In Eq. (15), the $\psi_\alpha(x)$'s are the Fermi fields describing electrons in the wires $\alpha = 1, \ldots, M$. $\gamma_\alpha = \gamma_\alpha^\dagger$ are Majorana fields constrained in a box connected with the wires and satisfying the Poisson algebra

$$\{\gamma_\alpha, \gamma_\beta\} = 2\delta_{\alpha,\beta} .$$

(16)

The symmetric matrix $\lambda_{\alpha,\beta} > 0$ is the analog of the coupling with the magnetic impurity in the usual Kondo problem. The couplings $h_{\alpha,\beta} = h_{\beta,\alpha}$ represent a direct coupling between only a pair of Majorana fermions. These can be made exponentially small when the Majorana zero modes are far enough from each other in real space. For this reason, we shall neglect them henceforth. A related, yet different model, with real spinless fermions in the bulk, has been analyzed in Ref. 67.

3.1. *Topological Kondo model from a Y-junction of one-dimensional spin systems*

To prove the mapping between a YSC of three critical chains and a topological Kondo Hamiltonian with $M = 3$, one starts from Ref. 56 where, by introducing the KF approach to the generalized JW transformation for a junction of quantum spin chains, it was shown that a Y-junction of quantum XX-spin chains in their gapless phase (corresponding to taking the limit $\gamma = 1$ in $H^{(0)} + H_\Delta$) hosts a remarkable realization of the four-channel, spin-$1/2$ Kondo effect.

In the low-energy, long-wavelength limit for the lead excitations, the Y-junction of quantum XX-spin chains exactly maps onto the Hamiltonian in Eq. (15) for $M = 3$. In order to see this, one first of all notices that, for $\gamma = 1$, one obtains

$$H_{\gamma=1} = H^{(0)}_{\gamma=1} + H_{\Delta,\gamma=1} =$$

(17)

$$\sum_{\alpha=1,2,3} \left\{ -J \sum_{j=1}^{\ell-1} [a^\dagger_{j,\alpha} a_{j+1,\alpha} + a^\dagger_{j+1,\alpha} a_{j,\alpha}] - H \sum_{j=1}^{\ell} a^\dagger_{j,\alpha} a_{j,\alpha} \right\} + 2J_\Delta [\vec{\Sigma}_1 + \vec{\Upsilon}_1] \cdot \vec{\mathcal{R}} \ .$$

In the "critical" region, $2J > |H|$, the lattice fermions in Eq. (18) show a gapless spectrum (in the $\ell \to \infty$ limit), thus behaving as gapless free fermions on the leads of the junction. This allows to expand the lattice fermion operators by only retaining low-energy modes around the Fermi momenta of the single-fermion spectrum in each lead, $\pm k_F = \pm \arccos \left(\frac{H}{2J}\right)$. Specifically, one sets

$$a_{j,\alpha} \approx \sqrt{a} \left\{ e^{ik_F(j-1)} \psi_{R,\alpha}(x_j) + e^{-ik_F(j-1)} \psi_{L,\alpha}(x_j) \right\} \ ,$$

(18)

with $x_j = ja$ and a being the lattice step (which we will set to 1 henceforth, unless specifically required to do otherwise for the sake of clarity). In terms of the continuum fermion operators at the right-hand side of Eq. (18), one obtains

$$H_{\gamma-1} \approx -i\hat{v}_F \sum_{\alpha=1}^{3} \int_0^\ell dx \, \{\psi^\dagger_{R,\alpha}(x)\partial_x \psi_{R,\alpha}(x) - \psi^\dagger_{L,\alpha}(x)\partial_x \psi_{L,\alpha}(x)\}$$

$$+ \sum_{\alpha \neq \beta} J_\Delta \sigma^\alpha \sigma^\beta [\psi^\dagger_{R,\alpha}(0) + \psi^\dagger_{L,\alpha}(0)][\psi_{R,\beta}(0) + \psi_{L,\beta}(0)] \ ,$$

(19)

with $\hat{v}_F = 2J \cos(k_F)$. To recover the Hamiltonian in Eq. (15), one should "unfold" the chiral fields by defining 3 right-handed field $\psi_\alpha(x)$ such that

$$\psi_\alpha(x) = \begin{cases} \psi_{R,\alpha}(x) \ , & (x > 0) \\ \psi_{L,\alpha}(-x) \ , & (x < 0) \end{cases} \ .$$

(20)

Once $H_{\gamma=1}$ is rewritten in terms of the unfolded fields in Eq. (20), by comparison with Eq. (15), one readily sees that $H_{\gamma=1}$ corresponds to an $M = 3$ TKM, provided one identifies the Majorana modes $\{\gamma_\alpha\}$ with the KF $\{\sigma^\alpha\}$ and μ with H; all the $\lambda_{\alpha\beta}$ should be independent of α and β and all equal to J_Δ.

To pertinently complement our discussion about the realization of the TKM with a YSC, it is worth recalling that a different spin-chain realization of the topological Kondo effect may be realized at a Y-junction of three quantum Ising chains. Indeed, using the approach highlighted above, in his paper on a YSC of three critical quantum Ising chains,[53] Alexei Tsvelik provided a remarkable realization of a two-channel version of the topological Kondo effect.[97] Upon setting $\gamma = 0$ in Eq. (11) and, at the same time, assuming that H is tuned at the quantum-critical point $H = J$, Eq. (11) becomes

$$H^{(0)}_{\gamma=0,\mathrm{crit}} = \sum_{\alpha=1,2,3} \left\{ -\frac{J}{2} \sum_{j=1}^{\ell-1} [a^\dagger_{j,\alpha} - a_{j,\alpha}][a^\dagger_{j+1,\alpha} + a_{j+1,\alpha}] - \frac{J}{2} \sum_{j=1}^{\ell} [a^\dagger_{j,\alpha} + a_{j,\alpha}][a^\dagger_{j,\alpha} - a_{j,\alpha}] \right\}.$$
(21)

Upon defining the real-fermion vector operator $\vec{\Psi}_j$ so that

$$\vec{\Psi}_j = \begin{bmatrix} a^\dagger_{j,1} + a_{j,1} \\ a^\dagger_{j,2} + a_{j,2} \\ a^\dagger_{j,3} + a_{j,3} \end{bmatrix} , \quad j \text{ odd}$$

$$\vec{\Psi}_j = \begin{bmatrix} -i(a^\dagger_{j,1} - a_{j,1}) \\ -i(a^\dagger_{j,2} - a_{j,2}) \\ -i(a^\dagger_{j,3} - a_{j,3}) \end{bmatrix} , \quad j \text{ even} ,$$
(22)

with $j = 1, \ldots, 2\ell$, one sees that the definition in Eqs. (22) allows for rewriting $H_{\gamma=0,\mathrm{crit}} = H^{(0)}_{\gamma=0,\mathrm{crit}} + H_{\Delta,\gamma=0}$ as

$$H_{\gamma=0,\mathrm{crit}} = -\frac{iJ}{2} \sum_{j=1}^{2\ell-1} \vec{\Psi}_j \cdot \vec{\Psi}_{j+1} - 2iJ_\Delta(\vec{\Psi}_1 \times \vec{\Psi}_1) \cdot \vec{\mathcal{R}} .$$
(23)

As outlined in Ref. 102, the right-hand side of Eq. (23) corresponds to the lattice version of the real-fermion realization of a two-channel Kondo model. As the details of the renormalization group flow of the running coupling, of the nature of the strongly coupled Kondo fixed point and of possible ways of experimentally probing the Kondo physics have been largely addressed in the literature (see, for instance Refs. 83, 102 and 103), here we just stress once more how this remarkable properties are determined by the emergence of the KF at the junction and by their interaction with the dynamical degrees of freedom of the junctions. At the same time, the topological nature of the spin operator $\vec{\mathcal{R}}$ is witnessed by the fact that its components are nonlocal functionals of operators related to different leads of the junction, so that once one lead is disconnected, the operator itself ceases to exist as an effective quantum spin-1/2 degree of freedom.[97]

For generic values of γ and H, each single-chain fermionic Hamiltonian at the right-hand side of Eq. (11) may be readily diagonalized, yielding the dispersion relation[83]

$$\pm \epsilon_k = \pm \sqrt{[J(1+\gamma)\cos(k) - H]^2 + J^2(1-\gamma)^2 \sin^2(k)} \quad . \tag{24}$$

For generic values of the parameters, the dispersion relation in Eq. (24) is fully gapped, with the gap $\Delta_w = \min \left\{ J(1-\gamma)\sqrt{1 - \frac{H^2}{\gamma^2 J^2}}, |J(1+\gamma) - H| \right\}$. Moving along the "critical" line $\Delta_w = 0$, one may continuously interpolate between the Y-junction of QICs and the Y-junction of XX spin chains studied in Ref. 56.

In order to pertinently interpolate between the two Kondo systems discussed above, both characterized by a gapless dispersion relation of the lead fermions, one has to move along a line in parameter space on which $\Delta_w = 0$. This is explicitly illustrated in Fig. 2. The line continuously connecting the two-channel Kondo system at $\gamma = 0$ with the four-channel one at $\gamma = 1$ is drawn in red and while the vertical line with $\gamma = 1$ is green.

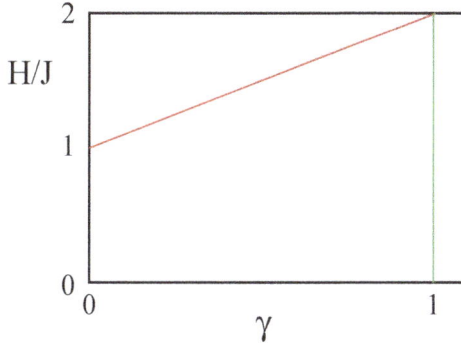

Fig. 2. Phase diagram of the XY-chain in a magnetic field in the $\gamma - (H/J)$ plane: the green line at $\gamma = 1$ corresponds to the XX line. The red line corresponds to the gap closure at $H/J = 1+\gamma$ (see text).

Specifically, the "critical" line connecting the Ising limit $\gamma = 0$ with the four-channel line, spanned at constant $\gamma = 1$ by varying the ratio $H/(2J)$ between -1 and 1, satisfies the equation $H = J(1+\gamma)$, which implies

$$\pm \epsilon_{k,\text{crit}} = \pm 2J \left| \sin\left(\frac{k}{2}\right) \right| \sqrt{(1+\gamma)^2 \sin^2\left(\frac{k}{2}\right) + (1-\gamma)^2 \cos^2\left(\frac{k}{2}\right)} \quad , \tag{25}$$

with the gap closing at $k = 0$.

An important observation is that, though, for any $0 \leq \gamma < 1$, H_Δ corresponds to a four-channel, spin-1/2 Kondo Hamiltonian, the condition $J_2/J_1 = \gamma < 1$ makes the coupling to be pairwise inequivalent. This implies that J_1 will flow towards strong coupling before J_2. Once J_1 has flown to strong coupling, the flow of the other running coupling stops and the system behaves as a two-channel Kondo model.

3.2. *Thermodynamics of the topological Kondo model*

The thermodynamics of the TKM with an arbitrary number of wires M is analyzed in Ref. 51, where it is provided the complete Bethe ansatz solution for $T \neq 0$. The results for $T = 0$ are given in Refs. 104 and 105. We refer to Ref. 51 for details of the finite temperature Bethe ansatz solution. Here, we limit ourselves to stress the most important results for the entropy and the specific heat.

The residual entropy at zero temperature shows that the degrees of freedom introduced by the Majorana modes, contribute as

$$S_J^{(0)} = \log \sqrt{\frac{M}{2}} ,\tag{26}$$

for even M and

$$S_J^{(0)} = \log \sqrt{M} ,\tag{27}$$

for odd M, in both cases are in agreement with the boundary conformal field theory results of Ref. 99.

The signature of the non-Fermi liquid nature of the strongly coupled fixed point is given by the next-to-leading term in the expansion of the junction free energy. As a result, the Majorana contribution to the specific heat behaves at low temperatures as[51]

$$C_J = -T\frac{\partial^2 F_J}{\partial T^2} \sim \left(\frac{T}{T_K}\right)^{\frac{2(M-2)}{M}} ,\tag{28}$$

where the (dimensionless) Kondo temperature T_K depends on the coupling between legs as

$$T_K \sim e^{-\frac{\pi}{\lambda(M-2)}} .\tag{29}$$

A non-integer power is a strong and experimentally detectable signature of the presence of a non-Fermi liquid fixed point. In particular, it is related to the operator content of the conformal field theory describing the fixed point at strong coupling, as explained in Ref. 106.

4. Josephson junction networks and junctions of spin chains

In Sec. 2 we have shown how real-fermion Majorana modes can emerge at a YSC. Spin chain models typically provide simplified, though effective, descriptions of strongly correlated, many-body systems. A basic question to answer, when proposing a YSC as a system to probe emerging Majorana modes, is how to realize it in practice with controlled parameters. A possible answer to this question is provided by the mapping between arrays of Josephson junction chains (JJCs) with pertinently chosen parameters and quantum spin chains.[107–109] In the following we recall how the mapping works for XY model.

4.1. *The XY model as an effective description of a Josephson junction chain*

Starting from the pioneering work by Bradley and Doniach,[107] it is by now well established how JJCs can be well described in terms of either classical, or quantum, spin chains, in various regions of values of their parameters.[107–109]

The first relevant example is provided by a quantum XX-chain realized as a JJC with finite charging energy E_Q.

One considers a chain of quantum Josephson junctions realized between superconducting grains, each one characterized by the value ϕ_i of the local superconducting order parameter. The charge operator at grain-i is canonically conjugated to ϕ_i and, in units of the Cooper pair charge $e^* = 2e$, it is given by $\hat{Q}_i = -i\frac{\partial}{\partial \phi_i}$. Additionally, one assumes that the charge at each grain is fixed by a gate voltage V_g. As a result, letting E_Q, E_J and $\mathcal{N}$ respectively be the charging energy at each grain, the Josephson coupling between nearest-neighbor grains, and the V_g-depending average charge at each grain, a uniform JJC with ℓ sites is described by the Hamiltonian[108,109]

$$H_{\mathrm{JJC}} = \frac{E_Q}{2} \sum_{j=1}^{\ell} \left[-i\frac{\partial}{\partial \phi_j} - \mathcal{N} \right]^2 - E_J \sum_{j=1}^{\ell-1} \cos[\phi_j - \phi_{j+1}] \quad . \tag{30}$$

To map H_{JJC} onto the model Hamiltonian for a quantum spin-1/2 XX-spin chain, one should assume that the JJC parameters are set so that $E_Q/E_J \gg 1$. When $\mathcal{N}$ is integer, or near so, this corresponds to the so-called "charging" regime of the JJC, in which, typically, the chain behaves as a (Mott) insulator, as the Coulomb blockade, due to the large charging energy, prevents charges from tunneling from grain to grain, forbidding current transport across the chain.[107] In fact, at integer $\mathcal{N}$, the Coulomb blockade is determined by the condition that the minimum energy state at each grain is nondegenerate, with total charge $\sim e^*\mathcal{N}$. At variance, when $\mathcal{N} = n + \frac{1}{2}$, with integer n, two different charge eigenstates, with charge equal to e^*n and to $e^*(n+1)$, are degenerate, at each site of the chain. This means that one may recover the low-energy dynamics of H_{JJC} in Eq. (30) by only retaining those two states at each site. Accordingly, one naturally resorts to a quantum spin-1/2 re-formulation of the JJC dynamics, by defining, at each site i, the two states $|\uparrow\rangle_i \equiv |n+1\rangle_i$, and $|\downarrow\rangle_i = |n\rangle_i$, with $|n+1\rangle_i$ and $|n\rangle_i$ being the two charge eigenstates corresponding to charge $n+1$ and n at site i.

Defining the low-energy subspace as $\mathcal{F} = \oplus_{\sigma_1,...,\sigma_\ell=\pm1} \{|\sigma_1\rangle_1 \otimes ... \otimes |\sigma_\ell\rangle_\ell\}$, one defines the spin-1/2 operators $\vec{S}_j$ by projecting on $\mathcal{F}$ the operators at the right-hand side of Eq. (30). In particular, letting $\mathcal{P}_\mathcal{F}$ be the projector on $\mathcal{F}$, one obtains[109]

$$S_j^z \equiv \mathcal{P}_\mathcal{F} \left[-i\frac{\partial}{\partial \phi_j} - n - \frac{1}{2} \right] \mathcal{P}_\mathcal{F}$$

$$S_j^\pm \equiv \mathcal{P}_\mathcal{F} e^{\pm i\phi_j} \mathcal{P}_\mathcal{F} \quad . \tag{31}$$

In terms of the spin-1/2 operators in Eq. (31) one then gets

$$H_{\rm JJC} = -\frac{E_J}{2} \sum_{j=1}^{\ell-1} \{S_j^+ S_{j+1}^- + S_{j+1}^+ S_j^-\} - h \sum_{j=1}^{\ell} S_j^z \quad, \tag{32}$$

where, in Eq. (32), $h = E_Q \delta$, and δ corresponds to a possible small offset in $\mathcal{N}$ from the exact degeneracy point, that is, $\mathcal{N} = n + \frac{1}{2} + \delta$.

One should observe that, though states with charge at a site i different from either n, or $n+1$, are ruled out from $\mathcal{F}$ as "high-energy" states, at finite E_Q they can still play a role as virtual states, entering the mapping leading to Eq. (32) by means of higher-order contributions in E_J/E_Q. For instance, to first-order in E_J/E_Q an additional term at the right-hand side of Eq. (32) is generated, that is $\propto -\frac{E_J^2}{E_Q} \sum_{j=1}^{\ell-1} S_j^z S_{j+1}^z$.[108,109] While such a term is of great interest in view of novel phase transitions it can trigger, its analysis goes beyond the scope of this work. The previous derivation shows how a JJC with appropriate parameters may be simulated by a quantum XX-spin chain.

At variance, engineering a Josephson junction network realizing a QIC is much more challenging. In fact, Cooper-pair tunneling between superconducting grains across each Josephson junction, naturally determines an XX planar coupling, once one resorts to the spin-chain description of the JJC. In order to recover an Ising-like coupling in spin space it is required to use a Josephson junction network, where the "elementary unit" is a rhombus, made out of a single, circular, four-junction chain.[103]

4.2. *Josephson junction network realization of Y-junctions of quantum spin chains*

As a general observation, it is worth stressing again how, due to the optimal level of control reached on their fabrication and control parameters, Josephson junction networks provide an excellent arena to engineer reliable and largely tunable quantum devices.[110] In particular, it is by now known that highly coherent two-level quantum systems may emerge at pertinently engineered Josephson junction rhombi chains.[111,112] More generally, it is well estabilished how a Josephson junction rhombi chain is able to induce charge $4e$ superconducting correlations,[113–115] in the bulk as well as in a tunneling process across a quantum impurity.[111,112,116] Once one has set the mapping between a Josephson junction rhombi chain and a QIC, three Josephson junction rhombi chains may be glued together into a Y-junction, as shown in Ref. 103, thus providing the JJC realization of Tsvelik's YSC.

Given the mapping between pertinent Josephson junction networks and quantum spin chains (XX-spin chains or QICs), one expects to translate all the observations concerning the emergence of Majorana modes at a YSC to Y-junctions realized with appropriate Josephson junction networks. To illustrate how the task can be achieved, one begins with a Y-junction of JJCs such as the ones described in Eq. (30). One considers three chains, described by the Hamiltonian $H_{\mathrm{JJC}}^{(3)}$, given by

$$H_{\mathrm{JJC}}^{(3)} = \sum_{\alpha=1,2,3} \left\{ \frac{E_Q}{2} \sum_{j=1}^{\ell} \left[-i\frac{\partial}{\partial \phi_{j,\alpha}} - \mathcal{N} \right]^2 - E_J \sum_{j=1}^{\ell-1} \cos[\phi_{j,\alpha} - \phi_{j+1,\alpha}] \right\} , \quad (33)$$

with $\phi_{j,a}$ being the phase of the superconducting grain j of chain α and $-i\frac{\partial}{\partial \phi_{j,\alpha}}$ being the corresponding charge operator. The three chains are connected at one of their endpoints, say $j = 1$, to the other two chains by means of a Josephson coupling term H_J, given by[56]

$$H_J = -J' \sum_{\alpha=1,2,3} \{ e^{i\phi_{1,\alpha}} e^{-i\phi_{1,\alpha+1}} + e^{i\phi_{1,\alpha+1}} e^{-i\phi_{1,\alpha}} \} . \quad (34)$$

Assuming that the parameters are all the same in the "bulk" of each chain, one may go through the same projection onto the joint low-energy subspace of the three chains, thus resorting to an effective spin-1/2 quantum spin chain description of the network. As a result, setting $\mathcal{N} = \bar{N} + \frac{1}{2} + h$, with $|h| \ll 1$, $\mathcal{F}_a a = \oplus_{\sigma_{1,\alpha},\ldots,\sigma_{\ell,\alpha}=\pm 1} \{ |\sigma_{1,\alpha}\rangle_{1,\alpha} \otimes \cdots \otimes |\sigma_{\ell,\alpha}\rangle_{\ell,\alpha} \}$ and $\mathcal{F} = \prod_{\alpha=1,2,3} \mathcal{F}_\alpha$, one defines $S_{j,\alpha}^z = \mathcal{P}_{\mathcal{F}} \left[-i\frac{\partial}{\partial \phi_{j,\alpha}} - \bar{N} - \frac{1}{2} \right] \mathcal{P}_{\mathcal{F}}$, $S_{j,\alpha}^{\pm} = \mathcal{P}_{\mathcal{F}} e^{\pm i\phi_{j,\alpha}} \mathcal{P}_{\mathcal{F}}$. Then one obtains that the whole junction (the set of the three JJCs plus the coupling at the endpoints of the chains) is described by the Hamiltonian H_{junction}, given by

$$H_{\mathrm{junction}} = \mathcal{P}_{\mathcal{F}}[H_{\mathrm{JJC}}^{(3)} + H_J]\mathcal{P}_{\mathcal{F}} = \quad (35)$$

$$\sum_{\alpha=1,2,3} \left\{ -\frac{E_J}{2} \sum_{j=1}^{\ell-1} [S_{j,\alpha}^+ S_{j+1,\alpha}^- + S_{j+1,\alpha}^+ S_{j,\alpha}^-] - h \sum_{j=1}^{\ell} S_{j,\alpha}^z \right\} - 2J_{\Delta} \sum_{\alpha=1,2,3} S_{1,\alpha}^x S_{1,\alpha+1}^x ,$$

with $J_{\Delta}/E_J < 1$, that is, the Hamiltonian of a YSC in the isotropic case $\gamma = 1$.

One may construct a YSC of quantum Ising chains, as illustrated in Ref. 103, by considering three equal rhombi chains described by the Hamiltonian $H_{\mathrm{m}}^{(3)}$ given by

$$H_{\mathrm{m}}^{(3)} = \sum_{\alpha=1,2,3} \left\{ -J \sum_{p=1}^{\ell} \sum_{r=1}^{4} \{ e^{-\frac{i}{4}\varphi} \sigma_{r,p,\alpha}^+ \sigma_{r+1,p,\alpha}^- + \mathrm{h.c.} \} \right.$$

$$\left. - h \sum_{p=1}^{\ell} \sum_{r=1}^{4} \sigma_{r,p,\alpha}^z - T \sum_{p=1}^{\ell-1} \{ \sigma_{3,p,\alpha}^+ \sigma_{1,p+1,\alpha}^- + \mathrm{h.c.} \} \right\} . \quad (36)$$

To effectively make a Y-junction, one assumes the existence of Josephson couplings with strength $\mathcal{J}$ between sites number 2 of the endpoint-rhombus of each chain, described by the Hamiltonian $H_{\mathrm{MB;J}}$ given by

$$H_{\text{MB};\text{J}} = -\mathcal{J}\{\sigma_{1,1,2}^{+}\sigma_{2,1,2}^{-} + \sigma_{2,1,2}^{+}\sigma_{3,1,2}^{-} + \sigma_{3,1,2}^{+}\sigma_{1,1,2}^{-} + \text{h.c.}\} \quad . \tag{37}$$

One then projects with $\mathcal{P}_{\mathcal{G}}$ the total Hamiltonian $H_m^{(3)} + H_{\text{MB};\text{J}}$. As a result, one gets the Hamiltonian H_Y, given by

$$H_Y = \sum_{\lambda=1,2,3} \left\{ J_x \sum_{p=1}^{\ell-1} S_{p,\lambda}^x S_{p+1,\lambda}^x - 2 \sum_{p=1}^{\ell} H_p S_{p,\lambda}^z \right\} + H_B \quad , \tag{38}$$

with

$$H_B = J_K \{ S_{1,1}^x S_{2,1}^x + S_{2,1}^x S_{3,1}^x + S_{3,1}^x S_{1,1}^x \} + \delta H_B \quad , \tag{39}$$

and $J_K = \frac{\mathcal{J}^2}{J}$. $\delta H_B = -\frac{\sqrt{2}}{16}\frac{\mathcal{J}^2}{J}\{ S_{1,1}^z + S_{2,1}^z + S_{3,1}^z \}$ is a boundary magnetic field term which does not affect the behavior of the system. Apart for this term (and for a change in sign of J_x), the Hamiltonian in Eqs. (38) and (39) is exactly the one describing the YSC of QICs introduced before.

5. Ultracold atoms and junctions of spin chains

In order to implement a Y-junction with cold atomic systems,[117] one has to recall that:

1. Tonks–Girardeau (TG) gas of one-dimensional bosons on a Y-junction is a physical realization of the XX model on the Y-junction itself — and therefore can be mapped via the JW transformation of Sec. 2 in the TKM Hamiltonian;

2. TG gases of ultracold bosons have been implemented in several experiments;[118]

3. Stable Y-configurations may be created by holographic techniques, as experimentally shown in Ref. 117.

For our purpose here, one should focus only on point 1) and show that (15) may be obtained from a model Hamiltonian for interacting bosons in the TG limit, confined to M one-dimensional waveguides arranged in a Y-junction.

In each waveguide $\alpha = 1, \cdots, M$ the Lieb–Lininger Hamiltonian describing interacting bosons in one-dimensional guides of length $\mathcal{L}$ reads:[119–121]

$$H^{(\alpha)} = \int_0^{\mathcal{L}} dx \left[\frac{\hbar^2}{2m} \partial_x \Psi_\alpha^\dagger(x) \partial_x \Psi_\alpha(x) + \frac{c}{2}\Psi_\alpha^\dagger(x)\Psi_\alpha^\dagger(x)\Psi_\alpha(x)\Psi_\alpha(x) \right] \quad . \tag{40}$$

The parameter m is the mass of the bosons and $c > 0$ is the repulsion strength, as determined by the s-wave scattering length.[122] The bosonic fields Ψ_α satisfy canonical commutation relations $[\Psi_\alpha(x), \Psi_\alpha^\dagger(y)] = \delta(x - y)$.

The coupling of the Lieb–Lininger Hamiltonian, denoted by γ, is proportional to c/n where $n \equiv \mathcal{N}/\mathcal{L}$ is the density of bosons and $\mathcal{N}$ is the number of bosons per waveguide. More specifically one has $\gamma = mc/\hbar^2 n$. The limit of vanishing γ corresponds to an ideal one-dimensional Bose gas, while the limit of infinite γ corresponds to the TG gas,[123,124] which generally has the expectation values and thermodynamic quantities of a one-dimensional ideal Fermi gas.[118,121,125,126] The experimental realization of the TG gas with cold atoms[127,128] triggered intense activity in the last decade, reviewed in Refs. 118, 125 and 126.

One considers M copies of this one-dimensional Bose gas and joins them together by the ends of the segments, in such a way that the bosons can tunnel from one waveguide to the others. The bosonic fields in different legs commute:

$$\left[\Psi_\alpha(x), \Psi_\beta^\dagger(y)\right] = \delta_{\alpha,\beta}\,\delta(x - y),$$

and the total Hamiltonian has the form $H = \sum_{\alpha=1}^{M} H^{(\alpha)} + H_J$ where the junction term H_J describes the tunneling process among legs.

As a tool for performing computations, as well as to give a precise meaning to the tunneling processes at the edges of the legs, in each leg we discretize space into a lattice of L sites with lattice spacing a (where $La = \mathcal{L}$ and the total number of sites N_S of the star lattice is $N_S \equiv LM$). This discretization can be physically realized by superimposing optical lattices on the legs.[129] One can then perform a tight-binding approximation[130,131] and write the bosonic fields as $\Psi_\alpha(x) = \sum_{\alpha,j} w_{\alpha,j}(x)\, b_{\alpha,j}$ where $b_{\alpha,j}$ is the operator destroying a particle at the site $j = 1, \cdots, L$ of the leg α and $w_{\alpha,j}(x)$ is the appropriate Wannier wavefunction localized at the same site.

The resulting lattice Bose–Hubbard Hamiltonian on each leg then reads[130,132]

$$H_U^{(\alpha)} = -t \sum_{j=1}^{L-1} \left(b_{\alpha,j}^\dagger b_{\alpha,j+1} + b_{\alpha,j+1}^\dagger b_{\alpha,j}\right) + \frac{U}{2}\sum_{j=1}^{L} b_{\alpha,j}^\dagger b_{\alpha,j}^\dagger b_{\alpha,j} b_{\alpha,j} \tag{41}$$

where the interaction coefficient is $U = c\int |w_\alpha(x)|^4\, dx$ ($\alpha = 1, \cdots, L$), the hopping coefficient is $t = -\int w_{\alpha,j}\hat{T} w_{\alpha,j+1}\, dx$ with $\alpha = 1, \cdots, L-1$, and $\hat{T} = -(\hbar^2/2m)\partial^2/\partial x^2$ is the kinetic energy operator.

The total lattice Hamiltonian for a Y-junction of atomic condensates is obtained by taking 3 copies of the system, connected to one another by a hopping term. The total Hamiltonian is then written as:

$$H_U = \sum_{\alpha=1}^{3} H_U^{(\alpha)} + H_J, \tag{42}$$

where the junction term has the form

$$H_J = -\lambda \sum_{1 \leq \alpha < \beta \leq 3} \left(b_{\alpha,1}^\dagger b_{\beta,1} + b_{\beta,1}^\dagger b_{\alpha,1}\right) \tag{43}$$

with λ being the hopping between the first site of a leg and the first sites of the others. Typically one has $\lambda > 0$, which corresponds to an *antiferromagnetic* Kondo model, as shown in the following. Nevertheless, we observe that the sign of t and λ could be changed by shaking the trap.[133]

The total number of bosons in the system, $N = \mathcal{N}\mathcal{L}$, is a conserved quantity in the lattice model and can be tuned in experiments. In the canonical ensemble $N = \sum_{\alpha,j} \langle b^\dagger_{\alpha,j} b_{\alpha,j} \rangle$. The phase diagram of the bulk Hamiltonian (41) in each leg undergoes quantum phase transitions between superfluid and Mott insulating phases:[134] notice that in the canonical ensemble the system is superfluid as soon as the filling N/N_S is not integer.

One is interested in the limit $U \to \infty$, so that after the continuous limit is taken back again, the TG gas is retrieved in the bulk. It is well known that this limit brings substantial simplifications in the computation: it was shown in Ref. 135 that, on each leg, the spectrum and the scattering matrix are equivalent to a system of spins in the $s = 1/2$ representation. As is customary, one may map the hard-core bosons to $1/2$ spins. The Hamiltonian (42) written in spin variables is given by

$$H^{(\alpha)}_\infty = -t \sum_{j=1}^{N-1} \left(S^+_{j,\alpha} \sigma^-_{j+1,\alpha} + S^+_{j+1,\alpha} \sigma^-_{j,\alpha} \right) \tag{44}$$

$$H_J = -\lambda \sum_{\alpha<\beta}^{3} \left(S^+_{1,\alpha} \sigma^-_{1,\beta} + \sigma^+_{1,\beta} \sigma^-_{1,\alpha} \right) \tag{45}$$

which coincides with a junction of XX-type spin chains.[56] Now one can proceed exactly as in Sec. 2, finding an Hamiltonian of the form (3) via the correct identification of the parameters.

From the Hamiltonian (44) and (45), one obtains

$$H = -t \sum_{j=1}^{N-1} \sum_{\alpha=1}^{3} \left(c^\dagger_{\alpha,j} c_{\alpha,j+1} + c^\dagger_{\alpha,j+1} c_{\alpha,j} \right) + H_J \tag{46}$$

$$H_J = -\lambda \sum_{1\leq\alpha<\beta\leq3} \gamma_\alpha \gamma_\beta \left(c^\dagger_{\alpha,1} c_{\beta,1} + c_{\alpha,1} c^\dagger_{\beta,1} \right). \tag{47}$$

In conclusion, one has mapped the Hamiltonian (45), acting on N_S spin variables, onto another one, defined in terms of N_S spinless fermionic degrees of freedom plus one Klein factor per leg. In other words, the hard-core boson Hamiltonian (42) in the limit $U \to \infty$ is mapped onto the fermionic Hamiltonian (46), given by the sum of non-interacting wires and the highly nontrivial junction term H_J. The Fermi energy of the non-interacting fermions in (each of) the external wires is denoted by E_F.

It should be remarked that the only mechanism in which the topological protection of the degree of freedom encoded in the Majorana modes can be spoiled is the loss of bosonic atoms by the trap. With our parameters, we found this probability negligible on the time scales needed for the experiment, since the energy barrier for the atom loss is ~ 500 nK, much larger than the typical Fermi energy of the TG gases.

6. Concluding remarks

We argued that a pertinent modification of a remarkable result obtained in the seventies by Roman Jackiw and collaborators[1,2] opened the avenue to consistently deal with non-local effects in field theories of relevant condensed matter systems.

In this paper we dealt with these non-local effects in spin models relevant for the study of the multi-channel Kondo effect. One advantage of these Kondo models is that the perfect symmetry of the couplings of the various channels with the spin impurity is guaranteed by a topological degeneracy and thus robust against decoherence. The topological Kondo effect stems from the interaction of localized Majorana modes with external 1D channels arranged in a pertinent graph geometry.

We considered the simple case of star-like geometries on which XX and XY spin models were defined. We showed how, from these models, the four- and two-channel Kondo models emerge.

We finally showed how these topological Kondo models may be realized in Josephson junction arrays and networks of Tonks–Girardeau gases.

Acknowledgements

We all benefited from discussions with Professor Roman Jackiw and from his lectures on Majorana fermions and charge fractionalization. We are very happy to contribute to his festschrift volume.

I (P.S.) enjoyed Roman's friendship since 1985 when I first joined his research group at the Centre for Theoretical Physics of MIT. The possibility to visit MIT as a postdoctoral fellow in Roman's group was offered to me by Professor Sergio Fubini who, after the many discussions we had at CERN, assisted me in getting the financial support for my first visit at MIT. There, it started a very memorable time of my scientific and personal life! Indeed, I had the opportunity to enjoy the many teachings from Roman on a variety of topics ranging from Chern–Simons field theories to field theories of condensed matter systems. Our frequent discussions shaped my scientific attitude through the years and drove my interest towards the study of the role of topology in relativistic field theories and condensed matter systems. The friendship with Roman gave to me the unique chance to appreciate also his sense of humor, his loyalty as a friend, and his patience as a teacher; in addition, he had a unique ability in choosing always the very good restaurants where, from time to time, we celebrated our encounters. I wish him a very happy birthday and many serene and interesting years to come.

It is also a pleasure to thank I. Affleck, H. Babujian, A. Bayat, S. Bose, F. Buccheri, M. Burrello, G. Campagnano, N. Crampé, D. Cassettari, R. Egger, R. Graham, V. Korepin, L. Lepori, D. Rossini, and A. Tagliacozzo for many enlightening discussions during our collaboration on some of the topics presented in this paper.

References

1. R. Jackiw and C. Rebbi, Solitons with fermion number 1/2, *Phys. Rev. D* **13**, 3398–3409 (1976). doi: 10.1103/PhysRevD.13.3398.
2. R. Jackiw and P. Rossi, Zero modes of the vortex–fermion system, *Nucl. Phys. B* **190**(4), 681–691 (1981). doi: 10.1016/0550-3213(81)90044-4.
3. E. Majorana, Teoria simmetrica dell'elettrone e del positrone, *Il Nuovo Cimento* **14** (4), 171 (1937). doi: 10.1007/BF02961314.
4. R. Rajaraman and J. S. Bell, On solitons with half integral charge, *Phys. Lett. B* **116**(2), 151–154 (1982). doi: 10.1016/0370-2693(82)90996-0.
5. A. Niemi and G. Semenoff, Fermion number fractionization in quantum field theory, *Phys. Rep.* **135**(3), 99–193 (1986). doi: 10.1016/0370-1573(86)90167-5.
6. R. Jackiw, A. Kerman, I. Klebanov, and G. Semenoff, Fluctuations of fractional charge in soliton anti-soliton systems, *Nucl. Phys. B* **225**(2), 233–246, (1983). doi: 10.1016/0550-3213(83)90051-2.
7. R. Jackiw and J. Schrieffer, Solitons with fermion number 1/2 in condensed matter and relativistic field theories, *Nucl. Phys. B* **190**(2), 253–265 (1981). doi: 10.1016/0550-3213(81)90557-5.
8. R. Jackiw, Fractional and Majorana fermions: The physics of zero-energy modes, *Physica Scripta.* **T146**, 014005 (2012). doi: 10.1088/0031-8949/2012/t146/014005.
9. R. Jackiw and S.-Y. Pi, State space for planar Majorana zero modes, *Phys. Rev. B* **85**, 033102 (2012). doi: 10.1103/PhysRevB.85.033102.
10. R. Jackiw, Emergent Majorana fermions and their restricted Clifford algebra, *Theor. Math. Phys.* **181**(1), 1164–1168 (2014). doi: 10.1007/s11232-014-0206-6.
11. S.-H. Ho, F.-L. Lin, and X.-G. Wen, Majorana zero-modes and topological phases of multi-flavored Jackiw–Rebbi model, *J. High Energy Phys.* **2012**(12), 74 (2012). doi: 10.1007/JHEP12(2012)074.
12. J. Lee and F. Wilczek, Algebra of Majorana doubling, *Phys. Rev. Lett.* **111**, 226402 (2013). doi: 10.1103/PhysRevLett.111.226402.
13. G. W. Semenoff and P. Sodano, Stretching the electron as far as it will go, *Electron. J. Theor. Phys.* **10**, 157–190 (2006).
14. J. Alicea, New directions in the pursuit of Majorana fermions in solid state systems, *Rep. Prog. Phys.* **75**(7), 076501 (2012). doi: 10.1088/0034-4885/75/7/076501.
15. C. Beenakker, Search for Majorana fermions in superconductors, *Ann. Rev. Cond. Mat. Phys.* **4**(1), 113–136 (2013). doi: 10.1146/annurev-conmatphys-030212-184337.
16. M. Franz, Majorana's wires, *Nature Nanotechnology.* **8**, 149–152 (2013). doi: 10.1038/nnano.2013.33.
17. M. Leijnse and K. Flensberg, Introduction to topological superconductivity and Majorana fermions, *Semiconductor Science and Technology.* **27**(12), 124003 (2012). doi: 10.1088/0268-1242/27/12/124003.
18. V. Mourik, K. Zuo, S. M. Frolov, S. R. Plissard, E. P. A. M. Bakkers, and L. P. Kouwenhoven, Signatures of Majorana fermions in hybrid superconductor-semiconductor nanowire devices, *Science.* **336**(6084), 1003–1007 (2012). doi: 10.1126/science.1222360.
19. A. Y. Kitaev, Unpaired Majorana fermions in quantum wires, *Physics-Uspekhi.* **44** (10S), 131–136 (2001). doi: 10.1070/1063-7869/44/10s/s29.
20. G. W. Semenoff and P. Sodano, Stretched quantum states emerging from a Majorana medium, *J. Phys. B.* **40**(8), 1479–1488 (2007). doi: 10.1088/0953-4075/40/8/002.
21. J. Nilsson, A. R. Akhmerov, and C. W. J. Beenakker, Splitting of a Cooper pair by a pair of Majorana bound states, *Phys. Rev. Lett.* **101**, 120403 (2008). doi:

10.1103/PhysRevLett.101.120403.

22. L. Fu, Electron teleportation via Majorana bound states in a mesoscopic superconductor, *Phys. Rev. Lett.* **104**, 056402 (2010). doi: 10.1103/PhysRevLett.104.056402.

23. L. Jiang, D. Pekker, J. Alicea, G. Refael, Y. Oreg, and F. von Oppen, Unconventional Josephson signatures of Majorana bound states, *Phys. Rev. Lett.* **107**, 236401 (2011). doi: 10.1103/PhysRevLett.107.236401.

24. J. D. Sau, E. Berg, and B. I. Halperin. On the possibility of the fractional ac Josephson effect in non-topological conventional superconductor-normal-superconductor junctions, (2012).

25. S. Tewari, S. Das Sarma, C. Nayak, C. Zhang, and P. Zoller, Quantum computation using vortices and Majorana zero modes of a $p_x + ip_y$ superfluid of fermionic cold atoms, *Phys. Rev. Lett.* **98**, 010506 (2007). doi: 10.1103/PhysRevLett.98.010506.

26. L. Fu and C. L. Kane, Superconducting proximity effect and Majorana fermions at the surface of a topological insulator, *Phys. Rev. Lett.* **100**, 096407 (2008). doi: 10.1103/PhysRevLett.100.096407.

27. V. Shivamoggi, G. Refael, and J. E. Moore, Majorana fermion chain at the quantum spin Hall edge, *Phys. Rev. B.* **82**, 041405 (2010). doi: 10.1103/PhysRevB.82.041405.

28. J. D. Sau, S. Tewari, R. M. Lutchyn, T. D. Stanescu, and S. Das Sarma, Non-Abelian quantum order in spin-orbit-coupled semiconductors: Search for topological Majorana particles in solid-state systems, *Phys. Rev. B.* **82**, 214509 (2010). doi: 10.1103/PhysRevB.82.214509.

29. J. D. Sau, R. M. Lutchyn, S. Tewari, and S. Das Sarma, Generic new platform for topological quantum computation using semiconductor heterostructures, *Phys. Rev. Lett.* **104**, 040502 (2010). doi: 10.1103/PhysRevLett.104.040502.

30. Y. Oreg, G. Refael, and F. von Oppen, Helical liquids and Majorana bound states in quantum wires, *Phys. Rev. Lett.* **105**, 177002 (2010). doi: 10.1103/PhysRevLett.105.177002.

31. R. M. Lutchyn, J. D. Sau, and S. Das Sarma, Majorana fermions and a topological phase transition in semiconductor-superconductor heterostructures, *Phys. Rev. Lett.* **105**, 077001 (2010). doi: 10.1103/PhysRevLett.105.077001.

32. A. M. Tsvelik. Zero energy Majorana modes in superconducting wires, (2011).

33. F. Hassler and D. Schuricht, Strongly interacting Majorana modes in an array of Josephson junctions, *New J. Phys.* **14**(12), 125018 (2012). doi: 10.1088/1367-2630/14/12/125018.

34. A. A. Nersesyan and A. M. Tsvelik, Zero-energy Majorana modes in spin ladders and a possible realization of the Kitaev model, *Europhys. Lett.* **96**(1), 17002 (2011). doi: 10.1209/0295-5075/96/17002.

35. C. Chamon, R. Jackiw, Y. Nishida, S.-Y. Pi, and L. Santos, Quantizing Majorana fermions in a superconductor, *Phys. Rev. B.* **81**, 224515 (2010). doi: 10.1103/PhysRevB.81.224515.

36. D. Sticlet, C. Bena, and P. Simon, Spin and Majorana polarization in topological superconducting wires, *Phys. Rev. Lett.* **108**, 096802 (2012). doi: 10.1103/PhysRevLett.108.096802.

37. I. Affleck and D. Giuliano, Screening clouds and Majorana fermions, *J. Stat. Phys.* **157**(4), 666–691 (2014). doi: 10.1007/s10955-014-1056-1.

38. C. Nayak, S. Simon, A. Stern, M. Freedman, and S. Das Sarma, Non-Abelian anyons and topological quantum computation, *Rev. Mod. Phys.* **80**, 1083–1159 (2008). doi: 10.1103/RevModPhys.80.1083.

39. S. Bose and P. Sodano, Nonlocal Hanbury–Brown–Twiss interferometry and entanglement generation from Majorana bound states, *New J. Phys.* **13**(8), 085002 (2011).

doi: 10.1088/1367-2630/13/8/085002.

40. G. Strübi, W. Belzig, M.-S. Choi, and C. Bruder, Interferometric and noise signatures of Majorana fermion edge states in transport experiments, *Phys. Rev. Lett.* **107**, 136403 (2011). doi: 10.1103/PhysRevLett.107.136403.

41. L. Fu and C. L. Kane, Probing neutral Majorana fermion edge modes with charge transport, *Phys. Rev. Lett.* **102**, 216403 (2009). doi: 10.1103/PhysRevLett.102.216403.

42. A. R. Akhmerov, J. Nilsson, and C. W. J. Beenakker, Electrically detected interferometry of Majorana fermions in a topological insulator, *Phys. Rev. Lett.* **102**, 216404 (2009). doi: 10.1103/PhysRevLett.102.216404.

43. A. Romito, J. Alicea, G. Refael, and F. von Oppen, Manipulating Majorana fermions using supercurrents, *Phys. Rev. B.* **85**, 020502 (2012). doi: 10.1103/PhysRevB.85.020502.

44. A. Kitaev, Fault-tolerant quantum computation by anyons, *Annals of Physics.* **303** (1), 2–30 (2003). doi: 10.1016/S0003-4916(02)00018-0.

45. A. Zazunov, P. Sodano, and R. Egger, Even–odd parity effects in Majorana junctions, *New J. Phys.* **15**(3), 035033 (2013). doi: 10.1088/1367-2630/15/3/035033.

46. A. Zazunov, A. L. Yeyati, and R. Egger, Coulomb blockade of Majorana-fermion-induced transport, *Phys. Rev. B.* **84**, 165440 (2011). doi: 10.1103/PhysRevB.84.165440.

47. R. Hützen, A. Zazunov, B. Braunecker, A. L. Yeyati, and R. Egger, Majorana single-charge transistor, *Phys. Rev. Lett.* **109**, 166403 (2012). doi: 10.1103/PhysRevLett.109.166403.

48. A. Altland and R. Egger, Multiterminal Coulomb–Majorana junction, *Phys. Rev. Lett.* **110**, 196401 (2013). doi: 10.1103/PhysRevLett.110.196401.

49. B. Béri and N. R. Cooper, Topological Kondo effect with Majorana fermions, *Phys. Rev. Lett.* **109**, 156803 (2012). doi: 10.1103/PhysRevLett.109.156803.

50. A. Altland, B. Beri, R. Egger, and A. M. Tsvelik, Majorana spin dynamics in the topological Kondo effect, *Phys. Rev. Lett.* **113**, 076401 (2014).

51. F. Buccheri, H. Babujian, V. E. Korepin, P. Sodano, and A. Trombettoni, Thermodynamics of the topological Kondo model, *Nucl. Phys. B.* **896**, 52–79 (2015). doi: 10.1016/j.nuclphysb.2015.04.016.

52. A. Bayat, S. Bose, and P. Sodano, Entanglement routers using macroscopic singlets, *Phys. Rev. Lett.* **105**, 187204 (2010). doi: 10.1103/PhysRevLett.105.187204.

53. A. M. Tsvelik, Majorana fermion realization of a two-channel Kondo effect in a junction of three quantum Ising chains, *Phys. Rev. Lett.* **110**, 147202 (2013). doi: 10.1103/PhysRevLett.110.147202.

54. A. Zazunov, F. Buccheri, P. Sodano, and R. Egger, 6π Josephson effect in Majorana box devices, *Phys. Rev. Lett.* **118**, 057001 (2017). doi: 10.1103/PhysRevLett.118.057001.

55. S. Plugge, A. Zazunov, P. Sodano, and R. Egger, Majorana entanglement bridge, *Phys. Rev. B.* **91**, 214507 (2015). doi: 10.1103/PhysRevB.91.214507.

56. N. Crampé and A. Trombettoni, Quantum spins on star graphs and the Kondo model, *Nucl. Phys. B.* **871**(3), 526–538 (2013). doi: 10.1016/j.nuclphysb.2013.03.001.

57. F. Buccheri, G. D. Bruce, A. Trombettoni, D. Cassettari, H. Babujian, V. E. Korepin, and P. Sodano, Holographic optical traps for atom-based topological Kondo devices, *New J. Phys.* **18**(7), 075012 (2016). doi: 10.1088/1367-2630/18/7/075012.

58. D. Giuliano, P. Sodano, and A. Trombettoni, Kondo length in bosonic lattices, *Phys. Rev. A.* **96**, 033603 (2017). doi: 10.1103/PhysRevA.96.033603.

59. C. Chamon, M. Oshikawa, and I. Affleck, Junctions of three quantum wires and the

dissipative Hofstadter model, *Phys. Rev. Lett.* **91**, 206403 (2003). doi: 10.1103/Phys-RevLett.91.206403.

60. M. Oshikawa, C. Chamon, and I. Affleck, Junctions of three quantum wires, *J. Stat. Mech.* **2006**(02), P02008 (2006). doi: 10.1088/1742-5468/2006/02/p02008.

61. D. Giuliano and P. Sodano, Y-junction of superconducting Josephson chains, *Nucl. Phys. B.* **811**(3), 395–419, (2009). doi: 10.1016/j.nuclphysb.2008.11.011.

62. A. Bayat, S. Bose, P. Sodano, and H. Johannesson, Entanglement probe of two-impurity Kondo physics in a spin chain, *Phys. Rev. Lett.* **109**, 066403 (2012). doi: 10.1103/PhysRevLett.109.066403.

63. B. Alkurtass, A. Bayat, I. Affleck, S. Bose, H. Johannesson, P. Sodano, E. S. Sørensen, and K. Le Hur, Entanglement structure of the two-channel Kondo model, *Phys. Rev. B.* **93**, 081106 (2016). doi: 10.1103/PhysRevB.93.081106.

64. A. Bayat, P. Sodano, and S. Bose, Negativity as the entanglement measure to probe the Kondo regime in the spin-chain Kondo model, *Phys. Rev. B.* **81**, 064429 (2010). doi: 10.1103/PhysRevB.81.064429.

65. A. Bayat, H. Johannesson, S. Bose, and P. Sodano, An order parameter for impurity systems at quantum criticality, *Nature Communications.* **5**(1), 3784 (2014). doi: 10.1038/ncomms4784.

66. L. A. Landau, S. Plugge, E. Sela, A. Altland, S. M. Albrecht, and R. Egger, Towards realistic implementations of a Majorana surface code, *Phys. Rev. Lett.* **116**, 050501 (2016). doi: 10.1103/PhysRevLett.116.050501.

67. A. M. Tsvelik, Topological Kondo effect in star junctions of Ising magnetic chains: Exact solution, *New J. Phys.* **16**(3), 033003 (2014). doi: 10.1088/1367-2630/16/3/033003.

68. P. Jordan and E. Wigner, Über das paulische äquivalenzverbot, *Zeitschrift für Physik.* **47**(9), 631–651 (1928). doi: 10.1007/BF01331938.

69. E. Lieb, T. Schultz, and D. Mattis, Two soluble models of an antiferromagnetic chain, *Annals of Physics.* **16**(3), 407–466 (1961). doi: 10.1016/0003-4916(61)90115-4.

70. A. Komnik and R. Egger, Nonequilibrium transport for crossed Luttinger liquids, *Phys. Rev. Lett.* **80**, 2881–2884 (1998). doi: 10.1103/PhysRevLett.80.2881.

71. S. Lal, S. Rao, and D. Sen, Junction of several weakly interacting quantum wires: A renormalization group study, *Phys. Rev. B.* **66**, 165327 (2002). doi: 10.1103/Phys-RevB.66.165327.

72. D. Giuliano and P. Sodano, Frustration of decoherence in Y-shaped superconducting Josephson networks, *New J. Phys.* **10**(9), 093023 (2008). doi: 10.1088/1367-2630/10/9/093023.

73. A. Agarwal, Time resolved transport properties of a Y junction of Tomonaga–Luttinger liquid wires, *Phys. Rev. B.* **90**, 195403 (2014). doi: 10.1103/Phys-RevB.90.195403.

74. S. Mardanya and A. Agarwal, Enhancement of tunneling density of states at a Y junction of spin-1/2 Tomonaga–Luttinger liquid wires, *Phys. Rev. B.* **92**, 045432 (2015). doi: 10.1103/PhysRevB.92.045432.

75. D. Giuliano and A. Nava, Dual fermionic variables and renormalization group approach to junctions of strongly interacting quantum wires, *Phys. Rev. B.* **92**, 125138 (2015). doi: 10.1103/PhysRevB.92.125138.

76. S. Yin and B. Béri, Universality and quantized response in bosonic mesoscopic tunneling, *Phys. Rev. B.* **93**, 245142 (2016). doi: 10.1103/PhysRevB.93.245142.

77. R. Burioni, D. Cassi, M. Rasetti, P. Sodano, and A. Vezzani, Bose–Einstein condensation on inhomogeneous complex networks, *J. Phys. B.* **34**(23), 4697–4710 (2001). doi: 10.1088/0953-4075/34/23/314.

78. I. Brunelli, G. Giusiano, F. P. Mancini, P. Sodano, and A. Trombettoni, Topology-induced spatial Bose–Einstein condensation for bosons on star-shaped optical networks, *J. Phys. B.* **37**(7), S275–S286 (2004). doi: 10.1088/0953-4075/37/7/072.

79. A. Tokuno, M. Oshikawa, and E. Demler, Dynamics of one-dimensional Bose liquids: Andreev-like reflection at y junctions and the absence of the Aharonov–Bohm effect, *Phys. Rev. Lett.* **100**, 140402 (2008). doi: 10.1103/PhysRevLett.100.140402.

80. A. Cirillo, M. Mancini, D. Giuliano, and P. Sodano, Enhanced coherence of a quantum doublet coupled to Tomonaga–Luttinger liquid leads, *Nucl. Phys. B.* **852**(1), 235–268 (2011). doi: 10.1016/j.nuclphysb.2011.06.014.

81. A. M. Tsvelik and W.-G. Yin, Possible realization of a multichannel Kondo model in a system of magnetic chains, *Phys. Rev. B.* **88**, 144401 (2013). doi: 10.1103/PhysRevB.88.144401.

82. J. von Delft and H. Schoeller, Bosonization for beginners — refermionization for experts, *Annalen der Physik.* **7**(4), 225–305 (1998). doi: 10.1002/(SICI)1521-3889(199811)7:4⟨225 :: $AID - ANDP225$⟩3.0.CO;2-L.

83. D. Giuliano, P. Sodano, A. Tagliacozzo, and A. Trombettoni, From four- to two-channel Kondo effect in junctions of XY spin chains, *Nucl. Phys. B.* **909**, 135–172 (2016). doi: 10.1016/j.nuclphysb.2016.05.003.

84. B. Béri, Majorana–Klein hybridization in topological superconductor junctions, *Phys. Rev. Lett.* **110**, 216803 (2013). doi: 10.1103/PhysRevLett.110.216803.

85. D. Giuliano and I. Affleck, Real fermion modes, impurity entropy, and non-trivial fixed points in the phase diagram of junctions of interacting quantum wires and topological superconductors, *Nucl. Phys. B.* **944**, 114645 (2019). doi: 10.1016/j.nuclphysb.2019.114645.

86. J. Kondo, Resistance minimum in dilute magnetic alloys, *Prog. Theor. Phys.* **32**(1), 37–49 (1964). doi: 10.1143/PTP.32.37.

87. A. C. Hewson, *The Kondo Problem to Heavy Fermions.* Cambridge Studies in Magnetism, (Cambridge University Press, 1993). doi: 10.1017/CBO9780511470752.

88. L. P. Kouwenhoven and L. Glazman, Revival of the Kondo effect, *Physics World.* **14**, 33–38 (2001).

89. P. Nozières, A "fermi-liquid" description of the Kondo problem at low temperatures, *J. Low Temp. Phys.* **17**(1), 31–42 (1974). doi: 10.1007/BF00654541.

90. A. P. Alivisatos, Semiconductor clusters, nanocrystals, and quantum dots, *Science.* **271**(5251), 933–937 (1996). doi: 10.1126/science.271.5251.933.

91. L. Kouwenhoven and C. Marcus, Quantum dots, *Physics World.* **11**(6), 35–40 (1998). doi: 10.1088/2058-7058/11/6/26.

92. D. Goldhaber-Gordon, H. Shtrikman, D. Mahalu, D. Abusch-Magder, U. Meirav, and M. A. Kastner, Kondo effect in a single-electron transistor, *Nature.* **391**, 156–159 (1998). doi: 10.1038/34373.

93. S. M. Cronenwett, T. H. Oosterkamp, and L. P. Kouwenhoven, A tunable Kondo effect in quantum dots, *Science.* **281**(5376), 540–544 (1998). doi: 10.1126/science.281.5376.540.

94. Y. Avishai, A. Golub, and A. D. Zaikin, Tunneling through an Anderson impurity between superconductors, *Phys. Rev. B.* **63**, 134515 (2001). doi: 10.1103/PhysRevB.63.134515.

95. M.-S. Choi, M. Lee, K. Kang, and W. Belzig, Kondo effect and Josephson current through a quantum dot between two superconductors, *Phys. Rev. B.* **70**, 020502 (2004). doi: 10.1103/PhysRevB.70.020502.

96. G. Campagnano, D. Giuliano, A. Naddeo, and A. Tagliacozzo, Josephson current in a quantum dot in the Kondo regime connected to two superconductors, *Physica C.*

406(1), 1–8 (2004). doi: 10.1016/j.physc.2004.03.002.

97. B. Béri and N. R. Cooper, Topological Kondo effect with Majorana fermions, *Phys. Rev. Lett.* **109**, 156803 (2012). doi: 10.1103/PhysRevLett.109.156803.

98. E. Eriksson, A. Nava, C. Mora, and R. Egger, Tunneling spectroscopy of Majorana–Kondo devices, *Phys. Rev. B.* **90**, 245417 (2014). doi: 10.1103/PhysRevB.90.245417.

99. A. Altland, B. Béri, R. Egger, and A. M. Tsvelik, Bethe ansatz solution of the topological Kondo model, *J. Phys. A.* **47**(26), 265001 (2014).

100. B. Béri, Majorana–Klein hybridization in topological superconductor junctions, *Phys. Rev. Lett.* **110**, 216803 (2013). doi: 10.1103/PhysRevLett.110.216803.

101. A. Zazunov, A. Altland, and R. Egger, Transport properties of the Coulomb–Majorana junction, *New J. Phys.* **16**(1), 015010 (2014).

102. P. Coleman, L. B. Ioffe, and A. M. Tsvelik, Simple formulation of the two-channel Kondo model, *Phys. Rev. B.* **52**, 6611–6627 (1995). doi: 10.1103/PhysRevB.52.6611.

103. D. Giuliano and P. Sodano, Realization of a two-channel Kondo model with Josephson junction networks, *Europhys. Lett.* **103**(5), 57006 (2013). doi: 10.1209/0295-5075/103/57006.

104. E. Ogievetsky, N. Reshetikhin, and P. Wiegmann, The principal chiral field in two dimensions on classical Lie algebras: The Bethe-ansatz solution and factorized theory of scattering, *Nucl. Phys. B.* **280**, 45–96 (1987). doi: 10.1016/0550-3213(87)90138-6.

105. H. J. De Vega and M. Karowski, Exact Bethe ansatz solution of o($2n$) symmetric theories, *Nucl. Phys. B.* **280**, 225–254 (1987). doi: 10.1016/0550-3213(87)90146-5.

106. I. Affleck and A. Ludwig, Critical theory of overscreened Kondo fixed points, *Nucl. Phys. B.* **360**(2-3), 641–696 (1991). doi: 10.1016/0550-3213(91)90419-X.

107. R. M. Bradley and S. Doniach, Quantum fluctuations in chains of Josephson junctions, *Phys. Rev. B.* **30**, 1138–1147 (1984). doi: 10.1103/PhysRevB.30.1138.

108. L. I. Glazman and A. I. Larkin, New quantum phase in a one-dimensional Josephson array, *Phys. Rev. Lett.* **79**, 3736–3739 (1997). doi: 10.1103/PhysRevLett.79.3736.

109. D. Giuliano and P. Sodano, Effective boundary field theory for a Josephson junction chain with a weak link, *Nucl. Phys. B.* **711**(3), 480–504 (2005). doi: 10.1016/j.nuclphysb.2005.01.037.

110. P. Ågren, K. Andersson, and D. B. Haviland, Kinetic inductance and Coulomb blockade in one dimensional Josephson junction arrays, *J. Low Temp. Phys.* **124**(1), 291–304 (2001). doi: 10.1023/A:1017594322332.

111. D. Giuliano and P. Sodano, Pairing of Cooper pairs in a Josephson junction network containing an impurity, *Europhys. Lett.* **88**(1), 17012 (2009). doi: 10.1209/0295-5075/88/17012.

112. D. Giuliano and P. Sodano, Competing boundary interactions in a Josephson junction network with an impurity, *Nucl. Phys. B.* **837**(3), 153–185 (2010). doi: 10.1016/j.nuclphysb.2010.04.022.

113. M. Rizzi, V. Cataudella, and R. Fazio, 4e-condensation in a fully frustrated Josephson junction diamond chain, *Phys. Rev. B.* **73**, 100502 (2006). doi: 10.1103/PhysRevB.73.100502.

114. I. V. Protopopov and M. V. Feigel'man, Anomalous periodicity of supercurrent in long frustrated Josephson-junction rhombi chains, *Phys. Rev. B.* **70**, 184519 (2004). doi: 10.1103/PhysRevB.70.184519.

115. I. V. Protopopov and M. V. Feigel'man, Coherent transport in Josephson-junction rhombi chain with quenched disorder, *Phys. Rev. B.* **74**, 064516 (2006). doi: 10.1103/PhysRevB.74.064516.

116. B. Doucot and J. Vidal, Pairing of Cooper pairs in a fully frustrated Josephson-junction chain, *Phys. Rev. Lett.* **88**, 227005 (2002). doi: 10.1103/Phys-

RevLett.88.227005.

117. F. Buccheri, G. D. Bruce, A. Trombettoni, D. Cassettari, H. Babujian, V. E. Korepin, and P. Sodano, Holographic optical traps for atom-based topological Kondo devices, *New J. Phys.* **18**(7), 075012 (2016). doi: 10.1088/1367-2630/18/7/075012.

118. M. A. Cazalilla, R. Citro, T. Giamarchi, E. Orignac, and M. Rigol, One dimensional bosons: From condensed matter systems to ultracold gases, *Rev. Mod. Phys.* **83**, 1405–1466 (2011). doi: 10.1103/RevModPhys.83.1405.

119. T. D. Schultz, D. C. Mattis, and E. H. Lieb, Two-dimensional Ising model as a soluble problem of many fermions, *Rev. Mod. Phys.* **36**, 856–871 (1964). doi: 10.1103/RevModPhys.36.856.

120. C. N. Yang and C. P. Yang, Thermodynamics of a one-dimensional system of bosons with repulsive delta-function interaction, *J. Math. Phys.* **10**(7), 1115–1122 (1969). doi: 10.1063/1.1664947.

121. V. Korepin, N. Bogoliubov, and A. Izergin, *Quantum Inverse Scattering Method and Correlation Functions*, Cambridge Monographs on Mathematical Physics (Cambridge University Press, 1997).

122. M. Olshanii, Atomic scattering in the presence of an external confinement and a gas of impenetrable bosons, *Phys. Rev. Lett.* **81**, 938–941 (1998). doi: 10.1103/PhysRevLett.81.938.

123. L. Tonks, The complete equation of state of one, two and three-dimensional gases of hard elastic spheres, *Phys. Rev.* **50**, 955–963 (1936). doi: 10.1103/PhysRev.50.955.

124. M. Girardeau, Relationship between systems of impenetrable bosons and fermions in one dimension, *J. Math. Phys.* **1**(6), 516–523 (1960). doi: 10.1063/1.1703687.

125. V. A. Yurovsky, M. Olshanii, and D. S. Weiss, Collisions, correlations, and integrability in atom waveguides, in eds. E. Arimondo, P. R. Berman, and C. C. Lin, *Advances in Atomic, Molecular, and Optical Physics*, Vol. 55, *Advances In Atomic, Molecular, and Optical Physics*, (Academic Press, 2008), pp. 61–138. doi: 10.1016/S1049-250X(07)55002-0.

126. I. Bouchoule, N. J. van Druten, and C. I. Westbrook, *Atom Chips and one-dimensional Bose gases*, in *Atom Chips* (Wiley-VCH Verlag GmbH & Co. KGaA, 2011), pp. 331–363. doi: 10.1002/9783527633357.ch11.

127. B. Paredes, A. Widera, V. Murg, O. Mandel, S. Fölling, I. Cirac, G. V. Shlyapnikov, T. W. Hänsch, and I. Bloch, Tonks–Girardeau gas of ultracold atoms in an optical lattice, *Nature* **429**, 277 (1996).

128. T. Kinoshita, T. Wenger, and D. Weiss, Observation of a one-dimensional Tonks–Girardeau gas, *Science* **305**(5687), 1125–1128 (2004). doi: 10.1126/science.1100700.

129. M. Lewenstein, A. Sanpera, and V. Ahufinger, *Ultracold Atoms in Optical Lattices: Simulating Quantum Many-Body Systems* (Oxford University Press, 2012).

130. D. Jaksch, C. Bruder, J. I. Cirac, C. W. Gardiner, and P. Zoller, Cold bosonic atoms in optical lattices, *Phys. Rev. Lett.* **81**, 3108–3111 (1998). doi: 10.1103/PhysRevLett.81.3108.

131. A. Trombettoni and A. Smerzi, Discrete solitons and breathers with dilute Bose–Einstein condensates, *Phys. Rev. Lett.* **86**, 2353–2356 (2001). doi: 10.1103/PhysRevLett.86.2353.

132. D. Jaksch and P. Zoller, The cold atom Hubbard toolbox, *Annals of Physics* **315**(1), 52–79 (2005). doi: 10.1016/j.aop.2004.09.010.

133. A. Eckardt, Colloquium: Atomic quantum gases in periodically driven optical lattices, *Rev. Mod. Phys.* **89**, 011004 (2017). doi: 10.1103/RevModPhys.89.011004.

134. M. P. A. Fisher, P. B. Weichman, G. Grinstein, and D. S. Fisher, Boson localization and the superfluid-insulator transition, *Phys. Rev. B.* **40**, 546–570 (1989). doi:

10.1103/PhysRevB.40.546.

135. R. Friedberg, T. Lee, and H. Ren, Equivalence between spin waves and lattice bosons with applications to the Heisenberg model, *Annals of Physics* **228**(1), 52–103 (1993). doi: 10.1006/aphy.1993.1088.

Chapter 18

Anomalies and Bose Symmetry

Daniel Kabat

Department of Physics and Astronomy
Lehman College, City University of New York, Bronx NY 10468, USA
`daniel.kabat@lehman.cuny.edu`

We point out a feature of the triangle diagram for three chiral currents which is perhaps not widely appreciated: Bose symmetry is not manifest and suffers from a momentum-routing ambiguity. Imposing Bose symmetry fixes the ambiguity and leads to the famous Adler–Bell–Jackiw anomaly.

Dedicated to Roman Jackiw on the occasion of his 80th birthday

1. Introduction

It was a privilege studying under Roman Jackiw at MIT in the early 90's. By that time the threat of midnight phone calls inquiring after the status of a calculation had somewhat subsided, but the benefits of Roman as an advisor remained. Never one to waste time with trivial discussion or unfounded speculation, he instead provided an unparalleled source of clear guidance and direction for his students.

Roman had an uncanny ability to formulate and solve mathematical systems of physical relevance. His work often laid the foundation for future developments, with an impact far beyond its original scope. Witness the resurgence of interest in Jackiw–Teitelboim gravity [1, 2] as a holographic system [3–5], or the Adler–Bell–Jackiw anomaly [6, 7] which had its origins in current algebra but then did so much to initiate the use of topological methods in gauge theories [8]. Another example may be his work on non-associative structures and 3-cocycles [9], a topic I believe will ultimately play an important role in quantizing gravity [10].

In these notes I'll consider a two-component fermion of definite chirality. Thus in Sec. 2 I'll work with a single chiral current, instead of the vector and axial combinations which can be built from a pair of chiral fermions, and I'll discuss a feature of the triangle diagram for three chiral currents which, although known, may not be widely appreciated: due to a momentum-routing ambiguity the diagram does not have manifest Bose symmetry. Bose symmetry may be imposed on the diagram by hand; this fixes the ambiguity and leads to a unique expression for the divergence of the chiral current. In Sec. 3 I'll make contact with the usual axial anomaly. In Sec. 4 I'll show that Bose symmetry can be restored with a local counterterm, while a different counterterm gives a covariant expression for the anomaly.

2. The chiral triangle

Consider a massless chiral fermion coupled to an external vector field. We'll describe the fermion using a Dirac spinor ψ, but with a projection condition so that the fermion is either right- or left-handed.[a] That is, we'll assume

$$\psi = \frac{1}{2}(1 \pm \gamma^5)\psi. \tag{1}$$

In what follows the upper sign will correspond to a right-handed spinor, the lower sign to left-handed. The coupling of ψ to the external (non-dynamical) vector field A_μ is described by the Lagrangian

$$\mathcal{L} = \bar{\psi} i \gamma^\mu \left(\partial_\mu + i q A_\mu \right) \psi. \tag{2}$$

With these ingredients the diagram for scattering three vector fields

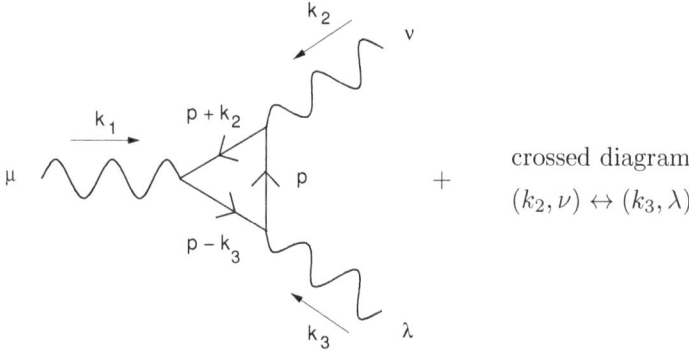

leads to the scattering amplitude

$$-i\mathcal{M}_{\mu\nu\lambda} = (-1) \int \frac{d^4p}{(2\pi)^4} \left[\text{Tr} \left\{ -iq\gamma_\mu \frac{i}{\not{p} + \not{k}_2} (-iq\gamma_\nu) \frac{i}{\not{p}} (-iq\gamma_\lambda) \frac{i}{\not{p} - \not{k}_3} \frac{1}{2}(1 \pm \gamma^5) \right\} \right.$$

$$\left. + \text{Tr} \left\{ -iq\gamma_\mu \frac{i}{\not{p} + \not{k}_3} (-iq\gamma_\lambda) \frac{i}{\not{p}} (-iq\gamma_\nu) \frac{i}{\not{p} - \not{k}_2} \frac{1}{2}(1 \pm \gamma^5) \right\} \right] \tag{3}$$

---

[a] Conventions: the metric $g_{\mu\nu} = \text{diag}(+---)$, the antisymmetric tensor $\epsilon_{0123} = +1$, and the Dirac matrices are

$$\gamma^0 = \begin{pmatrix} 0 & 1 \\ 1 & 0 \end{pmatrix}, \qquad \gamma^i = \begin{pmatrix} 0 & \sigma^i \\ -\sigma^i & 0 \end{pmatrix}, \qquad \gamma^5 \equiv i\gamma^0\gamma^1\gamma^2\gamma^3 = \begin{pmatrix} -1 & 0 \\ 0 & 1 \end{pmatrix}.$$

In this basis a Dirac spinor has left- and right-handed chiral components $\psi = \begin{pmatrix} \psi_L \\ \psi_R \end{pmatrix}$.

All external momenta are directed inward, with $k_1 + k_2 + k_3 = 0$. The projection operator $\frac{1}{2}(1 \pm \gamma^5)$ ensures that only a single chirality circulates in the loop.

We might expect the (necessarily chiral) currents $j^\mu = \bar{\psi}\gamma^\mu\psi$ to obey Bose statistics, so we might expect the amplitude to be invariant under permutations of the external lines. Indeed there's a simple argument which seems to show that Bose symmetry is satisfied. Invariance under exchange $(k_2, \nu) \leftrightarrow (k_3, \lambda)$ is manifest; given our labellings it just corresponds to exchanging the two diagrams. However we should check invariance under exchange of say (k_1, μ) with (k_2, ν). Making this exchange in (3) we get

$$-i\mathcal{M}_{\nu\mu\lambda} = (-1) \int \frac{d^4p}{(2\pi)^4} \left[\text{Tr} \left\{ -iq\gamma_\nu \frac{i}{\not{p} + \not{k}_1} (-iq\gamma_\mu) \frac{i}{\not{p}} (-iq\gamma_\lambda) \frac{i}{\not{p} - \not{k}_3} \frac{1}{2}(1 \pm \gamma^5) \right\} \right.$$
$$\left. + \text{Tr} \left\{ -iq\gamma_\nu \frac{i}{\not{p} + \not{k}_3} (-iq\gamma_\lambda) \frac{i}{\not{p}} (-iq\gamma_\mu) \frac{i}{\not{p} - \not{k}_1} \frac{1}{2}(1 \pm \gamma^5) \right\} \right]$$

Shifting the integration variable $p^\mu \to p^\mu + k_3^\mu$ in the first line, $p^\mu \to p^\mu - k_3^\mu$ in the second, and making some cyclic permutations inside the trace, we seem to recover the previous expression (3).

Famously, though, this argument for Bose symmetry is invalid. Instead a linearly divergent integral picks up a finite surface term when the integration variable is shifted.

$$\int \frac{d^4p}{(2\pi)^4} f(p + a) = \int \frac{d^4p}{(2\pi)^4} \left(f(p) + a^\mu \partial_\mu f(p) + \cdots \right)$$
$$= \int \frac{d^4p}{(2\pi)^4} f(p) - ia^\mu \frac{1}{8\pi^2} \lim_{p \to \infty} \langle p^2 p_\mu f(p) \rangle \tag{4}$$

Here angle brackets denote an average over the Lorentz group, and the limit is understood to mean large spacelike momentum.

We'll return below to see that the amplitude (3) indeed violates Bose symmetry. But for now, rather than study the violation of Bose symmetry in detail, we're simply going to demand that the scattering amplitude be symmetric. The most straightforward way to do this is to *define* the scattering amplitude to be given by averaging over all permutations of the external lines. Equivalently, we average over cyclic permutations of the internal momentum routing. That is, we define the Bose-symmetrized amplitude

$$
-i\mathcal{M}_{\mu\nu\lambda}^{\text{symm}} = \frac{1}{3}\left[\mu \quad\quad + \mu \quad\quad + \mu \quad\quad + \text{crossed diagrams} \atop (k_2,\nu)\leftrightarrow(k_3,\lambda) \right].
$$

Explicitly this gives

$$
-i\mathcal{M}_{\mu\nu\lambda}^{\text{symm}} = \mp\frac{1}{6}q^3\int\frac{d^4p}{(2\pi)^4}\,\text{Tr}\Big\{ \gamma_\mu\frac{1}{\not{p}}\gamma_\nu\frac{1}{\not{p}-\not{k}_2}\gamma_\lambda\frac{1}{\not{p}+\not{k}_1}\gamma^5 + \gamma_\mu\frac{1}{\not{p}+\not{k}_2}\gamma_\nu\frac{1}{\not{p}}\gamma_\lambda\frac{1}{\not{p}-\not{k}_3}\gamma^5
$$
$$
+\gamma_\mu\frac{1}{\not{p}-\not{k}_1}\gamma_\nu\frac{1}{\not{p}+\not{k}_3}\gamma_\lambda\frac{1}{\not{p}}\gamma^5 + \gamma_\mu\frac{1}{\not{p}}\gamma_\lambda\frac{1}{\not{p}-\not{k}_3}\gamma_\nu\frac{1}{\not{p}+\not{k}_1}\gamma^5
$$
$$
+\gamma_\mu\frac{1}{\not{p}+\not{k}_3}\gamma_\lambda\frac{1}{\not{p}}\gamma_\nu\frac{1}{\not{p}-\not{k}_2}\gamma^5 + \gamma_\mu\frac{1}{\not{p}-\not{k}_1}\gamma_\lambda\frac{1}{\not{p}+\not{k}_2}\gamma_\nu\frac{1}{\not{p}}\gamma^5 \Big\}.
$$

Here we've used the fact that only terms involving γ^5 contribute to the scattering amplitude (Furry's theorem).

Having enforced Bose symmetry, let's check current conservation by dotting this amplitude into k_1^μ. Using trivial identities such as

$$
\not{k}_1 = (\not{p}+\not{k}_1) - \not{p} \tag{5}
$$

to cancel the propagators adjacent to $\not{k}_1$, it turns out that most terms cancel, leaving only

$$
-ik_1^\mu\mathcal{M}_{\mu\nu\lambda}^{\text{symm}} = \pm\frac{1}{6}q^3\int\frac{d^4p}{(2\pi)^4}\,\text{Tr}\Big\{ \frac{1}{\not{p}+\not{k}_2}\gamma_\nu\frac{1}{\not{p}-\not{k}_1}\gamma_\lambda\gamma^5 - \frac{1}{\not{p}+\not{k}_1}\gamma_\nu\frac{1}{\not{p}-\not{k}_2}\gamma_\lambda\gamma^5
$$
$$
+\frac{1}{\not{p}-\not{k}_1}\gamma_\nu\frac{1}{\not{p}+\not{k}_3}\gamma_\lambda\gamma^5 - \frac{1}{\not{p}-\not{k}_3}\gamma_\nu\frac{1}{\not{p}+\not{k}_1}\gamma_\lambda\gamma^5 \Big\}.
$$

After shifting $p \to p + k_2 - k_1$ the second term seems to cancel the first, and after shifting $p \to p + k_3 - k_1$ the fourth term seems to cancel the third. This naive cancellation means that the whole expression is given just by a surface term.

$$
-ik_1^\mu\mathcal{M}_{\mu\nu\lambda}^{\text{symm}} = \pm\frac{1}{6}q^3\int\frac{d^4p}{(2\pi)^4}\Big[(k_2^\alpha - k_1^\alpha)\frac{\partial}{\partial p^\alpha}\text{Tr}\Big\{ \frac{1}{\not{p}}\gamma_\nu\frac{1}{\not{p}+\not{k}_3}\gamma_\lambda\gamma^5 \Big\}
$$
$$
+ (k_3^\alpha - k_1^\alpha)\frac{\partial}{\partial p^\alpha}\text{Tr}\Big\{ \frac{1}{\not{p}+\not{k}_2}\gamma_\nu\frac{1}{\not{p}}\gamma_\lambda\gamma^5 \Big\} \Big].
$$

Using the expression for the surface term (4), evaluating the Dirac traces with $\mathrm{Tr}\left(\gamma^\alpha\gamma^\beta\gamma\gamma^\delta\gamma^5\right) = 4i\epsilon^{\alpha\beta\gamma\delta}$, and averaging over the Lorentz group with $\langle p_\alpha p_\beta\rangle = \frac{1}{4}g_{\alpha\beta}\,p^2$ we see that the amplitude satisfies

$$-ik_1^\mu\mathcal{M}_{\mu\nu\lambda}^{\mathrm{symm}} = \mp\frac{q^3}{12\pi^2}\epsilon_{\nu\lambda\alpha\beta}k_2^\alpha k_3^\beta\ . \tag{6}$$

Famously current conservation is violated by the triangle diagram [6,7].

This consistent anomaly [11] can be encapsulated by writing down an effective action for the vector field $\Gamma[A]$ which incorporates the effect of the fermion triangle. The amplitude we've computed corresponds to the following non-local term in the effective action.[b]

$$\Gamma[A] = \cdots \pm \frac{q^3}{96\pi^2}\int d^4x d^4y\,\partial_\mu A^\mu(x)\,\Box^{-1}(x-y)\,\epsilon^{\alpha\beta\gamma\delta}F_{\alpha\beta}F_{\gamma\delta}(y) \tag{7}$$

In this expression $\Box^{-1}(x-y)$ should be thought of as a Green's function, the inverse of the operator $\partial_\mu\partial^\mu$, and $F_{\alpha\beta}$ is the field strength of A_μ. The fact that the current $j^\mu \equiv -\frac{1}{q}\frac{\delta\Gamma}{\delta A_\mu}$ is not conserved,

$$\partial_\mu j^\mu = -\frac{1}{q}\partial_\mu\frac{\delta\Gamma}{\delta A_\mu} = \pm\frac{q^2}{96\pi^2}\epsilon^{\alpha\beta\gamma\delta}F_{\alpha\beta}F_{\gamma\delta}\ , \tag{8}$$

manifests itself in the effective action as a breakdown of gauge invariance.

3. Recovering the axial anomaly

Suppose we have two spinors, one right-handed and one left-handed. Assembling them into a Dirac spinor ψ, the currents

$$j_R^\mu = \bar\psi\gamma^\mu\frac{1}{2}(1+\gamma^5)\psi\ , \qquad j_L^\mu = \bar\psi\gamma^\mu\frac{1}{2}(1-\gamma^5)\psi$$

have anomalous divergences

$$\partial_\mu j_R^\mu = \frac{q^2}{96\pi^2}\epsilon^{\alpha\beta\gamma\delta}R_{\alpha\beta}R_{\gamma\delta}\ , \qquad \partial_\mu j_L^\mu = -\frac{q^2}{96\pi^2}\epsilon^{\alpha\beta\gamma\delta}L_{\alpha\beta}L_{\gamma\delta}\ . \tag{9}$$

---

[b]To verify (7) note that the term we've written down in $\Gamma[A]$ corresponds to a vertex

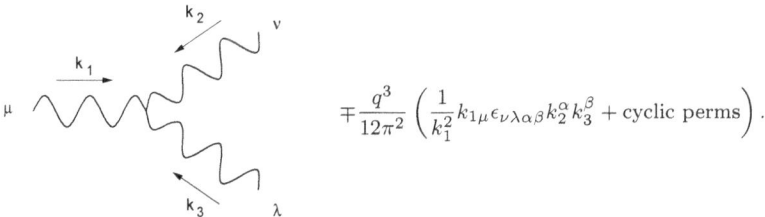

$$\mp\frac{q^3}{12\pi^2}\left(\frac{1}{k_1^2}k_{1\mu}\epsilon_{\nu\lambda\alpha\beta}k_2^\alpha k_3^\beta + \text{cyclic perms}\right).$$

When dotted into one of the external momenta this reproduces (6). The non-locality of Γ is crucial, as otherwise the anomaly could be cancelled with a local counterterm.

Here R_μ and L_μ are background vector fields which couple to the chiral components of ψ, and quantities with two indices are the corresponding field strengths. Note that we've taken the right- and left-handed components of ψ to have the same charge. The vector and axial currents

$$j^\mu = j_R^\mu + j_L^\mu = \bar{\psi}\gamma^\mu\psi , \qquad j^{\mu 5} = j_R^\mu - j_L^\mu = \bar{\psi}\gamma^\mu\gamma^5\psi$$

couple to the linear combinations

$$V_\mu = \frac{1}{2}\left(R_\mu + L_\mu\right) , \qquad A_\mu = \frac{1}{2}\left(R_\mu - L_\mu\right) .$$

As a consequence of (9) these currents have divergences

$$\partial_\mu j^\mu = \frac{q^2}{24\pi^2}\epsilon^{\alpha\beta\gamma\delta}V_{\alpha\beta}A_{\gamma\delta} ,$$

$$\partial_\mu j^{\mu 5} = \frac{q^2}{48\pi^2}\epsilon^{\alpha\beta\gamma\delta}\left(V_{\alpha\beta}V_{\gamma\delta} + A_{\alpha\beta}A_{\gamma\delta}\right) .$$

At first sight this seems no better than having a single chiral spinor. But consider adding the following local term to the effective action for V_μ and A_μ ,

$$\Delta\Gamma = \frac{cq^3}{6\pi^2}\int d^4x\,\epsilon^{\alpha\beta\gamma\delta}\partial_\alpha V_\beta V_\gamma A_\delta .$$

Here c is an arbitrary constant. This term violates both vector and axial gauge invariance, so it contributes to the divergences of the corresponding currents ,

$$\Delta\left(\partial_\mu j^\mu\right) = -\frac{1}{q}\partial_\mu\frac{\delta(\Delta\Gamma)}{\delta V_\mu} = -\frac{cq^2}{24\pi^2}\epsilon^{\alpha\beta\gamma\delta}V_{\alpha\beta}A_{\gamma\delta} ,$$

$$\Delta\left(\partial_\mu j^{\mu 5}\right) = -\frac{1}{q}\partial_\mu\frac{\delta(\Delta\Gamma)}{\delta A_\mu} = +\frac{cq^2}{24\pi^2}\epsilon^{\alpha\beta\gamma\delta}V_{\alpha\beta}V_{\gamma\delta} .$$

If we add this term to the effective action and set $c = 1$, we have a conserved vector current but an anomalous axial current:

$$\partial_\mu j^\mu = 0 \qquad \partial_\mu j^{\mu 5} = \frac{q^2}{16\pi^2}\epsilon^{\alpha\beta\gamma\delta}\left(V_{\alpha\beta}V_{\gamma\delta} + \frac{1}{3}A_{\alpha\beta}A_{\gamma\delta}\right) . \qquad (10)$$

Given a conserved vector current we can promote V_μ to a dynamical gauge field — usually the desired state of affairs. If we aren't interested in making V_μ dynamical then other choices for c are possible.

4. Restoring Bose symmetry

It's no surprise that by imposing Bose symmetry we've recovered the standard expression for the anomaly, as the importance of Bose symmetry for the result was emphasized from the very beginning [6,7]. But still, let's return to the momentum routing given in (3) and see how Bose symmetry and current conservation play out.

Using identities similar to (5) and the expression for the surface term (4) we find that

$$-ik_1^\mu \mathcal{M}_{\mu\nu\lambda} = 0$$

$$-ik_2^\nu \mathcal{M}_{\mu\nu\lambda} = \mp\frac{q^3}{8\pi^2}\epsilon_{\mu\lambda\alpha\beta}k_1^\alpha k_3^\beta \qquad (11)$$

$$-ik_3^\lambda \mathcal{M}_{\mu\nu\lambda} = \mp\frac{q^3}{8\pi^2}\epsilon_{\mu\nu\alpha\beta}k_1^\alpha k_2^\beta$$

Thus with the momentum routing (3) the current is conserved at one vertex but not at the other two, a peculiar state of affairs which shows that Bose symmetry is violated. To capture this in an effective action we introduce three distinct vector fields A, B, C, with field strengths denoted by the same letter, and take

$$\Gamma[A,B,C] = \pm\frac{q^3}{32\pi^2}\int d^4x d^4y \left(\partial_\mu B^\mu(x)\Box^{-1}(x-y)\,\epsilon^{\alpha\beta\gamma\delta}A_{\alpha\beta}C_{\gamma\delta}(y)\right.$$

$$\left. +\partial_\mu C^\mu(x)\Box^{-1}(x-y)\,\epsilon^{\alpha\beta\gamma\delta}A_{\alpha\beta}B_{\gamma\delta}(y)\right). \qquad (12)$$

Gauge invariance is respected for A but violated for B and C. The breakdown of Bose symmetry is manifest.

Since Bose symmetry could be restored by symmetrizing over momentum routings, it should also be possible to restore it with a local counterterm. Consider adding the following local term to the effective action,

$$\Delta\Gamma = \mp\frac{q^3}{48\pi^2}\int d^4x\,\epsilon^{\alpha\beta\gamma\delta}A_\alpha\left(B_{\beta\gamma}C_\delta + B_\beta C_{\gamma\delta}\right). \qquad (13)$$

When added to (12) the anomalous divergences become symmetric,

$$\partial_\mu j_A^\mu = -\frac{1}{q}\partial_\mu\frac{\delta(\Gamma+\Delta\Gamma)}{\delta A_\mu} = \pm\frac{q^2}{48\pi^2}\epsilon^{\alpha\beta\gamma\delta}B_{\alpha\beta}C_{\gamma\delta}\ ,$$

$$\partial_\mu j_B^\mu = -\frac{1}{q}\partial_\mu\frac{\delta(\Gamma+\Delta\Gamma)}{\delta B_\mu} = \pm\frac{q^2}{48\pi^2}\epsilon^{\alpha\beta\gamma\delta}A_{\alpha\beta}C_{\gamma\delta}\ , \qquad (14)$$

$$\partial_\mu j_C^\mu = -\frac{1}{q}\partial_\mu\frac{\delta(\Gamma+\Delta\Gamma)}{\delta C_\mu} = \pm\frac{q^2}{48\pi^2}\epsilon^{\alpha\beta\gamma\delta}A_{\alpha\beta}B_{\gamma\delta}\ ,$$

and can be captured by an effective action

$$\Gamma[A,B,C] = \pm\frac{q^3}{48\pi^2}\int d^4x d^4y \left(\partial_\mu A^\mu(x)\Box^{-1}\,\epsilon^{\alpha\beta\gamma\delta}B_{\alpha\beta}C_{\gamma\delta}(y) + \text{cyclic}\right). \qquad (15)$$

Then we're free to identify the three vector fields and, with a $1/3!$ for Bose symmetry, describe the anomaly with the effective action (7).

The procedure above gives a consistent anomaly. On the other hand consider adding to (12) the counterterm[c]

$$\Delta\Gamma = \mp\frac{q^3}{16\pi^2}\int d^4x\,\epsilon^{\alpha\beta\gamma\delta}A_\alpha\left(B_{\beta\gamma}C_\delta + B_\beta C_{\gamma\delta}\right) \tag{16}$$

Thanks to the larger coefficient this counterterm squeezes all of the anomaly into one of the legs:

$$\partial_\mu j_A^\mu = \pm\frac{q^2}{16\pi^2}\epsilon^{\alpha\beta\gamma\delta}B_{\alpha\beta}C_{\gamma\delta}, \qquad \partial_\mu j_B^\mu = \partial_\mu j_C^\mu = 0\ . \tag{17}$$

An effective action which captures this is

$$\Gamma[A,B,C] = \pm\frac{q^3}{16\pi^2}\int d^4x\,d^4y\,\partial_\mu A^\mu(x)\,\Box^{-1}\,\epsilon^{\alpha\beta\gamma\delta}B_{\alpha\beta}C_{\gamma\delta}(y)\ . \tag{18}$$

Since j_B and j_C are conserved, (18) respects gauge invariance for B and C, which means (17) can be identified as the covariant anomaly for A.[d] Evidently the covariant anomaly can be obtained by varying an effective action, despite the Wess–Zumino consistency condition [13], at the price of violating Bose symmetry. It would be interesting to see if a similar result holds in non-Abelian theories.

Happy birthday, Roman!

5. Acknowledgements

I'm grateful to V. Parameswaran Nair for comments on the manuscript and especially for suggesting the connection to covariant anomalies. DK is supported by U.S. National Science Foundation grant PHY-1820734.

References

1. R. Jackiw, "Lower dimensional gravity," *Nucl. Phys.* **B252** (1985) 343–356.
2. C. Teitelboim, "Gravitation and Hamiltonian structure in two space-time dimensions," *Phys. Lett.* **126B** (1983) 41–45.
3. A. Almheiri and J. Polchinski, "Models of AdS$_2$ backreaction and holography," *JHEP* **11** (2015) 014, arXiv:1402.6334 [hep-th].
4. J. Maldacena, D. Stanford, and Z. Yang, "Conformal symmetry and its breaking in two dimensional nearly anti-de-Sitter space," *PTEP* **2016** (2016) 12C104, arXiv:1606.01857 [hep-th].
5. J. Engelsöy, T. G. Mertens, and H. Verlinde, "An investigation of AdS$_2$ backreaction and holography," *JHEP* **07** (2016) 139, arXiv:1606.03438 [hep-th].

[c]Related counterterms appear in [12].
[d]The coefficient of (18) is larger than (7) by a Bose factor. Although we can't impose full Bose symmetry on (18), we could set $B = C$ and divide (18) by 2. This reproduces the relative normalizations of the consistent and covariant anomalies in [11].

6. S. L. Adler, "Axial vector vertex in spinor electrodynamics," *Phys. Rev.* **177** (1969) 2426–2438.

7. J. S. Bell and R. Jackiw, "A PCAC puzzle: $\pi^0 \to \gamma\gamma$ in the σ model," *Nuovo Cim.* **A60** (1969) 47–61.

8. R. Jackiw, "Topological investigations of quantized gauge theories." In *Relativity, Groups and Topology: Proceedings, 40th Summer School of Theoretical Physics, Les Houches, France, June 27 - August 4, 1983*, B. S. DeWitt and R. Stora, eds. (North-Holland, Amsterdam, 1984).

9. R. Jackiw, "3-cocycle in mathematics and physics," *Phys. Rev. Lett.* **54** (1985) 159–162.

10. D. Kabat and G. Lifschytz, "Does boundary quantum mechanics imply quantum mechanics in the bulk?," *JHEP* **03** (2018) 151, `arXiv:1801.08101 [hep-th]`.

11. W. A. Bardeen and B. Zumino, "Consistent and covariant anomalies in gauge and gravitational theories," *Nucl. Phys.* **B244** (1984) 421–453.

12. M. B. Paranjape, "Some aspects of anomalies of global currents," *Phys. Lett.* **156B** (1985) 376–380.

13. J. Wess and B. Zumino, "Consequences of anomalous Ward identities," *Phys. Lett.* **37B** (1971) 95–97.

Chapter 19

Sasakians and the Geometry of a Mass Term

V. P. Nair

Physics Department, City College of the CUNY
New York, NY 10031, USA
vpnair@ccny.cuny.edu

A gauge-invariant mass term for nonabelian gauge fields in two dimensions can be expressed as the Wess–Zumino–Witten (WZW) action. Hard thermal loops in the gauge theory in four dimensions at finite temperatures generate a screening mass for some components of the gauge field. This can be expressed in terms of the WZW action using the bundle of complex structures (for Euclidean signature) or the bundle of lightcones over Minkowski space. We show that a dynamically generated mass term in three dimensions can be put within the same general framework using the bundle of Sasakian structures.

1. Introduction

The early 1980s were a time of growing appreciation of the role of topology in quantum field theory, especially for gauge theories. Anomalies and Chern–Simons terms were very much in the air, so it was impossible for any graduate student to be unaware of the seminal contributions of Roman Jackiw. My own collaboration with Roman began somewhat later, during his sabbatical visit to Columbia University in 1990. At that time Roman was very much interested in solitons in Chern–Simons theories coupled to matter fields, both relativistic and nonrelativistic,[1] but we did talk about anomalies and anyons and representations of the Poincaró group in 2+1 dimensions. After he went back to MIT, we corresponded about anyons, and this evolved into our paper on the relativistic wave equation for anyons.[2]

Throughout the 1990s and early 2000s, we continued to collaborate on a number of projects of common interest, from finite-temperature field theories, nonabelian Clebsch parametrization,[3] a group-theory-based formulation of nonabelian magnetohydrodynamics,[4] etc. Particularly gratifying was my work with Efraty on an effective action for hard thermal loops,[5] and the subsequent work with Roman on developing it into a nonabelian version of the Kubo formula,[6] combining two of his favorite topics: field theory at finite temperature and Chern–Simons theory. Chromomagnetic screening masses and gap equations in 2+1 dimensional gauge theories was another topic on which I had many discussions with Roman and So-Young, although we never published any joint work on this.[7] Looking back, it is striking to me that we had overlap of interest on so many different topics. Yet, on second thought, it is perhaps not so remarkable since Roman has been a continuing influence on the development of field theory from the mid-1960s to the present, and hence anyone interested in field theory would be bound to have many points of overlap with his work.

To a man who has devoted decades to physics, appreciation must be shown in kind, not just in anecdotes and reminiscences alone. So, for my contribution to this Festschrift, I have decided to write on a novel aspect of something we have both worked on. I shall discuss dynamically generated mass terms in gauge theories, which brings together Chern–Simons actions and their eikonals, the Wess–Zumino–Witten actions, Dirac determinants, chromomagnetic screening effects and many facets of geometry and topology, which are all topics of interest to Roman. The key point is that while Kähler structures play an important role for physics in even dimensions, Sasakian structures should do so in odd dimensions. A mass term which can be dynamically generated in nonabelian gauge theories in odd dimensions, as I argue below, exemplifies this.

2. Masses for gauge theories in two and four dimensions

The prototypical example of a gauge-invariant mass term is given by the Wess–Zumino–Witten (WZW) action in two dimensions,[8] or, equivalently, by the logarithm of the Dirac determinant.[9]   This is the nonabelian generalization of Schwinger's result for the Abelian case.[10] Specifically, this mass term takes the form

$$\Gamma = -[\text{Tr}\log\left(-\bar{D}D\right) - \text{Tr}\log(-\bar{\partial}\partial)] = -A_R\, S_{\text{WZW}}(H)$$

$$S_{\text{WZW}}(H) = \frac{1}{2\pi}\int_{\mathcal{M}} d^2z\, \text{Tr}(\partial H\bar{\partial}H^{-1})$$

$$+\frac{i}{12\pi}\int_{\mathcal{M}^3} \text{Tr}(H^{-1}dH \wedge H^{-1}dH \wedge H^{-1}dH) \tag{1}$$

where we use complex coordinates in two dimensions, z, $\bar{z} = x_1 \mp ix_2$ and

$$D = \partial + A = \tfrac{1}{2}(\partial_1 + i\partial_2) + \tfrac{1}{2}(A_1 + iA_2) = \tfrac{1}{2}(\partial_1 + i\partial_2) + (-it_a)\tfrac{1}{2}(A_1^a + iA_2^a)$$
$$\bar{D} = \bar{\partial} + \bar{A} = \tfrac{1}{2}(\partial_1 - i\partial_2) + \tfrac{1}{2}(A_1 - iA_2) = \tfrac{1}{2}(\partial_1 - i\partial_2) + (-it_a)\tfrac{1}{2}(A_1^a - iA_2^a)\tag{2}$$

One can parametrize the gauge field in terms of a complex matrix M as $A = -\partial M M^{-1}$, $\bar{A} = M^{\dagger-1}\bar{\partial}M^{\dagger}$, which yields the second expression in (1) in terms of the WZW action with $H = M^{\dagger}M$. In (2), $\{t_a\}$ are a set of hermitian matrices forming a basis for the Lie algebra of the gauge group, $A_R = 1$ if the fields are in the fundamental representation (F), otherwise, for representation R, it is defined by $\text{Tr}(t_at_b)_R = A_R\,\text{Tr}(t_at_b)_F$. The integration is over the two-dimensional space $\mathcal{M}$ of interest; in the second term of S_{WZW} we extend the fields to a three-manifold $\mathcal{M}^3$ whose boundary is $\mathcal{M}$, as usual.

The WZW action we have written can also be expressed directly in terms of the gauge potentials, which is useful for explicit computations in a gauge theory. It reads

$$S_{\text{WZW}}(H) = \frac{1}{\pi}\left[\int d^2z\,\text{Tr}(A\bar{A}) - \pi\mathcal{I}(A) - \pi\bar{\mathcal{I}}(\bar{A})\right]$$

$$\mathcal{I}(A) = \sum_2^{\infty} \frac{(-1)^n}{n\,\pi^n}\int d^2z_1\cdots d^2z_n\,\frac{\text{Tr}\big(A(z_1,\bar{z}_1)A(z_2,\bar{z}_2)\cdots A(z_n,\bar{z}_n)\big)}{(\bar{z}_1-\bar{z}_2)(\bar{z}_2-\bar{z}_3)\cdots(\bar{z}_n-\bar{z}_1)} \quad (3)$$

and $\bar{\mathcal{I}}(\bar{A})$ is similar with $A \to \bar{A}$, $\bar{z}_1 - \bar{z}_2 \to z_1 - z_2$, etc. for the terms in the denominator of the expression in (3).

The fact that we have a complex structure for $\mathbb{R}^2$ is important in constructing this mass term. The WZW action, and hence the mass term, can be written for any Riemann surface, viewed as a complex manifold, by a simple generalization as

$$S_{\text{WZW}}(H) = \frac{1}{8\pi}\int_{\mathcal{M}} d^2x\,\sqrt{g}\,g^{ab}\,\text{Tr}(\partial_a H\partial_b H^{-1})$$

$$+ \frac{i}{12\pi}\int_{\mathcal{M}^3}\text{Tr}(H^{-1}dH \wedge H^{-1}dH \wedge H^{-1}dH) \quad (4)$$

where g_{ab} is the metric tensor for the two-manifold.

Consider now the extension of this to four dimensions. A mass term, similar to (1), can be written down if we can identify two complex coordinates out of the four real coordinates of $\mathbb{R}^4$. But there are many choices for a complex (or even a Kähler) structure. We can understand the inequivalent choices as follows. If we choose one set of complex combinations, say $w_1 = x_0 - ix_3$, $w_2 = x_2 - ix_1$, then a $U(2)$ transformation of (w_1, w_2) does not change the complex structure. In particular, a holomorphic function of $w = (w_1, w_2)$ remains holomorphic under the $U(2)$ transformation. However, we can do a rotation of all four coordinates, as $x_\mu \to x'_\mu = R_\mu{}^\nu x_\nu$, where $R_\mu{}^\nu$ is a rotation matrix, and this does lead to a different structure given by $x'_0 - ix'_3$, $x'_2 - ix'_1$. So the inequivalent ways of combining (x_0, x_1, x_2, x_3) into complex combinations are parametrized by $SO(4)/U(2) \sim S^2$. More explicitly, introduce a two-spinor (π_1, π_2) with the identification $\pi = (\pi_1, \pi_2) \sim \lambda(\pi_1, \pi_2)$, $\lambda \in \mathbb{C} - \{0\}$. The π's parametrize $\mathbb{CP}^1 \sim S^2$. The complex combinations can be taken as

$$\begin{pmatrix} w_1 \\ w_2 \end{pmatrix} = (x_0 - i\sigma_i x_i)\pi = \begin{bmatrix} x_0 - ix_3 & -x_2 - ix_1 \\ x_2 - ix_1 & x_0 + ix_3 \end{bmatrix}\begin{pmatrix} \pi_1 \\ \pi_2 \end{pmatrix} \quad (5)$$

where σ_i are the Pauli matrices. We can also define a real unit vector $Q_i = \bar{\pi}\sigma_i\pi/(\bar{\pi}\pi)$. We then find

$$\frac{1}{\bar{\pi}\pi}(\bar{\pi}w, \bar{w}\pi) = (x_0 - i\vec{Q}\cdot\vec{x},\ x_0 + i\vec{Q}\cdot\vec{x}) \equiv (z, \bar{z}) \quad (6)$$

These constitute two of the complex coordinates. The remaining two transverse coordinates are given by $\vec{x} \times \vec{Q}$. The unit vector Q_i gives an alternate way to parametrize S^2. For any fixed choice of $\vec{Q}$, we do lose rotational invariance but we can construct an invariant mass term as[5]

$$\Gamma = -k\int d\mu_{S^2}\,d^2x^T\,S_{\text{WZW}}(H) \quad (7)$$

where $H = M^\dagger M$ and

$$A = \tfrac{1}{2}(A_0 + i\vec{Q} \cdot \vec{A}) = -\tfrac{1}{2}(\partial_0 + i\vec{Q} \cdot \nabla)M\,M^{-1}$$
$$\bar{A} = \tfrac{1}{2}(A_0 - i\vec{Q} \cdot \vec{A}) = \tfrac{1}{2}M^{\dagger-1}(\partial_0 - i\vec{Q} \cdot \nabla)M^\dagger \qquad (8)$$

The integrations over all orientations of $\vec{Q}$, signified in (8) by $d\mu_{S^2}$, and over the transverse coordinates x^T, will make this mass term rotationally invariant. The integration over $z, \bar{z}$ is part of $S_{\mathrm{WZW}}(H)$, so the final result in (7) will have integration over all four coordinates with the measure d^4x. This mass term is also obviously gauge-invariant, in the same way as in two dimensions.

A number of comments are in order at this point. First of all, the key idea here is to use a pair of complex coordinates or more generally a two-dimensional complex subspace to construct the WZW action. This necessarily entails a lack of rotational symmetry. Symmetry is restored by integrating over all possible choices of complex coordinates. In other words, we may think of the total space of interest as the bundle of complex structures on $\mathbb{R}^4$. This is basically the way twistor space is defined.[11] In the present case, where the base space is flat, the bundle is trivial. One could do an analogous construction for other spaces, a notable example being S^4. For this latter case, we cannot have a global complex structure, so combinations of coordinates into complex ones correspond to local complex structures and one is considering the bundle of local complex structures over S^4. The bundle space is then $\mathbb{CP}^3$, with S^2 as the fiber and S^4 as the base, and the bundle is topologically nontrivial. In any case, the key point here is to consider the bundle of local complex structures, trivial or nontrivial.

Secondly, while the twistor space genesis of (7) may be mathematically gratifying, one may ask whether this mass term has anything to do with physics. The remarkable fact is that it does. Of course, its use in physics needs a continuation to Minkowski signature, not the Euclidean one we have used so far. This continuation can be done by the rules

$$A \to \tfrac{1}{2}(A_0 + \vec{Q} \cdot \vec{A}) = -\tfrac{1}{2}(\partial_0 + \vec{Q} \cdot \nabla)M\,M^{-1}$$
$$\bar{A} \to \tfrac{1}{2}(A_0 - \vec{Q} \cdot \vec{A}) = \tfrac{1}{2}M^{\dagger-1}(\partial_0 - \vec{Q} \cdot \nabla)M^\dagger \qquad (9)$$

If we consider a physical system described by a nonabelian gauge theory such as quantum chromodynamics (QCD), then, at finite temperature where we get a plasma of gluons (and quarks if they are included), the terms in the standard perturbative expansion have infrared divergences. There is an infinite sequence of terms, which are the leading infrared divergent terms, known as hard thermal loops (HTL).[12] These are special to the case of nonzero temperature and are in addition to the usual divergences (both in the ultraviolet and infrared) in the theory at zero temperature. These HTL terms have to be summed up and included at the lowest order to reformulate perturbation theory without infrared divergences. (This has to be done in a self-consistent way, the technology for this is well understood.) The summation of the HTL terms is a screening effect for the electric-type forces corresponding to the nonabelian gauge field. In fact it is the nonabelian generalization of the Debye screening effect well known for the Abelian plasma (and electrolytes). The sum of the HTL terms can be interpreted as a mass term, primarily for the A_0-component of the gauge field, with some contributions to the other components as well to satisfy the Gauss law. The HTL contributions can be calculated in the field theory at finite temperature and the result of the detailed calculations at finite temperature is exactly the mass term (7), with the continuation in (9), and with $k = (N + \frac{1}{2}N_F)T^2/6$.[5] This value of k is for the case of an $SU(N)$ gauge theory, with N_F massless fermion flavors and T denotes the temperature of the plasma. Thus, quite remarkably, what was defined purely as a mathematical generalization is indeed realized in explicit calculations in a very physical context, namely, the quark–gluon plasma.

A similar screening effect also occurs for a degenerate gas of quarks with a nonzero total baryon number, such as can occur deep inside a neutron star. This is the nonabelian generalization of the well-known Thomas–Fermi screening effect for electron gases. The mass term describing this is again of the same form, with $k = \mu_q^2/4\pi^2$, where μ_q is the chemical potential for the quark number $(= \frac{1}{3}$ of the baryon number).[13]

Finally, we may raise the question of Lorentz invariance. The Minkowski continuation of the mass term as written in (7) is not Lorentz-invariant. Physically, this is indeed as it should be, since thermal equilibrium and the specification of the temperature are obtained in the rest frame of the plasma without any overall drift velocity. For a Lorentz-invariant result, we need one more parameter, the overall drift velocity of the plasma, whose Lorentz transformation will lead to an invariant result. The relevant form of the mass term (i.e., the generalization for the moving plasma) was worked out many years ago and takes the form[14]

$$\Gamma = -k \int d\mu\, d^2 x^T\, S_{WZW}(H)$$

$$d\mu = 2i \frac{\pi \cdot d\pi\, \bar{\pi} \cdot d\bar{\pi}\, \xi \cdot d\xi\, \bar{\xi} \cdot d\bar{\xi}}{(\pi \cdot \xi)^2 (\bar{\pi} \cdot \bar{\xi})^2}\, \delta[\xi(e \cdot p)\bar{\pi}]\, \delta[\pi(e \cdot p)\bar{\xi}] \qquad (10)$$

We have introduced two sets of two-component $SL(2, \mathbb{C})$ spinors, π^A, ξ^A, $A = 1, 2$, with $\pi^{\dot{A}} = \overline{\pi^A}$ and $\xi^{\dot{A}} = \overline{\xi^A}$ as their complex conjugates. The components of the gauge fields used to define M and $M^\dagger$, and hence H, are given by

$$A_\pi = \tfrac{1}{2}\pi(e \cdot A)\bar{\pi}, \qquad A_\xi = \tfrac{1}{2}\xi(e \cdot A)\bar{\xi} \qquad (11)$$

Further $e^\mu = (1, \sigma^i)$, and p_μ in (10) denotes the drift 4-velocity of the plasma. The derivatives are defined in a way similar to the A's given above, namely by (11) with $A \to \partial$, with corresponding expressions for the coordinates. Notice the presence of the δ-functions in $d\mu$. Upon integration, they enforce a relation between the two spinors in a way which depends on p^μ. In the rest frame of the plasma, with $p^\mu = (1, 0, 0, 0)$, expression (10) reproduces the previous result (7).

Another feature worthy of remark is that the combinations of the gauge potentials in (9), as well as derivatives, can be written as $n \cdot A$, $n \cdot \partial$ and $n' \cdot A$, $n' \cdot \partial$, where

$n^\mu = (1, iQ_i)\, n'^\mu = (1, -iQ_i) = \bar{n}^\mu$ in Euclidean space. These are complex null vectors. Upon continuation to Minkowski space, we get $n^\mu = (1, Q_i)\, n'^\mu = (1, -Q_i)$, which are real null vectors. Thus the bundle space we are considering is the bundle of lightcones over Minkowski space.[11]

Turning now to three-dimensional space, obviously we cannot combine coordinates pairwise into complex combinations, so an immediate generalization seems difficult. However, there is a mathematical structure known as the Sasakian which can exist on certain odd-dimensional manifolds and which has been suggested as the closest we can get to a Kähler structure. We can try to utilize this to construct a mass term. In the next section, we give a general discussion of Sasakians for S^3 and $\mathbb{R}^3$ and write down this mass term. The final result agrees with what was suggested as the magnetic screening mass for the gluon plasma many years ago, although the Sasakian connection was not apparent at that time.

3. S^3 and $\mathbb{R}^3$ as Sasakian manifolds and a 3d mass term

We begin by briefly recalling the definition of a Sasakian manifold.[15] Let $\mathcal{M}$ be an odd-dimensional Riemannian manifold with the metric $ds^2_\mathcal{M}$. The Riemannian cone for $\mathcal{M}$ is $\mathcal{M} \times \mathbb{R}_+$ with the cone metric

$$ds^2 = dr^2 + r^2 \, ds^2_\mathcal{M} \tag{12}$$

where $r \in \mathbb{R}_+$ is the additional coordinate along the $\mathbb{R}_+$ direction. The manifold $\mathcal{M}$ is said to be a contact manifold if there is a one-form $\hat{\Theta}$ on $\mathcal{M}$ such that the two-form

$$\Omega = r^2 d\hat{\Theta} + 2r dr \, \hat{\Theta} \tag{13}$$

is symplectic. The manifold $\mathcal{M}$ endowed with $\hat{\Theta}$ is Sasakian if the two-form Ω and the metric ds^2 on the cone, i.e., (12), are Kähler. Since $\mathcal{M}$ is a transverse cross-section of the cone, it inherits many properties from the Kähler structure of the cone. In fact, it is generally considered that the Sasakian structure is the closest one can get to Kähler-type properties for an odd-dimensional space.

We can apply this specifically to S^3 by considering its embedding in $\mathbb{R}^4$ and taking r as the radial coordinate. Removing the origin, $\mathbb{R}^4 - \{0\}$ has the cone structure, with the metric on the cone being the flat Euclidean metric $ds^2 = dx_0^2 + dx_1^2 + dx_2^2 + dx_3^2$. To identify S^3 as a Sasakian space, we need to write this metric as a Kähler metric. As discussed in the last section, there are an infinity of inequivalent ways of doing this, the possible complex combinations being parametrized by π which form the homogeneous coordinates for $\mathbb{CP}^1$. Using the freedom of scaling, $\pi \sim l\pi$, $l \in \mathbb{C} - \{0\}$, we can bring it to the form

$$\begin{pmatrix} \pi_1 \\ \pi_2 \end{pmatrix} = \begin{pmatrix} -e^{-i\frac{\varphi}{2}} \cos\frac{\theta}{2} \\ -e^{i\frac{\varphi}{2}} \sin\frac{\theta}{2} \end{pmatrix}$$

$$= \begin{bmatrix} -e^{-i\frac{\varphi}{2}} \sin\frac{\theta}{2} & -e^{-i\frac{\varphi}{2}} \cos\frac{\theta}{2} \\ e^{i\frac{\varphi}{2}} \cos\frac{\theta}{2} & -e^{i\frac{\varphi}{2}} \sin\frac{\theta}{2} \end{bmatrix} \begin{pmatrix} 0 \\ 1 \end{pmatrix} \equiv g \begin{pmatrix} 0 \\ 1 \end{pmatrix} \tag{14}$$

The complex combinations for $\mathbb{R}^4$ can be taken as in (5). But as mentioned earlier, we are free to do a $U(2)$ rotation of the complex combinations without changing the complex structure. For our purpose here, it is useful to do this using $g^\dagger$, thus defining

$$\omega = g^\dagger (x_0 - i\vec{\sigma} \cdot \vec{x})\pi = g^\dagger (x_0 - i\vec{\sigma} \cdot \vec{x})g \begin{pmatrix} 0 \\ 1 \end{pmatrix}$$

$$= (x_0 - i\sigma_k R_{ki} x_i) \begin{pmatrix} 0 \\ 1 \end{pmatrix} \tag{15}$$

For the choice of π in (14), the components of the orthogonal matrix R_{ki} are given by

$$R_{ki} = \begin{bmatrix} -\cos\theta\cos\varphi & -\cos\theta\sin\varphi & \sin\theta \\ -\sin\varphi & \cos\varphi & 0 \\ -\sin\theta\cos\varphi & -\sin\theta\sin\varphi & -\cos\theta \end{bmatrix} \tag{16}$$

In terms of the complex coordinates ω, the Kähler forms and metric can be taken as

$$\Omega = i\,(d\bar{\omega}_1 \wedge d\omega_1 + d\bar{\omega}_2 \wedge d\omega_2) = d\mathcal{A}$$

$$\mathcal{A} = \frac{i}{2} [\bar{\omega}_1 \wedge d\omega_1 - \omega_1 \wedge d\bar{\omega}_1 + \bar{\omega}_2 \wedge d\omega_2 - \omega_2 \wedge d\bar{\omega}_2]$$

$$ds^2 = d\bar{\omega}_1 d\omega_1 + d\bar{\omega}_2 d\omega_2 \tag{17}$$

Using ω_1, ω_2 from (15), we can now separate out the radial coordinate, writing

$$\omega_1 = -i(R_{1i} - iR_{2i})x_i = r\left[-i(R_{1i} - iR_{2i})\phi_i\right]$$

$$\omega_2 = x_0 + iR_{3i}x_i = r\left[\phi_0 + iR_{3i}\phi_i\right] \tag{18}$$

with $\phi_0\phi_0 + \phi_i\phi_i = 1$. It is straightforward to simplify $\mathcal{A}$ to get $\mathcal{A} = r^2\,\hat{\Theta}$, with

$$\hat{\Theta} = R_{3i}(\phi_i d\phi_0 - \phi_0 d\phi_i) + \frac{i}{2}\left[(n \cdot \phi)\,d(\bar{n} \cdot \phi) - (\bar{n} \cdot \phi)\,d(n \cdot \phi)\right]$$

$$n_i = R_{1i} + iR_{2i} = (-\cos\theta\cos\varphi - i\sin\varphi, -\cos\theta\sin\varphi + i\cos\varphi, \sin\theta) \tag{19}$$

We have chosen a Kähler metric for the cone and we see that $\Omega = d\mathcal{A}$ does have the required structure (13). This is basically the Sasakian structure on S^3. Notice that the second term in $\hat{\Theta}$, namely,

$$\alpha = \frac{i}{2}\left[(n \cdot \phi)\,d(\bar{n} \cdot \phi) - (\bar{n} \cdot \phi)\,d(n \cdot \phi)\right] \tag{20}$$

defines a local Kähler structure for the two-dimensional subspace transverse to $R_{3i}\phi_i$. This is only local, since the separation of the third direction on S^3 can only be local. The existence of such a local transverse Kähler structure is a feature of Sasakian manifolds.

The vectors n_i, $\bar{n}_i$ define the choice of complex combinations on the cone. (R_{3i} is not independent, it is proportional to $(\vec{n} \times \bar{\vec{n}})_i$.) A particular Kähler structure on the cone corresponds to a particular choice of these vectors, each of them leading to a particular Sasakian structure for S^3. We have an S^2 worth of such choices, so the total space we are considering is the bundle of Sasakians over S^3. In this sense, it is the natural equivalent in three dimensions of the twistor space in four dimensions.[a] Notice that, since R_{ki} is an orthogonal matrix,

$$n_i n_i = \bar{n}_i \bar{n}_i = 0, \qquad n_i \bar{n}_i = 2 \qquad (21)$$

In other words, n_i, $\bar{n}_i$ are complex null vectors, normalized by the second relation in (21). Thus we may think of the bundle of Sasakians of S^3 as the bundle of complex null rays. In this case, the bundle is still trivial just as it was in $\mathbb{R}^4$.

To proceed further, we introduce stereographic coordinates y_i for S^3 by

$$\phi_0 = \frac{y^2 - R^2}{y^2 + R^2}, \qquad \phi_i = \frac{2y_i R}{y^2 + R^2} \qquad (22)$$

We also introduce the notation $z = n \cdot y$, $\bar{z} = \bar{n} \cdot y$, $R_{3i} y_i = v$. The Kähler potential for the transverse local Kähler one-form α in (20) is

$$K_T = (n \cdot \phi)(\bar{n} \cdot \phi) = \frac{4R^2 \bar{z} z}{(\bar{z}z + v^2 + R^2)^2} \qquad (23)$$

If we take the large R limit, which corresponds to blowing up the S^3 to get $\mathbb{R}^3$, we find

$$\hat{\Theta} \approx \frac{2dv}{R} + \frac{2i}{R^2}(zd\bar{z} - \bar{z}dz)$$

$$K_T \approx 4\frac{\bar{z}z}{R^2} \qquad (24)$$

This will be useful in applying our results to gauge fields in $\mathbb{R}^3$.

We can now write down the WZW action for the transverse space with the coordinates z, $\bar{z}$, with the (local) Kähler metric $(\partial\bar{\partial}K_T)\,dzd\bar{z}$. The factors involving $(\partial\bar{\partial}K_T)$ will drop out of S_{WZW} because of its conformal invariance. The action is thus the integral of a differential two-form given as

$$S_{\text{WZW}}(H) = -\frac{i}{4\pi}\int_{\mathcal{M}} \text{Tr}(\partial H \wedge \bar{\partial}H^{-1}) + \frac{i}{12\pi}\int_{\mathcal{M}^3} \text{Tr}(H^{-1}dH \wedge H^{-1}dH \wedge H^{-1}dH)$$

$$= -\frac{i}{4\pi}\left[\int_{\mathcal{M}} \text{Tr}(\partial H \wedge \bar{\partial}H^{-1})\right.$$

$$\left. - \int_{\mathcal{M}^3} \text{Tr}\left[H^{-1}dH \wedge (H^{-1}\partial H \wedge H^{-1}\bar{\partial}H - H^{-1}\bar{\partial}H \wedge H^{-1}\partial H)\right]\right] \quad (25)$$

The mass term of interest can now be written down as $\Gamma = m^2 S_m$, with[14]

$$S_m = -\int d\mu(S^2)\,\hat{\Theta}\,S_{\text{WZW}}(H)$$

---

[a]In Ref. 15, Boyer and Galicki study a particular version of what they name as the twistor space for Sasakian manifolds. They also mention that there could be another object which deserves the name of twistor space for Sasakians. The latter one is the trivial S^2 bundle for a Sasakian manifold inheriting the structure from the Kähler structure on the cone. It is this latter definition which applies to our case here.

$$= \frac{i}{4\pi} \int d\mu(S^2) \,\hat{\Theta} \wedge \left[\mathrm{Tr}(\partial H \wedge \bar{\partial} H^{-1}) \right.$$

$$\left. - \mathrm{Tr}\left[H^{-1} dH \wedge \left(H^{-1}\partial H \wedge H^{-1}\bar{\partial} H - H^{-1}\bar{\partial} H \wedge H^{-1}\partial H \right) \right] \right] \quad (26)$$

The same expression also applies to the large R (or $\mathbb{R}^3$) limit, where the one-form $\hat{\Theta}$ and the potential K_T simplify as given in (24).

The expression (26) may still seem rather cryptic, but it is straightforward to work out the expression as a series in terms of the gauge potentials, after Fourier transforming to momentum space. The first two terms are[16]

$$S_m = \frac{1}{2} \int \frac{d^3 k}{(2\pi)^3} A_i^a(-k) A_j^a(k) \left(\delta_{ij} - \frac{k_i k_j}{\vec{k}^2} \right)$$

$$+ \int \frac{d^3 k}{(2\pi)^3} \frac{d^3 q}{(2\pi)^3} A_i^a(k) A_j^b(q) A_k^c(-k-q) \, f^{abc} V_{ijk}(k, q, -k-q)$$

$$V_{ijk}(k, q, -(k+q)) = \frac{i}{6} \left[\frac{1}{k^2 q^2 - (q \cdot k)^2} \right] \left[\left\{ \frac{q \cdot k}{k^2} - \frac{q \cdot (q+k)}{(q+k)^2} \right\} k_i k_j k_k \right.$$

$$\left. + \frac{k \cdot (q+k)}{(q+k)^2} (q_i q_j k_k + q_k q_i k_j + q_j q_k k_i) - (q \leftrightarrow k) \right] \quad (27)$$

4. Properties of the 3d mass term

Our arguments in arriving at (26) show that it has a deep and interesting mathematical structure and that it is the most natural generalization to three dimensions of the results in two and four dimensions. But we can again ask the crucial question of whether it has anything to do with physics. Indeed that is the case, the motivation from physics is what led to this mass term, for $\mathbb{R}^3$, many years ago, although the Sasakian structure was not clear at that time.[14] The general expectation is that in nonabelian gauge theories a mass gap will be dynamically generated, so potentially, one can get a term like (26) in the effective action for such theories. This will be a highly nonperturbative result. One can attempt to demonstrate this, and calculate the coefficient m^2, via a gap equation approach where we add and subtract the same term to the standard Yang–Mills action,

$$S = S_{\mathrm{YM}} + m^2 S_m - \Delta S_m = \tilde{S} - \Delta S_m$$
$$\tilde{S} = S_{\mathrm{YM}} + m^2 S_m \quad (28)$$

The idea is to consider m^2 as the exact value of the mass generated by interactions while Δ is taken to have a loop expansion of the form $\Delta = \Delta^{(1)} + \Delta^{(2)} + \ldots$. Calculations can be done in a loop expansion, with the action $\tilde{S}$ used to construct the propagators and vertices at the tree level, but Δ starts at the one-loop level. Since m^2 is taken to be the exact dynamically generated mass, the pole of the propagator must remain at $k_0^2 - \vec{k}^2 = m^2$ as loop corrections are added. This requires choosing $\Delta^{(1)}$ to cancel the one-loop shift of the pole, $\Delta^{(2)}$ to cancel the two-loop shift of the pole, etc., as is usually done for mass renormalization. After this is done, the Δ so obtained must still equal m^2, since the theory is defined by just the YM action. Thus we must impose the condition

$$\Delta = \Delta^{(1)} + \Delta^{(2)} + \ldots = m^2 \tag{29}$$

This statement of equating the corrections to the mass term is the gap equation which determines m^2.[16] (This strategy can be continued to arbitrary orders of calculation.[7,16]) Notice that the approach is completely gauge-invariant. The calculation of m^2 along these lines was carried out in Ref. 16 and gave the value $m \approx 1.19 \, (e^2 c_A / 2\pi)$, where e is the gauge coupling and c_A is the quadratic Casimir value for the adjoint representation of the gauge group. (For other related approaches to the magnetic screening mass, see Refs. 17, 7.)

The gap equation can be viewed as the result of the summation of an infinite number of Feynman diagrams, a particular sequence being chosen by the form of the mass term. A very different approach is to use the Schrödinger equation in the Hamiltonian formulation of the theory and to solve it for the ground–state wave functional in a low energy approximation. Such an approach, which has been developed in a series of articles,[18] leads to a prediction for the string tension (which is in good agreement with lattice simulations[19]) and a value for m as $e^2 c_A / 2\pi$. This is close to the value obtained from the gap equation analysis. Yet another independent validation of the result comes from using the same Hamiltonian approach to calculate the Casimir energy for two parallel plates.[20] One can then obtain a direct and independent numerical estimate of the value of the mass by a lattice simulation of the parallel plate geometry for the Yang–Mills theory. Such a simulation yields the same value of $e^2 c_A / 2\pi$ to within a fraction of a percent.[21]

There are two other observations regarding this mass term which might be interesting. The first is about the one-loop correction generated by the mass term which determines the gap equation. Let us denote, in any gauge theory, the correction to the two-point function as $\int A_i^a(-k)\Pi_{ij}(k)A_j^a(k)$. Then one can analyze some of the excitations which can occur in intermediate states of the one-loop graph, corresponding to a unitarity cut of the diagram, by the singularity structure of $\Pi_{ij}(k)$. As an example, if we use a mass term

$$S_m = \int d^3x \, \mathrm{Tr}\left[F_i \left(\frac{1}{D^2} \right) F_i \right], \qquad F_i = \frac{1}{2}\epsilon_{ijk}F_{jk}, \tag{30}$$

one can show that there are singularities at $k^2 = 0$ in $\Pi_{ij}(k)$ indicating that there are still zero-mass excitations present in the spectrum.[7] One way to understand this is to note that if we write (30) in terms of local monomials with an auxiliary field, we get

$$S_m = -\int d^3x \, \mathrm{Tr}\left[\frac{1}{2}\phi_i (-D^2)\phi_i + \phi_i F_i \right] \tag{31}$$

This would give propagating massless solutions corresponding to $\Box \phi_i = 0$ in the absence of A_i. In contrast to this, $\Pi_{ij}(k)$ resulting from the mass term (26) or (27) has no zero-mass threshold singularities, at least to the one-loop order the calculations have been carried out. One way to understand this may be to note that the mass term rewritten in terms of local monomials with auxiliary fields is

$$
\begin{aligned}
S_m = \int d\mu(S^2) \Bigg[& \int dx^T S_{\text{WZW}}(G) \\
& + \frac{1}{\pi} \int d^3x \, \text{Tr}\left(G^{-1}\bar{\partial}G\, A - \bar{A}\,\partial G G^{-1} + A G^{-1}\bar{A}\, G - A\bar{A} \right) \Bigg]
\end{aligned}
\tag{32}
$$

The equations of motion for the group element G leads to the solution

$$
G = M^{\dagger -1}\tilde{V}(z)\, V(\bar{z})M^{-1}
\tag{33}
$$

Since the matrices M, $M^\dagger$ are only defined up to the ambiguity $M \to MV^{-1}(\bar{z})$, $M^\dagger \to \tilde{V}(z)M^\dagger$ by the equations $A = -\partial M\, M^{-1}$, $\bar{A} = M^{\dagger -1}\bar{\partial}M^\dagger$, we can absorb $\tilde{V}(z)\, V(\bar{z})$ in (33) into the definition of M, $M^\dagger$. Thus there are no independent solutions, or independent degrees of freedom, for the auxiliary field.

Our second observation is about the use of this mass term in the context of the quark–gluon plasma. The hard thermal loops generate a screening mass (7) or (10) for the chromoelectric forces, the mass term (26) can describe the magnetic screening or the magnetic mass of the plasma. In carrying out calculations at finite temperature, one can see that, even after taking account of the hard thermal loops and the corresponding chromoelectric screening effects, there are still infrared divergences left over in the nonabelian theory. These are cured by the screening mass for the chromomagnetic interactions, so the dynamical generation of such a mass term is an important feature. At very high temperatures, the 3+1 dimensional theory can be approximated by the same theory in three Euclidean dimensions, i.e., there is a dimensional reduction, the coupling of the 3d theory being $e^2 = g^2 T$, where g is the 4d coupling. The dynamically generated mass of the 3d theory can thus be interpreted as the magnetic screening mass of the high temperature limit of the 4d theory.[22] For this idea to be implemented in the full four-dimensional theory, we again need a 4d- Lorentz- invariant form of the mass term. It is indeed possible to construct such an invariant mass term.[14] The result is exactly of the form given in (10) with one change. Instead of the components of A_μ given in (11), we must use

$$
A_\pi = \tfrac{1}{2}\pi(e \cdot A)\xi, \qquad A_\xi = \tfrac{1}{2}\xi(e \cdot A)\bar{\pi}
\tag{34}
$$

Notice that in each of these combinations there is mixing of the spinors π, ξ, unlike the situation in (11).

Acknowledgements

I thank A. P. Balachandran and Denjoe O'Connor for discussions on the Sasakian structure. This research was supported in part by the U.S. National Science Foundation grant PHY-1820721 and by PSC-CUNY awards.

References

1. R. Jackiw and S.-Y. Pi, *Phys. Rev. Lett.* **64**, 2969 (1990).
2. R. Jackiw and V. P. Nair, *Phys. Rev.* **D43**, 1933 (1991); *Phys. Rev. Lett.* **73**, 2007 (1994); *Phys. Lett.* **B480**, 237 (2000); *Phys. Lett.* **B551**, 166 (2003).
3. R. Jackiw, V. P. Nair and S.-Y. Pi, *Phys. Rev.* **D62**, 085018 (2000).
4. B. Bistrovic *et al.*, *Phys. Rev.* **D67**, 025013 (2003); R. Jackiw, V. P. Nair, S.-Y. Pi and A. P. Polychronakos, *J. Phys. A: Math. Gen.* **37**, R327 (2004).
5. R. Efraty and V. P. Nair, *Phys. Rev. Lett.* **68**, 2891 (1992); *Phys. Rev.* **D47**, 5601 (1993).
6. R. Jackiw and V. P. Nair, *Phys. Rev.* **D48**, 4991 (1993).
7. R. Jackiw and S.-Y. Pi, *Phys. Lett.* **B368**, 131 (1996); *ibid.* **B403**, 297 (1997).
8. E. Witten, *Commun. Math. Phys.* **92**, 455 (1984); S. P. Novikov, *Usp. Mat. Nauk* **37**, 3 (1982).
9. A. M. Polyakov and P. B. Wiegmann, *Phys. Lett.* **B141**, 223 (1984).
10. J. S. Schwinger, *Phys. Rev.* **128**, 2425 (1962).
11. R. Penrose, *J. Math. Phys.* **8**, 345 (1967). The subject is by now very well developed, see R. Penrose and M. A. H. MacCallum, *Phys. Rep.* **6**, 241 (1972); R. Penrose and W. Rindler, *Spinors and Spacetime*, 2 volumes (Cambridge University Press, 1984 & 1987); S. A. Hugget and K. P. Tod, *An Introduction to Twistor Theory* (Cambridge University Press, 1993).
12. R. Pisarski, *Physica* **A158**, 246 (1989); *Phys. Rev. Lett.* **63**, 1129 (1989); E. Braaten and R. Pisarski, *Phys. Rev.* **D42**, 2156 (1990); *Nucl. Phys.* **B337**, 569 (1990); *ibid.* **B339**, 310 (1990); *Phys. Rev.* **D45**, 1827 (1992); J. Frenkel and J. C. Taylor, *Nucl. Phys.* **B334**, 199 (1990); J. C. Taylor and S. M. H. Wong, *Nucl. Phys.* **B346**, 115 (1990).
13. G. Alexanian and V. P. Nair, *Phys. Lett.* **B390**, 370 (1997).
14. V. P. Nair, *Phys. Lett.* **B352**, 117 (1995).
15. S. Sasaki, *Tohoku Math. J.* **2**, 459 (1960); For further developments and reviews, see C. Boyer and K. Galicki, *Surveys Diff. Geom.* **7**, 123 (1999), arXiv:hep-th/9810250.
16. G. Alexanian and V. P. Nair, *Phys. Lett.* **B352**, 435 (1995).
17. J. M. Cornwall, *Phys. Rev.* **D10**, 500 (1974); *ibid.* **D26**, 1453 (1982); *ibid.* **D57**, 3694 (1998); W. Buchmuller and O. Philipsen, *Nucl. Phys.* **B443** (1995) 47; O. Philipsen, in *TFT-98: Thermal Field Theories and their Applications*, U. Heinz (ed.), arXiv:hep-ph/9811469; J. M. Cornwall and B. Yan, *Phys. Rev.* **D53**, 4638 (1996); J. M. Cornwall, *Phys. Rev.* **D76**, 025012 (2007).
18. D. Karabali and V. P. Nair, *Nucl. Phys.* **B464**, 135 (1996); *Phys. Lett.* **B379**, 141 (1996); D. Karabali, C. Kim and V. P. Nair, *Nucl. Phys.* **B524**, 661 (1998); *Phys. Lett.* **B434**, 103 (1998); D. Karabali, V. P. Nair and A. Yelnikov, *Nucl. Phys.* **B824**, 387 (2010); For a short review, see V. P. Nair, in *Workshop on QCD Green's Functions, Confinement and Phenomenology, QCD-TNT*, September 2009, Trento, Italy, *Proceedings of Science*, POS (QCD-TNT09) 030, arXiv:0910.3252.
19. M. Teper, *Phys. Rev.* **D59**, 014512 (1999) and references therein; B. Lucini and M. Teper, *Phys. Rev.* **D66**, 097502 (2002); B. Bringoltz and M. Teper, *Phys. Lett.* **B645**, 383 (2007).
20. D. Karabali and V. P. Nair, *Phys. Rev.* **D98**, 105009 (2018).
21. M. N. Chernodub, V. A. Goy, A. V. Molochkov and H. H. Nguyen, *Phys. Rev. Lett.* **121**, 191601 (2018).
22. See, for example, D. Gross, R. Pisarski and L. Yaffe, *Rev. Mod. Phys.* **53**, 43 (1981) and references therein.

Chapter 20

Celestial Operator Products of Gluons and Gravitons
Dedicated to Roman Jackiw on his 80th birthday

Monica Pate[*‡], Ana-Maria Raclariu[*], Andrew Strominger[*] and Ellis Ye Yuan[*†]

Center for the Fundamental Laws of Nature, Harvard University, Cambridge, MA, USA

†*Zhejiang Institute of Modern Physics, Zhejiang University, Hangzhou, Zhejiang, China*

‡*Society of Fellows, Harvard University, Cambridge, MA, USA*

The operator product expansion (OPE) on the celestial sphere of conformal primary gluons and gravitons is studied. Asymptotic symmetries imply recursion relations between products of operators whose conformal weights differ by half-integers. It is shown, for tree-level Einstein–Yang–Mills theory, that these recursion relations are so constraining that they completely fix the leading celestial OPE coefficients in terms of the Euler beta function. The poles in the beta functions are associated with conformally soft currents.

1. Introduction

The subleading soft graviton theorem implies that any quantum theory of gravity in an asymptotically flat four-dimensional (4D) spacetime has an infinite-dimensional 2D conformal symmetry.[1,2] This symmetry acts on the celestial sphere at null infinity, with Lorentz transformations generating the global $SL(2,\mathbb{C})$ subgroup.[3] 4D scattering amplitudes in a conformal basis transform like a collection of correlators in a 2D 'celestial conformal field theory'. Properties of the so defined celestial CFTs have been extensively studied and differ in ways which are not yet fully understood from those of conventional CFTs. Celestial operator spectra were studied in Refs. 4–11 and celestial scattering amplitudes in Refs. 12–25.

In a general celestial CFT, the operator spectrum is continuous, with one continuum for every *stable* species of particles. Unstable particles decay before reaching infinity and are not part of the data on the celestial sphere. For a stable particle of spin s, a complete basis is given by celestial conformal primaries with conformal weights $(h, \bar{h}) = (\frac{\Delta+s}{2}, \frac{\Delta-s}{2})$ and $\text{Re}(\Delta) = 1$.[5]

In this paper we study the operator product expansion (OPE) of these celestial primaries. Poles in the celestial OPE for massless particles turn out to be Mellin transforms of collinear singularities in momentum space which can be computed with Feynman diagrams. The OPEs follow from the three-point vertices coupling the stable particles. We derive a simple and universal formula (12) relating the conformal weights in the operator product expansion to the bulk scaling dimension of the three-point vertex.

Celestial CFTs are subject to multiple infinities of asymptotic symmetry constraints beyond the familiar ones following from 2D conformal symmetry. These constraints have no analogs in conventional CFTs. They follow from the leading and subsubleading soft graviton theorems and, if there are gauge bosons, the subleading soft photon/gluon theorem. On the face of it, it would seem impossible for a collection of celestial amplitudes to satisfy additional infinities of constraints, but of course we know this seemingly overconstrained problem must have a solution as many celestial amplitudes have been explicitly constructed. So far there has been little study of the implications of these constraints.

In this paper we show that the additional symmetry constraints have remarkable implications for the operator product expansion. They imply recursion relations between products of celestial operators whose conformal weights differ by a half-integer. We analyze in detail tree-level Einstein–Yang–Mills (EYM) theory and find that the recursion relations, together with some analyticity assumptions, are so powerful that they completely determine (at least) all the conformal primary OPE coefficients of the leading poles in the operator product expansion. They are given by Euler beta functions (ratios of Gamma functions) with arguments given by the conformal weights. We check that the direct but lengthier Feynman-diagrammatic computation yields the same beta functions.

Inclusion of quantum, stringy or other corrections would introduce higher dimension terms into the effective action. These may alter both the three-point vertices and the (sub)subleading soft theorems, and hence the subleading terms in the OPEs in accord with the general formula (13) below. It will be interesting to study the symmetry constraints on OPEs in this more general context, as well as to extend the analysis beyond the leading poles.

In a conventional (unitary, discrete) CFT, the operator spectrum and the conformal primary OPEs fully determine the theory. Should an analogous result hold in celestial CFT, it would suggest that complete quantum theories of gravity are determined by these symmetry-constrained OPE coefficients. These are far fewer in number than the number of possible terms in the effective Lagrangian. This resonates with similar findings in the amplitudes program.[26–30] It would be interesting to study further constraints among these OPEs from crossing symmetry.

This paper is organized as follows. Section 2 contains conventions and useful formulae. Section 3 begins with a general derivation of the relation between the bulk dimension of the three-point couplings and the conformal weights of the OPE. Section 3.1 considers the gluon OPE poles in tree-level Yang–Mills (YM) theory. The subleading soft gluon theorem is shown to imply recursion relations among the OPE coefficients, with the overall normalization fixed by the leading soft gluon theorem. For the collinear pole terms, these are uniquely solved — subject to certain falloffs at large operator dimension — by Euler beta functions. Section 3.2 derives similar results, invoking the subsubleading soft graviton theorem, for the graviton OPEs in Einstein gravity, while 3.3 derives the EYM gluon–graviton OPEs. In Sec. 4, building on previous analyses of collinear limits of gravitons and gluons, we directly compute the collinear singularities in momentum space and then the OPE poles via a Mellin transformation. This direct analysis fully agrees with the symmetry-derived results. We generalize our results for operators associated to incoming and outgoing particles in Sec. 5. The EYM OPEs are all summarized in Sec. 5.3. Appendix A.1 details the relation between the bulk scaling dimension of a three-point vertex and the conformal weights entering the OPE. Appendix A.2 presents the list of all OPE coefficients which can be generated by higher-dimension operators. In Appendices A.3 and A.4 we review the unbroken global symmetries which are related to the subleading soft gluon theorem and subsubleading soft graviton theorem and used to derive the recursion relations. In Appendix A.5 we solve the recursion relations for the beta function and spell out the regularity conditions which make the solution unique.

2. Preliminaries

In this section we give our conventions for celestial scattering amplitudes and collect some useful formulae.

Celestial amplitudes $\tilde{A}$ of massless particles are obtained from momentum-space amplitudes A (including the momentum-conserving delta function) by performing Mellin transformations with respect to the particle energies[5,12]

$$\tilde{A}_{s_1 \cdots s_n}(\Delta_1, z_1, \bar{z}_1, \cdots, \Delta_n, z_n, \bar{z}_n) = \left(\prod_{k=1}^{n} \int_0^\infty d\omega_k \, \omega_k^{\Delta_k - 1} \right) \tag{1}$$
$$\times A_{s_1 \cdots s_n}(\epsilon_1 \omega_1, z_1, \bar{z}_1, \cdots, \epsilon_n \omega_n, z_n, \bar{z}_n),$$

where the helicity $s_k = \pm 1$ for gluons and $s_k = \pm 2$ for gravitons. In order to write momentum-space amplitudes as functions of $(\epsilon_k \omega_k, z_k, \bar{z}_k)$, we parametrize the Cartesian coordinate massless 4-momenta components as

$$p_k^\mu = \frac{\epsilon_k \omega_k}{\sqrt{2}} (1 + z_k \bar{z}_k, z_k + \bar{z}_k, -i(z_k - \bar{z}_k), 1 - z_k \bar{z}_k), \tag{2}$$

with $\mu = 0, 1, 2, 3$, $\epsilon_k = \pm 1$ for outgoing and incoming momenta respectively and helicities are defined with respect to outgoing momenta. In the following two sections we compute OPEs of outgoing states with $\epsilon_k = 1$. We finally explain how to generalize the analysis to mixed incoming and outgoing OPEs in Sec. 5. Color indices and in/out labels on celestial amplitudes are suppressed. We later use A to denote color-ordered partial amplitudes. We note that

$$p_1 \cdot p_2 = -\epsilon_1 \epsilon_2 \omega_1 \omega_2 z_{12} \bar{z}_{12}, \tag{3}$$

where

$$z_{12} = z_1 - z_2, \quad \bar{z}_{12} = \bar{z}_1 - \bar{z}_2. \tag{4}$$

For coordinates

$$x^\mu = u\partial_z \partial_{\bar{z}} q^\mu(z, \bar{z}) + r q^\mu(z, \bar{z}),$$
$$q^\mu(z, \bar{z}) = \frac{1}{\sqrt{2}}\left(1 + z\bar{z}, z + \bar{z}, -i(z - \bar{z}), 1 - z\bar{z}\right), \tag{5}$$

the flat metric is

$$ds^2 = dx^\mu dx_\mu = -2dudr + 2r^2 dz d\bar{z}, \tag{6}$$

the celestial sphere is conformally mapped to the celestial plane and z_k is the spatial location at which a particle of momentum p_k crosses $\mathcal{I}^+$. $\tilde{\mathcal{A}}$ transforms as a correlator of n weight $(h_k, \bar{h}_k) = (\frac{\Delta_k + s_k}{2}, \frac{\Delta_k - s_k}{2})$ primaries under conformal transformations of the celestial plane. In the next two sections we consider only OPEs between outgoing particles, and use $\mathcal{O}_{\Delta,s}$ to denote a generic such primary, $O_\Delta^{\pm a}$ for a primary gluon where $s = \pm 1$ (with a an adjoint group index) and $G_\Delta^\pm$ for a primary graviton with $s = \pm 2$. In Sec. 5 we reintroduce the additional label ϵ to distinguish between incoming and outgoing operators $\mathcal{O}_{\Delta,s}^\epsilon$. (Whenever the label is absent, the operator is taken to be outgoing.) Group structure constants $f^{ab}_{\;\;c}$ obey the Jacobi identity

$$f^{ab}_{\;\;d} f^{dce} + f^{bc}_{\;\;d} f^{dae} + f^{ca}_{\;\;d} f^{dbe} = 0, \tag{7}$$

and generators are normalized such that

$$\text{Tr}(T^a T^b) = g_{YM}^2 \delta^{ab}, \tag{8}$$

where T^a are in the fundamental representation. We work with the following polarization vectors for massless spin-1 particles

$$\varepsilon_k^{+\mu} = \frac{1}{\epsilon_k \omega_k} \partial_{z_k} p_k^\mu, \qquad \varepsilon_k^{-\mu} = \frac{1}{\epsilon_k \omega_k} \partial_{\bar{z}_k} p_k^\mu, \tag{9}$$

and polarization tensors $\varepsilon_k^{\pm\mu\nu} = \varepsilon_k^{\pm\mu} \varepsilon_k^{\pm\nu}$ for massless spin-2 particles. These obey

$$p_1 \cdot \varepsilon_2^- = \epsilon_1 \omega_1 z_{12}, \qquad p_1 \cdot \varepsilon_2^+ = \epsilon_1 \omega_1 \bar{z}_{12}. \tag{10}$$

Generically, the Mellin transform ω_k-integrals converge only for restricted values of Δ_k. For example in gauge theory they converge on the unitary principle series with $\text{Re}(\Delta) = 1$. However we will be interested in the celestial amplitudes for other complex values of Δ_k, where we define them by analytic continuation.

3. OPEs from asymptotic symmetries

In this section we study OPEs of conformal primary gluon and graviton operators on the celestial plane labeled by $(z, \bar{z})$. z and $\bar{z}$ will be varied independently. (These variables are independent in (2,2) signature, for which the celestial plane becomes Lorentzian.) Moreover we consider only the 'holomorphic limit' $z_{12} \to 0$ with $\bar{z}_1, \bar{z}_2$ fixed. Symmetry implies similar OPEs for $\bar{z}_{12} \to 0$ with z_1, z_2 fixed. However, order-of-limits subtleties arise when both $z_{12} \to 0$ and $\bar{z}_{12} \to 0$.[31-33] These are likely important for the structure of celestial amplitudes but are beyond the scope of this paper.

Singularities in the celestial OPEs are the Mellin transforms of collinear divergences in the momentum-space scattering amplitudes. This allows us to deduce some simple properties of the OPEs without any detailed computations. Collinear singularities arise when $p_1 \| p_2$ for massless particles which couple via a three-point vertex to form a nearly on-shell internal particle. The resulting propagator is proportional to $\frac{1}{p_1 \cdot p_2}$ which, according to (3), diverges as $\frac{1}{z_{12}}$ for $z_{12} \to 0$. Hence two-operator OPE singularities are at most simple poles in z_{12}. Schematically the OPE of conformal primaries $\mathcal{O}_{\Delta,s}$ with conformal weights $(h, \bar{h}) = (\frac{\Delta+s}{2}, \frac{\Delta-s}{2})$ takes the form

$$\mathcal{O}_{\Delta_1, s_1}(z_1, \bar{z}_1) \mathcal{O}_{\Delta_2, s_2}(z_2, \bar{z}_2) \propto \frac{1}{z_{12}} \mathcal{O}_{\Delta_3, s_3}(z_2, \bar{z}_2) + \text{order}(z_{12}^0). \quad (11)$$

Contributions to the OPE (11) arise from the three-point interaction vertices in the expansion of terms in the bulk effective Lagrangian around flat space. Since gravitons and gluons have bulk scaling dimension one, these are characterized by bulk dimension $d_V = 3 + m$, where m is the number of spacetime derivatives. For example the most relevant gluon–gluon–graviton vertex $h\partial A\partial A$ has $d_V = 5$, while the gluon–gluon–gluon vertex $A\partial AA$ has $d_V = 4$. The conformal weight Δ_3 of the operator on the right hand side of (11) can be inferred from d_V. Each derivative leads to one extra factor of ω inside the Mellin transform (1), and therefore shifts Δ_3 up by one. Accounting for all the factors of ω (including two in the internal propagator), one finds that the OPE of two operators of conformal weight Δ_1 and Δ_2 which couple via a three-point vertex of bulk dimension d_V can only produce an operator with conformal weight

$$\Delta_3 = \Delta_1 + \Delta_2 + d_V - 5. \quad (12)$$

Details are in Appendix A.1. Further, insisting on conformal invariance, one finds that the contribution to the OPE from a vertex of fixed d_V must take the form[a]

$$\mathcal{O}_{\Delta_1, s_1}(z_1, \bar{z}_1) \mathcal{O}_{\Delta_2, s_2}(z_2, \bar{z}_2)$$

$$\sim \sum_{n=0}^{d_V - 4} c_{n, d_V}(\Delta_1, s_1; \Delta_2, s_2) z_{12}^{n-1} \bar{z}_{12}^{d_V - 4 - n} \mathcal{O}_{\Delta_1 + \Delta_2 + d_V - 5, s_1 + s_2 + 3 + 2n - d_V}(z_2, \bar{z}_2). \quad (13)$$

---

[a]Since there are no gauge and coordinate invariant $d_V < 4$ relevant operators in a theory with only gluons or gravitons (except of course the cosmological constant, which we assume vanishes!), there are no $\frac{1}{z_{12}\bar{z}_{12}}$ singularities.

Although in the most general case (13) is an infinite series when summing over d_V, many terms are eliminated when the spins range over limited values. For example in a theory with only $s = \pm 1$ gluons the O^+O^+ OPE in (13) reduces to the two terms[b]

$$
O_{\Delta_1}^{+a}(z_1, \bar{z}_1)O_{\Delta_2}^{+b}(z_2, \bar{z}_2) \sim if^{ab}{}_c \left(c_{\frac{d_V}{2}-2,d_V} z_{12}^{d_V/2-3} \bar{z}_{12}^{d_V/2-2} O_{\Delta_1+\Delta_2+d_V-5}^{+c}(z_2, \bar{z}_2) \right.
$$
$$
\left. + c_{\frac{d_V}{2}-3,d_V} z_{12}^{d_V/2-4} \bar{z}_{12}^{d_V/2-1} O_{\Delta_1+\Delta_2+d_V-5}^{-c}(z_2, \bar{z}_2) \right).
$$
$$(14)$$

In this paper we consider in detail only symmetry constraints on the leading ($n = 0$) pole terms in EYM theory, for which there are only seven nonzero coefficients c_{0,d_V} with $d_V = 4, 5$. These are all completely fixed by asymptotic symmetries and summarized in Sec. 5.3. Equally powerful symmetry constraints apply to all terms in the expansion (13), but the more intricate higher-order analysis is left to future investigation.

3.1. *Gluons*

In this section we consider pure renormalizable glue theory with $d_V = 4$. In this case,[c]

$$
O_{\Delta_1}^{+a}(z_1, \bar{z}_1)O_{\Delta_2}^{+b}(z_2, \bar{z}_2) \sim -\frac{if^{ab}{}_c}{z_{12}} C(\Delta_1, \Delta_2) O_{\Delta_1+\Delta_2-1}^{+c}(z_2, \bar{z}_2), \tag{15}
$$

$$
O_{\Delta_1}^{+a}(z_1, \bar{z}_1)O_{\Delta_2}^{-b}(z_2, \bar{z}_2) \sim -\frac{if^{ab}{}_c}{z_{12}} D(\Delta_1, \Delta_2) O_{\Delta_1+\Delta_2-1}^{-c}(z_2, \bar{z}_2), \tag{16}
$$

for some to-be-determined coefficients $C(\Delta_1, \Delta_2) = C(\Delta_2, \Delta_1)$ and $D(\Delta_1, \Delta_2)$. O^-O^- is nonsingular in z_{12}. For gluons, the conformal primaries with $\mathrm{Re}(\Delta) = 1$ are a complete basis of square-integrable wave packets.[5] We see that in the renormalizable theory the OPEs (15) and (16) close on such operators.

The OPE coefficients are subject to a number of symmetry constraints. The simplest is translations $\mathcal{P}$ towards the 'north pole' of the celestial sphere, which involves a factor of ω in momentum space. In a conformal basis, this symmetry shifts the operator dimension:[7,19]

$$
\delta_{\mathcal{P}} O_\Delta^{\pm a}(z, \bar{z}) = O_{\Delta+1}^{\pm a}(z, \bar{z}). \tag{17}
$$

Acting on both sides of (15) and (16) with $\delta_{\mathcal{P}}$ gives the recursion relations

$$
C(\Delta_1, \Delta_2) = C(\Delta_1 + 1, \Delta_2) + C(\Delta_1, \Delta_2 + 1), \tag{18}
$$

$$
D(\Delta_1, \Delta_2) = D(\Delta_1 + 1, \Delta_2) + D(\Delta_1, \Delta_2 + 1). \tag{19}
$$

Such relations were also found in Ref. 11.

---

[b]Due to the lower limit in (13), the second term is absent for $d_V = 4$.
[c]Additional terms on the right hand side in the presence of gravitons are determined in Sec. 3.3. An F^3 term with $d_V = 6$ would lead to an O^- term on the right hand side of (15).

Next, the leading conformally soft theorem is[20-22]

$$\lim_{\Delta_1 \to 1} O_{\Delta_1}^{+a}(z_1, \bar{z}_1) O_{\Delta_2}^{\pm b}(z_2, \bar{z}_2) \sim -\frac{if^{ab}_{c}}{(\Delta_1 - 1)z_{12}} O_{\Delta_2}^{\pm c}(z_2, \bar{z}_2). \tag{20}$$

This implies poles in C and D with residues

$$\lim_{\Delta_1 \to 1}(\Delta_1 - 1)C(\Delta_1, \Delta_2) = \lim_{\Delta_1 \to 1}(\Delta_1 - 1)D(\Delta_1, \Delta_2) = 1. \tag{21}$$

Further, less familiar, constraints come from the subleading soft symmetry parametrized by $(Y^{za}, Y^{\bar{z}a})$. Under these symmetries, the gauge field on $\mathcal{I}^+$ shifts by[34]

$$\delta_Y A_z^a = u\partial_z^2 Y^{za}, \qquad \bar{\delta}_Y A_{\bar{z}}^a = u\partial_{\bar{z}}^2 Y^{\bar{z}a}. \tag{22}$$

If the right hand side is nonzero, the symmetry is spontaneously broken. The unbroken symmetries are the most useful for present purposes. These correspond to $Y^{za} = z\epsilon^a, \epsilon^a$ and $Y^{\bar{z}a} = \bar{z}\epsilon^a, \epsilon^a$ for constant ϵ^a. As shown in Appendix A.3 (see also Ref. 34), for the global symmetry $Y^{za} = z\epsilon^a$ conformal primary gluons transform as

$$\delta_b O_\Delta^{\pm a}(z, \bar{z}) = -(\Delta - 1 \pm 1 + z\partial_z)if^a_{bc}O_{\Delta-1}^{\pm c}(z, \bar{z}). \tag{23}$$

Similarly for $Y^{\bar{z}a} = \bar{z}\epsilon^a$ we have

$$\bar{\delta}_b O_\Delta^{\pm a}(z, \bar{z}) = -(\Delta - 1 \mp 1 + \bar{z}\partial_{\bar{z}})if^a_{bc}O_{\Delta-1}^{\pm c}(z, \bar{z}). \tag{24}$$

Since they are unbroken, the Ward identities for these symmetries involve no soft insertions

$$\sum_{k=1}^{n}\langle O_1 \cdots \delta O_k \cdots O_n\rangle = 0, \qquad \sum_{k=1}^{n}\langle O_1 \cdots \bar{\delta} O_k \cdots O_n\rangle = 0. \tag{25}$$

We now extract the consequences of this global symmetry for the OPE (15). This is complicated by the appearance of derivatives in the transformation laws (23) and (24) which mix up primaries and descendants, and therefore do not map the leading OPE relations (15) and (16) to themselves. These bothersome terms can be eliminated in $\bar{\delta}$ by considering the special case $\bar{z}_1 = \bar{z}_2 = 0$, where (15) still holds. (The z-analog of this trick cannot be used to analyze the implications of δ symmetry because (15) blows up for $z_1 = z_2 = 0$.) Acting with $\bar{\delta}_d$ on both sides of (15) we get

$$(\Delta_1 - 2)if^a_{dc}O_{\Delta_1-1}^{+c}(z_1, 0)O_{\Delta_2}^{+b}(z_2, 0) + (\Delta_2 - 2)O_{\Delta_1}^{+a}(z_1, 0)if^b_{dc}O_{\Delta_2-1}^{+c}(z_2, 0)$$
$$\sim \frac{\Delta_1 + \Delta_2 - 3}{z_{12}}C(\Delta_1, \Delta_2)f^{ab}_{c}f^c_{de}O_{\Delta_1+\Delta_2-2}^{+e}(z_2, 0). \tag{26}$$

Using the OPE again on the left hand side we obtain the consistency condition

$$(\Delta_1 - 2)C(\Delta_1 - 1, \Delta_2)f^a_{dc}f^{cb}_{e} + (\Delta_2 - 2)C(\Delta_1, \Delta_2 - 1)f^b_{dc}f^{ac}_{e}$$
$$= (\Delta_1 + \Delta_2 - 3)C(\Delta_1, \Delta_2)f^{ab}_{c}f^c_{de}. \tag{27}$$

Applying the Jacobi identity (7) this implies

$$(\Delta_1 - 2)C(\Delta_1 - 1, \Delta_2) = (\Delta_1 + \Delta_2 - 3)C(\Delta_1, \Delta_2). \qquad (28)$$

Under suitable assumptions spelled out in Appendix A.5 about boundedness and analyticity in Δ_1, Δ_2 (basically that there are no poles other than those implied by the soft theorems), (28) together with the normalization condition (21) have the unique solution[d]

$$C(\Delta_1, \Delta_2) = B(\Delta_1 - 1, \Delta_2 - 1), \qquad (29)$$

where B is the Euler beta function

$$B(x, y) = \frac{\Gamma(x)\Gamma(y)}{\Gamma(x + y)}. \qquad (30)$$

Acting with $\bar{\delta}_d$ on both sides of (16) gives a slightly different result because of the $\pm$ in (24). Instead of (28) we find two different recursion relations

$$(\Delta_1 - 2)D(\Delta_1 - 1, \Delta_2) = (\Delta_1 + \Delta_2 - 1)D(\Delta_1, \Delta_2),$$
$$\Delta_2 D(\Delta_1, \Delta_2 - 1) = (\Delta_1 + \Delta_2 - 1)D(\Delta_1, \Delta_2). \qquad (31)$$

Again, (31) together with the normalization condition (21), have the unique solution

$$D(\Delta_1, \Delta_2) = B(\Delta_1 - 1, \Delta_2 + 1). \qquad (32)$$

(29) and (32) agree with the expressions previously obtained in Ref. 20 by direct Mellin transform of the collinear singularities in momentum space. Here we see the OPE is entirely fixed by symmetries.

In fact there are further consistency conditions, which we did not need to use to fix C and D, but it can be checked that they are satisfied. One of these is that the OPEs have properly normalized poles at $\Delta_1 \to 0$ corresponding to the subleading soft theorem. This is indeed manifest in (29) and (32). We have used here only a few global symmetries. There are infinitely many more constraints from the infinity of soft symmetries. However these may all be obtained by commuting the global symmetries with the local conformal symmetry, which is manifestly built in to our construction and so their satisfaction is guaranteed.

3.2. *Gravitons*

For gravitons in Einstein gravity the three-point vertex has $d_V = 5$. According to (13) this leads to an OPE of the form[e]

$$G^+_{\Delta_1}(z_1, \bar{z}_1)G^\pm_{\Delta_2}(z_2, \bar{z}_2) \sim \frac{\bar{z}_{12}}{z_{12}} E_\pm(\Delta_1, \Delta_2)G^\pm_{\Delta_1 + \Delta_2}(z_2, \bar{z}_2), \qquad (33)$$

---

[d]Symmetry of $C(\Delta_1, \Delta_2)$ under $\Delta_1 \leftrightarrow \Delta_2$ together with the subleading soft symmetry constraint (28) in fact imply the translation invariance relation (18), which therefore does not further constrain C.

[e]A contribution of the form $\frac{\bar{z}_{12}^5}{z_{12}} E'_+(\Delta_1, \Delta_2)G^-_{\Delta_1 + \Delta_2 + 4}$ to the $G^+_{\Delta_1} G^+_{\Delta_2}$ OPE might for example be generated by an R^3 correction to the Einstein–Hilbert action.

for some to-be-determined coefficients $E_+(\Delta_1, \Delta_2) = E_+(\Delta_2, \Delta_1)$ and $E_-(\Delta_1, \Delta_2)$, while G^-G^- is nonsingular in the $z_{12} \to 0$ limit. As for the case of gluons, translation invariance implies the recursion relation

$$E_\pm(\Delta_1, \Delta_2) = E_\pm(\Delta_1 + 1, \Delta_2) + E_\pm(\Delta_1, \Delta_2 + 1). \tag{34}$$

The residue of a pole at $\Delta_1 \to 1$ is fixed by the leading soft graviton theorem[f,24]

$$\lim_{\Delta_1 \to 1} E_\pm(\Delta_1, \Delta_2) \sim -\frac{\kappa}{2(\Delta_1 - 1)}, \qquad \kappa = \sqrt{32\pi G}. \tag{35}$$

The subleading soft symmetry corresponds to 2D conformal transformations, which are generated by the shadow of G_0^+.[7,10,35,36] However, by working in a conformal basis, we have already ensured that the OPE is conformally invariant, and no further constraints on $E_\pm$ are obtained from the subleading soft symmetry.

The role of the subleading soft gluon theorem in constraining gauge theory OPEs is here played by the *subsubleading* soft graviton theorem, which implies further global symmetries. We show in Appendix A.4 that the relevant gravitational analog of the gauge theory relation (24) is

$$\bar{\delta}G_\Delta^\pm(z, \bar{z}) = -\frac{\kappa}{4}\left[(\Delta \mp 2)(\Delta \mp 2 - 1) + 4(\Delta \mp 2)\bar{z}\partial_{\bar{z}} + 3\bar{z}^2\partial_{\bar{z}}^2\right] G_{\Delta-1}^\pm(z, \bar{z}). \tag{36}$$

However, to study the consequences of this symmetry on the OPE, we cannot directly set $\bar{z}_1 = \bar{z}_2 = 0$ in (33) because that will set the right hand side to zero and no useful relation would be obtained. To avoid this we first differentiate with respect to $\bar{z}_1$, and then set $\bar{z}_1 = \bar{z}_2 = 0$. The positive-helicity graviton OPE in (33) is then

$$\partial_{\bar{z}_1}G_{\Delta_1}^+(z_1, 0)G_{\Delta_2}^+(z_2, 0) \sim \frac{E_+(\Delta_1, \Delta_2)}{z_{12}}G_{\Delta_1+\Delta_2}^+(z_2, 0). \tag{37}$$

Equation (36) becomes

$$\bar{\delta}G_\Delta^+(z, 0) = -\frac{\kappa}{4}(\Delta - 2)(\Delta - 3)G_{\Delta-1}^+(z, 0), \tag{38}$$

and in addition implies

$$\bar{\delta}\partial_{\bar{z}}G_\Delta^+(z, 0) = -\frac{\kappa}{4}(\Delta - 2)(\Delta + 1)\partial_{\bar{z}}G_{\Delta-1}^+(z, 0). \tag{39}$$

Invariance of the OPE (33) then holds if and only if

$$(\Delta_1 + 1)(\Delta_1 - 2)E_\pm(\Delta_1 - 1, \Delta_2) + (\Delta_2 \mp 2 - 1)(\Delta_2 \mp 2)E_\pm(\Delta_1, \Delta_2 - 1)$$
$$= (\Delta_1 + \Delta_2 \mp 2)(\Delta_1 + \Delta_2 \mp 2 - 1)E_\pm(\Delta_1, \Delta_2). \tag{40}$$

The two recursion relations (34) and (40), together with the normalization condition (35) are again solved by Euler beta functions

$$E_\pm(\Delta_1, \Delta_2) = -\frac{\kappa}{2}B(\Delta_1 - 1, \Delta_2 \mp 2 + 1). \tag{41}$$

In Sec. 4.1 (see Eqs. (57) and (58)) we directly compute the Mellin transform of the near-collinear graviton amplitudes and find complete agreement with (41).

Additionally, the OPE coefficients $E_\pm$ must have properly normalized poles at $\Delta_1 \to 0$ and $\Delta_1 \to -1$ associated to the subleading and subsubleading soft graviton symmetries, respectively. As in the gauge theory case, we did not impose such conditions in our derivation, but find that our results are consistent with these conditions.

---

[f]Supertranslations are generated by the current $P_{\bar{z}} = -\frac{2}{\kappa}\lim_{\Delta \to 1}(\Delta - 1)\partial_{\bar{z}}G_\Delta^+$.

3.3. *Gravitons and Gluons*

In this section we consider OPEs involving both gravitons and gluons. The Einstein–Yang–Mills interaction (schematically hF^2) has $d_V = 5$. The relevant term in (13) is

$$G_{\Delta_1}^+(z_1, \bar{z}_1)O_{\Delta_2}^{\pm a}(z_2, \bar{z}_2) \sim \frac{\bar{z}_{12}}{z_{12}}F_\pm(\Delta_1, \Delta_2)O_{\Delta_1+\Delta_2}^{\pm a}(z_2, \bar{z}_2). \tag{42}$$

Translation invariance again implies the recursion relation (18) for $F_\pm$. A second set of relations is determined from the global symmetry associated to subsubleading soft graviton theorem, whose action on gluons is shown in Appendix A.4 to be

$$\bar{\delta}O_\Delta^{\pm a}(z, \bar{z}) = -\frac{\kappa}{4}\left[(\Delta \mp 1 - 1)(\Delta \mp 1) + 4(\Delta \mp 1)\bar{z}\partial_{\bar{z}} + 3\bar{z}^2\partial_{\bar{z}}^2\right]O_{\Delta-1}^{\pm a}(z, \bar{z}). \tag{43}$$

Consistency of the OPE with this symmetry requires

$$(\Delta_1 + 1)(\Delta_1 - 2)F_\pm(\Delta_1 - 1, \Delta_2) + (\Delta_2 \mp 1 - 1)(\Delta_2 \mp 1)F_\pm(\Delta_1, \Delta_2 - 1)$$
$$= (\Delta_1 + \Delta_2 \mp 1 - 1)(\Delta_1 + \Delta_2 \mp 1)F_\pm(\Delta_1, \Delta_2), \tag{44}$$

where these relations are derived by studying the OPE of $\partial_{\bar{z}_1}G_{\Delta_1}^+(z_1, 0)O_{\Delta_2}^{\pm a}(z_2, 0)$ as in the previous section. Fixing the normalization with the leading soft graviton theorem one finds

$$F_\pm(\Delta_1, \Delta_2) = -\frac{\kappa}{2}B(\Delta_1 - 1, \Delta_2 \mp 1 + 1). \tag{45}$$

In the presence of gravitons, the right hand side of the gluon OPE (16) can also receive a correction of the form

$$O_{\Delta_1}^{+a}(z_1, \bar{z}_1)O_{\Delta_2}^{-b}(z_2, \bar{z}_2) \sim -\frac{if^{ab}_{\quad c}}{z_{12}}B(\Delta_1 - 1, \Delta_2 + 1)O_{\Delta_1+\Delta_2-1}^{-c}(z_2, \bar{z}_2)$$
$$+ \delta^{ab}\frac{\bar{z}_{12}}{z_{12}}H(\Delta_1, \Delta_2)G_{\Delta_1+\Delta_2}^-(z_2, \bar{z}_2), \tag{46}$$

corresponding to the fact that two gluons can make a graviton. This new term might seem to violate the subleading soft gluon theorem. Indeed, we will find shortly that symmetry constrains H to have a pole associated with the subleading soft gluon symmetry at $\Delta_1 = 0$. However, as shown in Refs. 37 and 38, this theorem is corrected at tree-level in Einstein–Yang–Mills theory by the hF^2 coupling! The known form of the correction in fact can be used to fix the constant normalization of H.

Translation invariance implies H obeys a recursion relation of the form (18), while the subsubleading soft graviton theorem implies H obeys the recursion relation

$$(\Delta_1 + 2)(\Delta_1 - 1)H(\Delta_1 - 1, \Delta_2) + \Delta_2(\Delta_2 + 1)H(\Delta_1, \Delta_2 - 1)$$
$$= (\Delta_1 + \Delta_2 + 2)(\Delta_1 + \Delta_2 + 1)H(\Delta_1, \Delta_2). \tag{47}$$

The properly normalized solution is

$$H(\Delta_1, \Delta_2) = \frac{\kappa}{2}B(\Delta_1, \Delta_2 + 2). \tag{48}$$

The symmetry-derived results (45) and (48) agree with the Mellin transforms of direct Feynman diagram computations found in the next section.

The appearance of a graviton in the OPE of two gluons is presumably the boundary manifestation of the still-enigmatic double-copy relation,[39–41] in which gravity is the square of gauge theory. A remarkable discovery due to Stieberger and Taylor[32,42] is that a pair of collinear gluons in a scattering amplitude can be replaced by a single graviton. If we take $\Delta_1 = \Delta_2 = 0$ in (46), the right hand side contains G_0^- which is the shadow of the boundary stress tensor. This is a Sugawara-like construction of the stress tensor from a pair of subleading soft currents. We leave these fascinating connections to future exploration.

4. OPEs from collinear singularities

In this section we directly compute the celestial OPEs among gravitons and gluons in EYM by Mellin transforms of Feynman diagrams. We begin by reviewing the collinear limits of gauge and gravity amplitudes. The various OPEs are derived by Mellin transforming the corresponding amplitudes in the collinear limit and found in all cases to agree with the symmetry-inferred results summarized later in Sec. 5.3. The OPEs among gluons were already derived in this manner in Ref. 20. Their computation confirms (29) and (32) and will not be repeated here.

4.1. *Gravitons*

The collinear limits of gravity amplitudes were first derived in Ref. 43 and further developments are in Refs. 44 and 45. The leading divergence is generically protected against loop corrections.[43] Here we specialize to a holomorphic collinear limit.

Consider a tree-level n-graviton scattering amplitude. In the limit when $z_{ij} \to 0$ for fixed $\bar{z}_i, \bar{z}_j$, the amplitude contains a universal piece which factorizes as

$$\lim_{z_{ij} \to 0} \mathcal{A}_{s_1 \cdots s_n}(p_1, \cdots, p_n) \longrightarrow \sum_{s=\pm 2} \text{Split}^s_{s_i s_j}(p_i, p_j) \mathcal{A}_{s_1 \cdots s \cdots s_n}(p_1, \cdots, P, \cdots, p_n),$$

(49)

where in the collinear limit[g]

$$P^\mu = p_i^\mu + p_j^\mu, \qquad \omega_P = \omega_i + \omega_j.$$

(50)

The collinear factor $\text{Split}^s_{s_i s_j}(p_i, p_j)$ then takes the form[43,h]

$$\text{Split}^2_{22}(p_i, p_j) = -\frac{\kappa}{2} \frac{\bar{z}_{ij}}{z_{ij}} \frac{\omega_P^2}{\omega_i \omega_j}, \qquad \text{Split}^{-2}_{2-2}(p_i, p_j) = -\frac{\kappa}{2} \frac{\bar{z}_{ij}}{z_{ij}} \frac{\omega_j^3}{\omega_i \omega_P^2},$$

(51)

---

[g]At subleading order in z_{ij}, (50) receives corrections, but these do not affect the leading singularities considered here. For a discussion of subleading terms see Ref. 32.
[h]We work with the Einstein-Hilbert action normalized as $S = \frac{2}{\kappa^2} \int d^4x \sqrt{-g} R$, $g_{\mu\nu} = \eta_{\mu\nu} + \kappa h_{\mu\nu}$.
This yields the following leading soft factor $S^\pm_{(0)} = \frac{\kappa}{2} \sum_k \frac{(p_k \cdot \varepsilon^\pm)^2}{p_k \cdot q}$.

with all other combinations of helicities vanishing. In the collinear limit, the celestial gravity amplitude $\tilde{\mathcal{A}}$ becomes

$$\tilde{\mathcal{A}}_{s_1\cdots s_n}(\Delta_1, z_1, \bar{z}_1, \cdots, \Delta_n, z_n, \bar{z}_n) \xrightarrow{i||j}$$

$$\prod_{k=1}^{n} \int_0^\infty dw_k \, w_k^{\Delta_k-1} \sum_{s=\pm 2} \mathrm{Split}^s_{s_i s_j}(p_i, p_j) \mathcal{A}_{s_1\cdots s\cdots s_n}(p_1, \cdots, P, \cdots, p_n) + \cdots. \tag{52}$$

To simplify, we make the following change of variables,

$$w_i = tw_P, \qquad w_j = (1-t)w_P, \tag{53}$$

so that for example

$$\int_0^\infty dw_i w_i^{\Delta_i-1} \int_0^\infty dw_j w_j^{\Delta_j-1} \mathrm{Split}^2_{22}(p_i, p_j) = -\frac{\kappa}{2} \frac{\bar{z}_{ij}}{z_{ij}} \int_0^1 dt \, t^{\Delta_i-2}(1-t)^{\Delta_j-2}$$

$$\times \int_0^\infty dw_P \, w_P^{\Delta_i+\Delta_j-1}. \tag{54}$$

The t integral is immediately recognizable as the integral representation of the Euler beta function,

$$B(x, y) = \int_0^1 dt \, t^{x-1}(1-t)^{y-1}, \tag{55}$$

whose origin is hence a splitting factor for the conformal weight between the two collinear external particles. Since the only t dependence on the right hand side of (52) comes from $\mathrm{Split}^s_{s_i s_j}(p_i, p_j)$, one finds

$$\lim_{z_{ij}\to 0} \tilde{\mathcal{A}}_{s_1\cdots 2\cdots 2\cdots}(\Delta_1, z_1, \bar{z}_1, \cdots, \Delta_i, z_i, \bar{z}_i, \cdots, \Delta_j, z_j, \bar{z}_j, \cdots) \longrightarrow$$

$$-\frac{\kappa}{2} \frac{\bar{z}_{ij}}{z_{ij}} B(\Delta_i - 1, \Delta_j - 1)\tilde{\mathcal{A}}_{s_1\cdots 2\cdots}(\Delta_1, z_1, \bar{z}_1, \cdots, \Delta_i + \Delta_j, z_j, \bar{z}_j, \cdots) + \mathrm{order}(z_{ij}^0). \tag{56}$$

Since this holds in any celestial amplitude, it implies the leading OPE between two positive-helicity gravitons is

$$G^+_{\Delta_1}(z_1, \bar{z}_1)G^+_{\Delta_2}(z_2, \bar{z}_2) \sim -\frac{\kappa}{2} \frac{\bar{z}_{12}}{z_{12}} B(\Delta_1 - 1, \Delta_2 - 1)G^+_{\Delta_1+\Delta_2}(z_2, \bar{z}_2), \tag{57}$$

in agreement with (41). By similar arguments, one also finds the following leading OPE between opposite-helicity gravitons

$$G^+_{\Delta_1}(z_1, \bar{z}_1)G^-_{\Delta_2}(z_2, \bar{z}_2) \sim -\frac{\kappa}{2} \frac{\bar{z}_{12}}{z_{12}} B(\Delta_1 - 1, \Delta_2 + 3)G^-_{\Delta_1+\Delta_2}(z_2, \bar{z}_2), \tag{58}$$

again in agreement with (41).

4.2. *Gravitons and gluons*

In order to derive graviton–gluon OPEs from collinear limits of EYM amplitudes, we here derive the collinear limits of conventional momentum-space amplitudes.

We start with the general Stieberger–Taylor formula which relates a momentum-space amplitude of n gluons and one graviton to a sum over color-ordered partial amplitudes of $n+1$ gluons[46,i]

$$A_{s_1 \cdots s_n; \pm 2}(p_1, \cdots, p_n; p)$$

$$= -\frac{\kappa}{2} \sum_{\ell=1}^{n-1} (\varepsilon^{\pm}(p) \cdot \chi_\ell) A_{s_1 \cdots s_\ell \pm 1 \, s_{\ell+1} \cdots s_n}(p_1, \cdots, p_\ell, p, p_{\ell+1}, \cdots, p_n), \tag{59}$$

where p_i, $i = 1, ..., n$ are the momenta of the gluons, p is the momentum of the graviton, $\varepsilon(p)$ is the polarization of a gluon of momentum p and

$$\chi_\ell = \sum_{k=1}^{\ell} p_k. \tag{60}$$

This formula allows us to determine collinear graviton–gluon limits from collinear gluon limits. The known leading collinear behavior of gluon amplitudes arises from adjacent gluons in color-ordered partial amplitudes[47]

$$\lim_{z_{ij} \to 0} A_{s_1 \cdots s_n}(p_1, \cdots, p_i, p_j, \cdots, p_n) \longrightarrow$$

$$\sum_{s = \pm 1} \mathrm{Split}^s_{s_i s_j}(p_i, p_j) A_{s_1 \cdots s \cdots s_n}(p_1, \cdots, P, \cdots, p_n), \tag{61}$$

where P was defined in (50) and the non-vanishing $\mathrm{Split}^s_{s_i s_j}(p_i, p_j)$ for collinear gluons are given by

$$\mathrm{Split}^1_{11}(p_i, p_j) = \frac{1}{z_{ij}} \frac{\omega_P}{\omega_i \omega_j}, \qquad \mathrm{Split}^{-1}_{1-1}(p_i, p_j) = \frac{1}{z_{ij}} \frac{\omega_j}{\omega_i \omega_P}. \tag{62}$$

Consider the collinear limit between a positive-helicity gluon of momentum p_i and a positive-helicity graviton. In the collinear limit, the leading-order contributions from the right hand side of (59) are just the two terms where the gluon of momentum p, which replaces the graviton, is adjacent to the ith gluon:

$$\lim_{z_i - z \to 0} A_{s_1 \cdots 1 \cdots s_n; 2}(p_1, \cdots, p_i, \cdots, p_n; p)$$

$$\longrightarrow -\frac{\kappa}{2} \left[\left(\varepsilon^+(p) \cdot \chi_{i-1} \right) A_{s_1 \cdots s_{i-1} \, 1 \, s_i \cdots s_n}(p_1, \cdots, p_{i-1}, p, p_i, \cdots, p_n) \right.$$

$$\left. + \left(\varepsilon^+(p) \cdot \chi_i \right) A_{s_1 \cdots s_i \, 1 \, s_{i+1} \cdots s_n}(p_1, \cdots, p_i, p, p_{i+1}, \cdots, p_n) \right] \tag{63}$$

$$\longrightarrow -\frac{\kappa}{2} \varepsilon^+(p) \cdot (\chi_i - \chi_{i-1}) \frac{1}{z_i - z} \frac{\omega_P}{\omega_i \omega}$$

$$\times A_{s_1 \cdots s_{i-1} \, 1 \, s_{i+1} \cdots s_n}(p_1, \cdots, p_{i-1}, P, p_{i+1}, \cdots, p_n).$$

---

[i]Note that $\varepsilon^{\pm}(p) \cdot \chi_n = -\varepsilon^{\pm}(p) \cdot p = 0$ by momentum conservation, hence the sum in (59) can be taken from 1 to n.

We use (60) to further simplify

$$\chi_i - \chi_{i-1} = p_i \tag{64}$$

and using (10),

$$\varepsilon^+(p) \cdot (-\chi_{i-1} + \chi_i) = \varepsilon^+(p) \cdot p_i = \omega_i(\bar{z}_i - \bar{z}). \tag{65}$$

Putting it all together, we obtain the following collinear limit for a positive-helicity gluon and graviton

$$\lim_{z_i - z \to 0} A_{s_1 \cdots 1 \cdots s_n ; 2}(p_1, \cdots, p_i, \cdots, p_n; p) \longrightarrow$$

$$-\frac{\kappa}{2} \frac{\bar{z}_i - \bar{z}}{z_i - z} \frac{\omega P}{\omega} A_{s_1 \cdots s_{i-1} \, 1 \, s_{i+1} \cdots s_n}(p_1, \cdots p_{i-1}, P, p_{i+1}, \cdots, p_n). \tag{66}$$

By similar arguments, keeping only singular terms in $z_i - z$, we obtain the following collinear graviton–gluon limit for the mixed-helicity case

$$\lim_{z_i - z \to 0} A_{s_1 \cdots -1 \cdots s_n ; 2}(p_1, \cdots, p_i, \cdots, p_n; p) \longrightarrow$$

$$-\frac{\kappa}{2} \frac{\bar{z}_i - \bar{z}}{z_i - z} \frac{\omega_i^2}{\omega \omega_P} A_{s_1 \cdots s_{i-1} \, -1 \, s_{i+1} \cdots s_n}(p_1, \cdots p_{i-1}, P, p_{i+1}, \cdots, p_n). \tag{67}$$

Taking Mellin transforms, we find the leading OPEs

$$G^+_{\Delta_1}(z_1, \bar{z}_1) O^{+a}_{\Delta_2}(z_2, \bar{z}_2) \sim -\frac{\kappa}{2} \frac{\bar{z}_{12}}{z_{12}} B(\Delta_1 - 1, \Delta_2) O^{+a}_{\Delta_1 + \Delta_2}(z_2, \bar{z}_2),$$

$$G^+_{\Delta_1}(z_1, \bar{z}_1) O^{-a}_{\Delta_2}(z_2, \bar{z}_2) \sim -\frac{\kappa}{2} \frac{\bar{z}_{12}}{z_{12}} B(\Delta_1 - 1, \Delta_2 + 2) O^{-a}_{\Delta_1 + \Delta_2}(z_2, \bar{z}_2), \tag{68}$$

which agree with Eq. (45).

Now we compute the graviton contribution to the mixed-helicity gluon OPE. Since we are interested in the contribution from G^- to the O^+O^- OPE, consider the on-shell vertex

$$V(p_1, p_2, p_3) = -i\kappa\delta^{a_1 a_2} \left[(\varepsilon_1^+ \cdot \varepsilon_2^-)(\varepsilon_3^+ \cdot p_1)(\varepsilon_3^+ \cdot p_2) - (\varepsilon_1^+ \cdot p_2)(\varepsilon_3^+ \cdot p_1)(\varepsilon_2^- \cdot \varepsilon_3^+) \right]. \tag{69}$$

Here $\varepsilon_1, \varepsilon_2$ are the polarizations of the positive and negative-helicity gluons of momenta p_1, p_2 and colors a_1, a_2 respectively. $\varepsilon^{\pm}_{3\mu\nu} = \varepsilon^{\pm}_{3\mu}\varepsilon^{\pm}_{3\nu}$ is the graviton polarization. Evaluating in our parametrization (2) and (9), the on-shell vertex becomes

$$V(p_1, p_2, p_3) = -i\kappa\delta^{a_1 a_2} \omega_1 \omega_2 \bar{z}_{13}^2. \tag{70}$$

In $(2,2)$ signature, the result is non-vanishing and upon taking $z_1 = z_2 = z_3$, momentum conservation reduces to

$$\omega_1 + \omega_2 + \omega_3 = 0,$$

$$\omega_1 \bar{z}_1 + \omega_2 \bar{z}_2 + \omega_3 \bar{z}_3 = 0. \tag{71}$$

Solving for $\bar{z}_3$, we find

$$\bar{z}_3 = \frac{\omega_1}{\omega_1 + \omega_2}\bar{z}_1 + \frac{\omega_2}{\omega_1 + \omega_2}\bar{z}_2 \quad \Rightarrow \quad \bar{z}_{13} = \frac{\omega_2}{\omega_1 + \omega_2}\bar{z}_{12}. \tag{72}$$

Then, accounting for the graviton propagator, we find that the collinear singularity for opposite-helicity gluons due to the EYM vertex (69) is

$$\text{Split}_{1-1}^{-2}(p_1, p_2) = \frac{\kappa}{2} \frac{\bar{z}_{12}}{z_{12}} \frac{\omega_2^2}{\omega_P^2}. \tag{73}$$

Taking a Mellin transform, we deduce that the $O_{\Delta_1}^{+a}(z_1)O_{\Delta_2}^{-b}(z_2)$ OPE contains a term of the form

$$\delta^{ab} \frac{\kappa}{2} \frac{\bar{z}_{12}}{z_{12}} B(\Delta_1, \Delta_2 + 2) G_{\Delta_1 + \Delta_2}^-(z_2, \bar{z}_2), \tag{74}$$

in agreement with the symmetry-derived result (48).

5. Celestial incoming and outgoing OPEs

In this section we generalize our results to account for the presence of both incoming and outgoing particles. We introduce celestial operators

$$\mathcal{O}_{\Delta_k, s_k}^{\epsilon_k}(z_k, \bar{z}_k) = \int_0^\infty d\omega_k \, \omega_k^{\Delta_k - 1} \mathcal{O}_{s_k}(\epsilon_k \omega_k, z_k, \bar{z}_k) \tag{75}$$

carrying an additional label $\epsilon_k = \pm 1$ which distinguishes between outgoing and incoming states respectively. $\mathcal{O}_{s_k}(\epsilon_k \omega_k, z_k, \bar{z}_k)$ are operators associated to the standard 'out' and 'in' momentum eigenstates through the parametrization (2). Since the action of the translation operator on 'in' and 'out' momentum eigenstates differs by a sign, the action of $\mathcal{P}$ on the celestial operators generalizes to

$$\delta_\mathcal{P} \mathcal{O}_{\Delta, s}^\epsilon(z, \bar{z}) = \epsilon \mathcal{O}_{\Delta+1, s}^\epsilon(z, \bar{z}). \tag{76}$$

Note, since the 'in' and 'out' labels of asymptotic states are directly related to charges of the corresponding operators under a global symmetry of the celestial CFT, these labels are naturally a part of the celestial CFT data.

Likewise, since the inverse of $\mathcal{P}$ appears in the relevant subleading gluon and subsubleading graviton symmetry actions (see Appendices A.3 and A.4), the actions of these symmetries (A.17) and (A.25) generalize to

$$\bar{\delta}^a \mathcal{O}_{\Delta_k, s_k}^{\epsilon_k}(z_k, \bar{z}_k) = \left[-\epsilon_k \left(\Delta_k - s_k - 1 + \bar{z}_k \partial_{\bar{z}_k} \right) T_k^a \mathcal{P}_k^{-1} \right.$$
$$\left. - \frac{\kappa}{2} \bar{z}_k \mathcal{F}_k^{+a} + \frac{\kappa}{2} \bar{z}_k \mathcal{G}_k^{+a} \right] \mathcal{O}_{\Delta_k, s_k}^{\epsilon_k}(z_k, \bar{z}_k),$$
$$\bar{\delta} \mathcal{O}_{\Delta_k, s_k}^{\epsilon_k}(z_k, \bar{z}_k) = -\frac{\kappa}{4} \epsilon_k \left[(\Delta_k - s_k)(\Delta_k - s_k - 1) \right.$$
$$\left. +4(\Delta_k - s_k)\bar{z}_k \partial_{\bar{z}_k} + 3\bar{z}_k^2 \partial_{\bar{z}_k}^2 \right] \mathcal{O}_{\Delta_k - 1, s_k}^{\epsilon_k}(z_k, \bar{z}_k), \tag{77}$$

where $\mathcal{F}$ and $\mathcal{G}$ are defined in Appendix A.3.

5.1. *Gluon OPEs from asymptotic symmetries*

We now determine the OPE coefficients among outgoing and incoming gluons from (77). The case when both operators are incoming is mostly identical to the previously studied case with both operators outgoing since the symmetry constraints remain unchanged. That is, up to normalization, these OPE coefficients are solved by the Euler beta functions (29) and (32) for gluons of identical and opposite helicity respectively. We therefore consider the OPEs of outgoing and incoming gluons where as we will see, the constraints from symmetry differ.

Generalizing (15) and (16), we begin with the ansatz

$$
O_{\Delta_1}^{+a,\epsilon}(z_1,\bar{z}_1)O_{\Delta_2}^{+b,-\epsilon}(z_2,\bar{z}_2) \sim -\epsilon\frac{if^{ab}_{\ \ c}}{z_{12}}\left[C'(\Delta_1,\Delta_2)O_{\Delta_1+\Delta_2-1}^{+c,\epsilon}(z_2,\bar{z}_2)\right.
$$
$$
\left.+C''(\Delta_1,\Delta_2)O_{\Delta_1+\Delta_2-1}^{+c,-\epsilon}(z_2,\bar{z}_2)\right], \tag{78}
$$

$$
O_{\Delta_1}^{+a,\epsilon}(z_1,\bar{z}_1)O_{\Delta_2}^{-b,-\epsilon}(z_2,\bar{z}_2) \sim -\epsilon\frac{if^{ab}_{\ \ c}}{z_{12}}\left[D'(\Delta_1,\Delta_2)O_{\Delta_1+\Delta_2-1}^{-c,\epsilon}(z_2,\bar{z}_2)\right.
$$
$$
\left.+D''(\Delta_1,\Delta_2)O_{\Delta_1+\Delta_2-1}^{-c,-\epsilon}(z_2,\bar{z}_2)\right]. \tag{79}
$$

Using the generalized action of the translation operator (76), we find the OPE coefficients must obey

$$
C'(\Delta_1+1,\Delta_2)-C'(\Delta_1,\Delta_2+1)=C'(\Delta_1,\Delta_2),
$$
$$
C''(\Delta_1+1,\Delta_2)-C''(\Delta_1,\Delta_2+1)=-C''(\Delta_1,\Delta_2), \tag{80}
$$

and

$$
D'(\Delta_1+1,\Delta_2)-D'(\Delta_1,\Delta_2+1)=D'(\Delta_1,\Delta_2),
$$
$$
D''(\Delta_1+1,\Delta_2)-D''(\Delta_1,\Delta_2+1)=-D''(\Delta_1,\Delta_2). \tag{81}
$$

As before, these recursion relations do not fully constrain the answer, so we turn to the subleading soft gluon symmetry. Constraining (78) with the symmetry in (77) and following the logic in Sec. 3.1, we obtain the following relations

$$
(\Delta_1-2)C'(\Delta_1-1,\Delta_2)f^a_{\ dc}f^{cb}_{\ \ e}-(\Delta_2-2)C'(\Delta_1,\Delta_2-1)f^b_{\ dc}f^{ac}_{\ \ e}
$$
$$
=(\Delta_1+\Delta_2-3)C'(\Delta_1,\Delta_2)f^{ab}_{\ \ c}f^c_{\ de},
$$
$$
(\Delta_1-2)C''(\Delta_1-1,\Delta_2)f^a_{\ dc}f^{cb}_{\ \ e}-(\Delta_2-2)C''(\Delta_1,\Delta_2-1)f^b_{\ dc}f^{ac}_{\ \ e}
$$
$$
=-(\Delta_1+\Delta_2-3)C''(\Delta_1,\Delta_2)f^{ab}_{\ \ c}f^c_{\ de}, \tag{82}
$$

which using the Jacobi identity reduce to

$$
(\Delta_1-2)C'(\Delta_1-1,\Delta_2)=(\Delta_1+\Delta_2-3)C'(\Delta_1,\Delta_2),
$$
$$
-(\Delta_2-2)C'(\Delta_1,\Delta_2-1)=(\Delta_1+\Delta_2-3)C'(\Delta_1,\Delta_2), \tag{83}
$$

and

$$
(\Delta_1-2)C''(\Delta_1-1,\Delta_2)=-(\Delta_1+\Delta_2-3)C''(\Delta_1,\Delta_2),
$$
$$
(\Delta_2-2)C''(\Delta_1,\Delta_2-1)=(\Delta_1+\Delta_2-3)C''(\Delta_1,\Delta_2). \tag{84}
$$

By shifting the arguments and taking a linear combination of the two constraints for each OPE coefficient, one can verify that these new recursion relations imply the modified recursion relation (80) from translation symmetry. (83) and (84) are solved by

$$
\begin{aligned}
C'(\Delta_1, \Delta_2) &= -B(\Delta_2 - 1, 3 - \Delta_1 - \Delta_2), \\
C''(\Delta_1, \Delta_2) &= B(\Delta_1 - 1, 3 - \Delta_1 - \Delta_2),
\end{aligned}
\tag{85}
$$

where we have used the celestial soft gluon theorem, generalized for incoming and outgoing operators,

$$
\lim_{\Delta_1 \to 1} O_{\Delta_1}^{+a,\epsilon}(z_1, \bar{z}_1) O_{\Delta_2}^{+b,-\epsilon}(z_2, \bar{z}_2) = -\epsilon \frac{i f^{ab}{}_c}{z_{12}} \frac{1}{\Delta_1 - 1} O_{\Delta_2}^{+c,-\epsilon}(z_2, \bar{z}_2)
\tag{86}
$$

to fix the normalization. Note that both C' and C'' are fixed by (86) due to the symmetry of (78) under exchange of labels which implies that they have soft poles at $\Delta_2, \Delta_1 = 1$ respectively, as seen explicitly in (85). As we will see now, this will not usually be the case and a more general argument will be needed.

For opposite-helicity gluons, the soft gluon symmetry constraints on (79) reduce to

$$
\begin{aligned}
(\Delta_1 - 2) D'(\Delta_1 - 1, \Delta_2) &= (\Delta_1 + \Delta_2 - 1) D'(\Delta_1, \Delta_2), \\
-\Delta_2 D'(\Delta_1, \Delta_2 - 1) &= (\Delta_1 + \Delta_2 - 1) D'(\Delta_1, \Delta_2),
\end{aligned}
\tag{87}
$$

$$
\begin{aligned}
(\Delta_1 - 2) D''(\Delta_1 - 1, \Delta_2) &= -(\Delta_1 + \Delta_2 - 1) D''(\Delta_1, \Delta_2), \\
\Delta_2 D''(\Delta_1, \Delta_2 - 1) &= (\Delta_1 + \Delta_2 - 1) D''(\Delta_1, \Delta_2).
\end{aligned}
\tag{88}
$$

The leading soft gluon theorem implies

$$
\lim_{\Delta_1 \to 1} O_{\Delta_1}^{+a,\epsilon}(z_1, \bar{z}_1) O_{\Delta_2}^{-b,-\epsilon}(z_2, \bar{z}_2) = -\epsilon \frac{i f^{ab}{}_c}{z_{12}} \frac{1}{\Delta_1 - 1} O_{\Delta_2}^{-c,-\epsilon}(z_2, \bar{z}_2),
\tag{89}
$$

which together with the recursion relation (88) uniquely fixes

$$
D''(\Delta_1, \Delta_2) = B(\Delta_1 - 1, 1 - \Delta_1 - \Delta_2).
\tag{90}
$$

On the other hand, (87) is solved by

$$
D'(\Delta_1, \Delta_2) = \alpha B(\Delta_2 + 1, 1 - \Delta_1 - \Delta_2)
\tag{91}
$$

for some yet-to-be determined constant α. To fix α, consider the mixed-helicity gluon OPE, evaluated at $\Delta_1 = \Delta_2 \equiv \Delta$,

$$
O_\Delta^{+a,\epsilon}(z_1, \bar{z}_1) O_\Delta^{-b,-\epsilon}(z_2, \bar{z}_2)
$$

$$
\sim \frac{-i f^{ab}{}_c}{z_{12}} \epsilon \left[\alpha B(\Delta + 1, 1 - 2\Delta) O_{2\Delta-1}^{-c,\epsilon}(z_2, \bar{z}_2) + B(\Delta - 1, 1 - 2\Delta) O_{2\Delta-1}^{-c,-\epsilon}(z_2, \bar{z}_2) \right].
\tag{92}
$$

Taking $\Delta \to 1$, which corresponds to a double soft limit of a scattering amplitude, we obtain an OPE among celestially soft operators

$$
\lim_{\Delta \to 1} (\Delta - 1)^2 O_\Delta^{+a,\epsilon}(z_1, \bar{z}_1) O_\Delta^{-b,-\epsilon}(z_2, \bar{z}_2)
$$

$$
\sim \frac{-i f^{ab}{}_c}{z_{12}} \epsilon \frac{1}{2} \lim_{\Delta \to 1} \left[\alpha (\Delta - 1) O_{2\Delta-1}^{-c,\epsilon}(z_2, \bar{z}_2) + (\Delta - 1) O_{2\Delta-1}^{-c,-\epsilon}(z_2, \bar{z}_2) \right].
\tag{93}
$$

The above OPE is related to another OPE for celestially soft operators

$$\lim_{\Delta \to 1} (\Delta - 1)^2 O_\Delta^{+a,\epsilon}(z_1, \bar{z}_1) O_\Delta^{-b,\epsilon}(z_2, \bar{z}_2) \sim -if^{ab}{}_c \frac{\epsilon}{z_{12}} \lim_{\Delta \to 1} (\Delta - 1) O_{2\Delta-1}^{-c,\epsilon}(z_2, \bar{z}_2)$$

(94)

by the crossing relation for soft modes,[31] which on the celestial sphere takes the form

$$\lim_{\Delta \to 1} (\Delta - 1) O_\Delta^{\pm a,\epsilon}(z, \bar{z}) = -\lim_{\Delta \to 1} (\Delta - 1) O_\Delta^{\pm a,-\epsilon}(z, \bar{z}).$$

(95)

Comparing the two, we find

$$\alpha = -1.$$

(96)

5.2. Gluon OPEs from collinear singularities

We now confirm the symmetry-derived results from a momentum-space amplitude calculation. As before, the OPE coefficients can be derived by Mellin transforming the collinear splitting functions. For incoming and outgoing gluons these take the general form

$$\mathrm{Split}_{1s_2}^{s_2}(p_1, p_2) = \frac{1}{z_{12}} (\epsilon_1 \omega_1)^\alpha (\epsilon_2 \omega_2)^\beta (\epsilon_1 \omega_1 + \epsilon_2 \omega_2)^\gamma.$$

(97)

To evaluate

$$\int_0^\infty d\omega_1 \int_0^\infty d\omega_2 \, \omega_1^{\Delta_1-1} \omega_2^{\Delta_2-1} \mathrm{Split}_{1s_2}^{s_2}(p_1, p_2),$$

(98)

it is convenient to make the following change of variables

$$\omega_1 = (1 - \epsilon_1 \epsilon_2 t) \omega_P, \qquad \omega_2 = t\omega_P,$$

(99)

where

$$\omega_1 + \epsilon_1 \epsilon_2 \omega_2 = \omega_P.$$

(100)

For $\epsilon_1 \epsilon_2 = -1$, (98) splits into two integrals such that the celestial OPE takes the form

$$O_{\Delta_1}^{+a,\epsilon_1}(z_1, \bar{z}_1) O_{\Delta_2}^{\pm b,\epsilon_2}(z_2, \bar{z}_2)$$

$$\sim \frac{-if^{ab}{}_c}{z_{12}} \epsilon_1^{\alpha+\gamma} \epsilon_2^\beta \left[\int_0^\infty d\omega_P \int_0^\infty dt \, (1+t)^{\Delta_1-1+\alpha} t^{\Delta_2-1+\beta} \right.$$

$$\times \omega_P^{\Delta_1+\Delta_2+\alpha+\beta+\gamma-1} O^{\pm c}(\epsilon_1 \omega_P, z_2, \bar{z}_2)$$

$$- \int_{-\infty}^0 d\omega_P \int_{-\infty}^{-1} dt \, (1+t)^{\Delta_1-1+\alpha} t^{\Delta_2-1+\beta}$$

$$\left. \times \omega_P^{\Delta_1+\Delta_2+\alpha+\beta+\gamma-1} O^{\pm c}(\epsilon_1 \omega_P, z_2, \bar{z}_2) \right]$$

$$= \frac{-if^{ab}{}_c}{z_{12}} \epsilon_1^{\alpha+\gamma} \epsilon_2^\beta \left[\int_0^\infty dt \, (1+t)^{\Delta_1-1+\alpha} t^{\Delta_2-1+\beta} O_{\Delta_1+\Delta_2+\alpha+\beta+\gamma}^{\pm c,\epsilon_1}(z_2, \bar{z}_2) \right.$$

$$\left. +(-1)^\gamma \int_0^\infty dt \, t^{\Delta_1-1+\alpha} (1+t)^{\Delta_2-1+\beta} O_{\Delta_1+\Delta_2+\alpha+\beta+\gamma}^{\pm c,-\epsilon_1}(z_2, \bar{z}_2) \right],$$

(101)

where to obtain the second line, we performed the change of variables $\omega_P \to -\omega_P$ and $t \to -(1+t)$ on the second term. Upon making a further change of variables $t = \dfrac{u}{1-u}$, we find the remaining t-integrals once again take the form (55) so that the OPE coefficients are given by Euler beta functions

$$O^{+a,\epsilon_1}_{\Delta_1}(z_1,\bar{z}_1)O^{\pm b,\epsilon_2}_{\Delta_2}(z_2,\bar{z}_2)$$

$$\sim \frac{-if^{ab}{}_c}{z_{12}}\epsilon_1^{\alpha+\gamma}\epsilon_2^{\beta}\left[B(\Delta_2+\beta,1-\Delta_1-\Delta_2-\alpha-\beta)O^{\pm c,\epsilon_1}_{\Delta_1+\Delta_2+\alpha+\beta+\gamma}(z_2,\bar{z}_2)\right.$$

$$\left.+(-1)^\gamma\, B(\Delta_1+\alpha,1-\Delta_1-\Delta_2-\alpha-\beta)O^{\pm c,-\epsilon_1}_{\Delta_1+\Delta_2+\alpha+\beta+\gamma}(z_2,\bar{z}_2)\right].$$

$$(102)$$

For equal-helicity gluons $\alpha=\beta=-\gamma=-1$ and so the in/out OPE is

$$O^{+a,\epsilon}_{\Delta_1}(z_1,\bar{z}_1)O^{+b,-\epsilon}_{\Delta_2}(z_2,\bar{z}_2) \sim if^{ab}{}_c\frac{\epsilon}{z_{12}}\left[B(3-\Delta_1-\Delta_2,\Delta_2-1)O^{+c,\epsilon}_{\Delta_1+\Delta_2-1}(z_2,\bar{z}_2)\right.$$

$$\left.-B(\Delta_1-1,3-\Delta_1-\Delta_2)O^{+c,-\epsilon}_{\Delta_1+\Delta_2-1}(z_2,\bar{z}_2)\right].$$

$$(103)$$

For opposite-helicity gluons $\alpha=-\beta=\gamma=-1$ and we find

$$O^{+a,\epsilon}_{\Delta_1}(z_1,\bar{z}_1)O^{-b,-\epsilon}_{\Delta_2}(z_2,\bar{z}_2) \sim if^{ab}{}_c\frac{\epsilon}{z_{12}}\left[B(-\Delta_1-\Delta_2+1,\Delta_2+1)O^{-c,\epsilon}_{\Delta_1+\Delta_2-1}(z_2,\bar{z}_2)\right.$$

$$\left.-B(\Delta_1-1,1-\Delta_1-\Delta_2)O^{-c,-\epsilon}_{\Delta_1+\Delta_2-1}(z_2,\bar{z}_2)\right],$$

$$(104)$$

which agree with the symmetry-derived OPEs. Analogous computations yield the graviton and gluon–graviton in/out OPEs. We summarize the results in the following section.

5.3. *Summary of OPE coefficients*

In summary, all the nonzero leading z_{12} poles for all possible configurations of incoming and outgoing gluon and graviton OPEs are determined by the asymptotic symmetries in tree-level EYM. The equal-helicity gluon OPEs are

$$O^{+a,\epsilon}_{\Delta_1}(z_1,\bar{z}_1)O^{+b,\epsilon}_{\Delta_2}(z_2,\bar{z}_2) \sim \frac{-if^{ab}{}_c}{z_{12}}\epsilon B(\Delta_1-1,\Delta_2-1)O^{+c,\epsilon}_{\Delta_1+\Delta_2-1}(z_2,\bar{z}_2),$$

$$O^{+a,\epsilon}_{\Delta_1}(z_1,\bar{z}_1)O^{+b,-\epsilon}_{\Delta_2}(z_2,\bar{z}_2) \sim \frac{-if^{ab}{}_c}{z_{12}}\epsilon\left[-B(\Delta_2-1,3-\Delta_1-\Delta_2)O^{+c,\epsilon}_{\Delta_1+\Delta_2-1}(z_2,\bar{z}_2)\right.$$

$$\left.+B(\Delta_1-1,3-\Delta_1-\Delta_2)O^{+c,-\epsilon}_{\Delta_1+\Delta_2-1}(z_2,\bar{z}_2)\right].$$

$$(105)$$

The mixed-helicity gluon OPEs are

$$O^{+a,\epsilon}_{\Delta_1}(z_1,\bar{z}_1)O^{-b,\epsilon}_{\Delta_2}(z_2,\bar{z}_2) \sim \frac{-if^{ab}_{c}}{z_{12}}\epsilon B(\Delta_1-1,\Delta_2+1)O^{-c,\epsilon}_{\Delta_1+\Delta_2-1}(z_2,\bar{z}_2)$$

$$+\frac{\kappa}{2}\frac{\bar{z}_{12}}{z_{12}}\delta^{ab}B(\Delta_1,\Delta_2+2)G^{-,\epsilon}_{\Delta_1+\Delta_2}(z_2,\bar{z}_2),$$

$$O^{+a,\epsilon}_{\Delta_1}(z_1,\bar{z}_1)O^{-b,-\epsilon}_{\Delta_2}(z_2,\bar{z}_2) \sim \frac{-if^{ab}_{c}}{z_{12}}\epsilon\left[-B(\Delta_2+1,1-\Delta_1-\Delta_2)O^{-c,\epsilon}_{\Delta_1+\Delta_2-1}(z_2,\bar{z}_2)\right.$$

$$\left.+B(\Delta_1-1,1-\Delta_1-\Delta_2)O^{-c,-\epsilon}_{\Delta_1+\Delta_2-1}(z_2,\bar{z}_2)\right]$$

$$+\frac{\kappa}{2}\frac{\bar{z}_{12}}{z_{12}}\delta^{ab}\left[B(\Delta_2+2,-1-\Delta_1-\Delta_2)G^{-,\epsilon}_{\Delta_1+\Delta_2}(z_2,\bar{z}_2)\right.$$

$$\left.+B(\Delta_1,-1-\Delta_1-\Delta_2)G^{-,-\epsilon}_{\Delta_1+\Delta_2}(z_2,\bar{z}_2)\right].$$

$$(106)$$

The graviton OPEs are

$$G^{+,\epsilon}_{\Delta_1}(z_1,\bar{z}_1)G^{\pm,\epsilon}_{\Delta_2}(z_2,\bar{z}_2) \sim -\frac{\kappa}{2}\frac{\bar{z}_{12}}{z_{12}}B(\Delta_1-1,\Delta_2+1\mp 2)G^{\pm,\epsilon}_{\Delta_1+\Delta_2}(z_2,\bar{z}_2),$$

$$G^{+,\epsilon}_{\Delta_1}(z_1,\bar{z}_1)G^{\pm,-\epsilon}_{\Delta_2}(z_2,\bar{z}_2) \sim \frac{\kappa}{2}\frac{\bar{z}_{12}}{z_{12}}\left[B(\Delta_2+1\mp 2,1\pm 2-\Delta_1-\Delta_2)G^{\pm,\epsilon}_{\Delta_1+\Delta_2}(z_2,\bar{z}_2)\right.$$

$$\left.+B(\Delta_1-1,1\pm 2-\Delta_1-\Delta_2)G^{\pm,-\epsilon}_{\Delta_1+\Delta_2}(z_2,\bar{z}_2)\right].$$

$$(107)$$

The gluon–graviton OPEs are

$$G^{+,\epsilon}_{\Delta_1}(z_1,\bar{z}_1)O^{\pm a,\epsilon}_{\Delta_2}(z_2,\bar{z}_2) \sim -\frac{\kappa}{2}\frac{\bar{z}_{12}}{z_{12}}B(\Delta_1-1,\Delta_2+1\mp 1)O^{\pm a,\epsilon}_{\Delta_1+\Delta_2}(z_2,\bar{z}_2),$$

$$G^{+,\epsilon}_{\Delta_1}(z_1,\bar{z}_1)O^{\pm a,-\epsilon}_{\Delta_2}(z_2,\bar{z}_2) \sim -\frac{\kappa}{2}\frac{\bar{z}_{12}}{z_{12}}\left[B(\Delta_2+1\mp 1,1\pm 1-\Delta_1-\Delta_2)O^{\pm a,\epsilon}_{\Delta_1+\Delta_2}(z_2,\bar{z}_2)\right.$$

$$\left.-B(\Delta_1-1,1\pm 1-\Delta_1-\Delta_2)O^{\pm a,-\epsilon}_{\Delta_1+\Delta_2}(z_2,\bar{z}_2)\right].$$

$$(108)$$

From (8), we recall a factor of g_{YM} is absorbed in f^{ab}_{c}. The $\bar{z}_{12} \to 0$ celestial OPEs are obtained in a similar way by imposing the δ symmetry instead.

The presence of higher-dimension operators due to quantum, stringy or other corrections is expected to augment this list with the finite number of additions allowed by the general formula (13). A list of all possible corrections in theories with only gluons and gravitons is given in Appendix A.2.

Acknowledgements

We are grateful to Nima Arkani-Hamed, Prahar Mitra, Sabrina Pasterski, Andrea Puhm, Shu-Heng Shao, Mark Spradlin and Tom Taylor for useful conversations. This work was supported by NSF grant 1205550 and the John Templeton Foundation. M.P. acknowledges the support of a Junior Fellowship at the Harvard Society of Fellows.

A.S. is deeply indebted to Roman Jackiw for explaining to him the operator product expansion, conformal symmetry and a pragmatic approach to physics in general.

A.1. Celestial OPEs from bulk three-point vertices

In this appendix we relate the conformal weights of the operators which are allowed to appear in the OPE of two conformal primaries to the bulk dimensions of the corresponding three-point vertices. We consider a bulk three-point vertex among gluons and gravitons which schematically takes the form

$$V = \partial^m \Phi_1(x)\Phi_2(x)\Phi_3(x), \tag{A.1}$$

where $\Phi_1, \Phi_2, \Phi_3 \in \{A_\mu, h_{\mu\nu}\}$, and we omitted Lorentz indices which should be contracted accordingly. m is the total number of derivatives in the interaction, which are appropriately distributed among Φ_1, Φ_2, Φ_3. Since both gluons and gravitons have dimension 1, the net dimension of the vertex is

$$d_V = 3 + m. \tag{A.2}$$

Suppose Φ_1, Φ_2 are taken to be outgoing external legs (on-shell states). In momentum space, each derivative is associated with a factor of momentum. Upon parametrizing momenta as in (2), Mellin transforming with respect to ω_1 and ω_2 and taking the collinear limit $z_{12} \to 0$, the celestial amplitude takes the general form

$$\tilde{A} = \sum_{\alpha,\beta} \int_0^\infty d\omega_1 \int_0^\infty d\omega_2\, \omega_1^{\Delta_1-1}\omega_2^{\Delta_2-1} \frac{\omega_1^{m+\alpha}\omega_2^\beta}{\omega_P^{\alpha+\beta}} \frac{1}{\omega_1\omega_2} F_{\alpha,\beta}(z_1, \bar{z}_1, z_2, \bar{z}_2; \cdots), \tag{A.3}$$

where we used momentum conservation and accounted for the Φ_3 propagator. Since we're working in a collinear expansion, $F_{\alpha,\beta}$ depends only on ω_P, but not ω_1 or ω_2 independently. In general, the amplitude involves a sum over terms with different α, β. The details depend on the precise form of the interaction but turn out to be irrelevant in determining the scaling dimension of the allowed operators. Setting

$$\omega_1 = \omega_P t, \qquad \omega_2 = \omega_P(1-t), \tag{A.4}$$

the celestial amplitude becomes

$$\tilde{A} = \sum_{\alpha,\beta} B(\Delta_1 + m + \alpha - 1, \Delta_2 + \beta - 1) \int_0^\infty d\omega_P \omega_P^{\Delta_1 + \Delta_2 - 3 + m} F_{\alpha,\beta}(z_1, \bar{z}_1, z_2, \bar{z}_2; \omega_P, \cdots). \tag{A.5}$$

This allows one to read off the scaling dimension of the associated operator in the OPE expansion

$$\Delta_3 - 1 = \Delta_1 + \Delta_2 - 3 + m \implies \Delta_3 = \Delta_1 + \Delta_2 + d_V - 5, \tag{A.6}$$

where in the last equation we used (A.2). We therefore conclude that the primaries in the Φ_1, Φ_2 OPE can be classified according to the dimension of the possible corresponding bulk three-point vertices as in (13).

A.2. Higher order OPEs

There is a finite number of primaries which contribute to the OPE (13) to any finite order in the z_{12} expansion. To get a flavor of this, in this appendix we collect all possible single-pole or finite terms. For the gluon–gluon OPEs these are

$$O_{\Delta_1}^{+a}(z_1, \bar{z}_1) O_{\Delta_2}^{+b}(z_2, \bar{z}_2) : \frac{1}{z_{12}} O_{\Delta_1+\Delta_2-1}^{+c}(z_2, \bar{z}_2), \quad \frac{\bar{z}_{12}^2}{z_{12}} O_{\Delta_1+\Delta_2+1}^{-c}(z_2, \bar{z}_2),$$

$$\bar{z}_{12} O_{\Delta_1+\Delta_2+1}^{+c}(z_2, \bar{z}_2), \quad \bar{z}_{12}^3 O_{\Delta_1+\Delta_2+3}^{-c}(z_2, \bar{z}_2),$$

$$G_{\Delta_1+\Delta_2}^{+}(z_2, \bar{z}_2), \quad \frac{\bar{z}_{12}^3}{z_{12}} G_{\Delta_1+\Delta_2+2}^{-}(z_2, \bar{z}_2), \quad \bar{z}_{12}^4 G_{\Delta_1+\Delta_2+4}^{-}(z_2, \bar{z}_2),$$

$$O_{\Delta_1}^{+a}(z_1, \bar{z}_1) O_{\Delta_2}^{-b}(z_2, \bar{z}_2) : \frac{1}{z_{12}} O_{\Delta_1+\Delta_2-1}^{-c}(z_2, \bar{z}_2), \quad \bar{z}_{12} O_{\Delta_1+\Delta_2+1}^{-c}(z_2, \bar{z}_2),$$

$$\frac{\bar{z}_{12}}{z_{12}} G_{\Delta_1+\Delta_2}^{-}(z_2, \bar{z}_2), \quad \bar{z}_{12}^2 G_{\Delta_1+\Delta_2+2}^{-}(z_2, \bar{z}_2).$$

$$(A.7)$$

Operators on the right hand side of dimension $\Delta_1 + \Delta_2 - 1$ arise from the pure YM three-point vertices while those of dimension $\Delta_1 + \Delta_2$ from three-point vertices in EYM (excluding the three-gluon vertex). All other operators of dimensions $\Delta_1 + \Delta_2 + n$, $n = 1, ..., 4$ correspond to the following higher derivative vertices in order: $F^3, RF^2, \partial^2 F^3, \partial^2 RF^2$.

Similarly, the finite or single-pole terms in the graviton–graviton OPE are

$$G_{\Delta_1}^{+}(z_1, \bar{z}_1) G_{\Delta_2}^{+}(z_2, \bar{z}_2) : \frac{\bar{z}_{12}}{z_{12}} G_{\Delta_1+\Delta_2}^{+}(z_2, \bar{z}_2), \quad \bar{z}_{12}^2 G_{\Delta_1+\Delta_2+2}^{+}(z_2, \bar{z}_2),$$

$$\frac{\bar{z}_{12}^5}{z_{12}} G_{\Delta_1+\Delta_2+4}^{-}(z_2, \bar{z}_2), \quad \bar{z}_{12}^6 G_{\Delta_1+\Delta_2+6}^{-}(z_2, \bar{z}_2), \quad (A.8)$$

$$G_{\Delta_1}^{+}(z_1, \bar{z}_1) G_{\Delta_2}^{-}(z_2, \bar{z}_2) : \frac{\bar{z}_{12}}{z_{12}} G_{\Delta_1+\Delta_2}^{-}(z_2, \bar{z}_2), \quad \bar{z}_{12}^2 G_{\Delta_1+\Delta_2+2}^{-}(z_2, \bar{z}_2).$$

Operators of dimensions $\Delta_1 + \Delta_2 + n$, $n = 2, 4, 6$ arise from the following higher derivative vertices in order: $R^2, R^3, \partial^2 R^3$. The coefficient of the R^2 term can be eliminated by field redefinition.[48]

The finite or single-pole terms in the gluon–graviton OPEs are

$$G_{\Delta_1}^{+}(z_1, \bar{z}_1) O_{\Delta_2}^{+a}(z_2, \bar{z}_2) : \frac{\bar{z}_{12}}{z_{12}} O_{\Delta_1+\Delta_2}^{+a}(z_2, \bar{z}_2), \quad \frac{\bar{z}_{12}^3}{z_{12}} O_{\Delta_1+\Delta_2+2}^{-a}(z_2, \bar{z}_2),$$

$$\bar{z}_{12}^2 O_{\Delta_1+\Delta_2+2}^{+a}(z_2, \bar{z}_2), \quad \bar{z}_{12}^4 O_{\Delta_1+\Delta_2+4}^{-a}(z_2, \bar{z}_2),$$

$$G_{\Delta_1}^{+}(z_1, \bar{z}_1) O_{\Delta_2}^{-a}(z_2, \bar{z}_2) : \frac{\bar{z}_{12}}{z_{12}} O_{\Delta_1+\Delta_2}^{-a}(z_2, \bar{z}_2), \quad O_{\Delta_1+\Delta_2}^{+a}(z_2, \bar{z}_2), \quad \bar{z}_{12}^2 O_{\Delta_1+\Delta_2+2}^{-a}(z_2, \bar{z}_2).$$

$$(A.9)$$

Operators of dimensions $\Delta_1 + \Delta_2 + n$, $n = 2, 4$ correspond to the higher derivative vertices RF^2 and $\partial^2 RF^2$ respectively.

A.3. Subleading soft gluon symmetry

Tree-level gauge theory amplitudes obey the soft relation (see Ref. 34 and references therein)

$$A^a_{n+1}(p_1, ..., p_n; q) = \left(J^a_{(0)} + J^a_{(1)} \right) A_n(p_1, ..., p_n) + \mathcal{O}(q), \tag{A.10}$$

where $J^a_{(0)}, J^a_{(1)}$ are the leading and subleading gluon soft factors and we suppressed all color indices except for a, the one associated with the soft gluon. In this section we derive the action of the subleading soft gluon symmetry on outgoing gluons in a conformal basis. The subleading soft gluon operators are

$$J^{\pm a}_{(1)} = \sum_{k=1}^{n} i \frac{\varepsilon^{\pm}_{\mu} q_{\nu} J^{\mu\nu}_k}{q \cdot p_k} T^a_k, \tag{A.11}$$

where T^a_k are the generators of the non-abelian gauge group in representation k. In the parametrization (2) and (9), (A.11) takes the form

$$J^{-a}_{(1)} = \sum_{k=1}^{n} \frac{1}{\bar{z} - \bar{z}_k} \left(-\frac{s_k}{\omega_k} + \partial_{\omega_k} + \frac{z - z_k}{\omega_k} \partial_{z_k} \right) T^a_k,$$
$$J^{+a}_{(1)} = \sum_{k=1}^{n} \frac{1}{z - z_k} \left(\frac{s_k}{\omega_k} + \partial_{\omega_k} + \frac{\bar{z} - \bar{z}_k}{\omega_k} \partial_{\bar{z}_k} \right) T^a_k. \tag{A.12}$$

Upon performing a Mellin transform we find

$$J^{-a}_{(1)} = \sum_{k=1}^{n} \frac{1}{\bar{z} - \bar{z}_k} \left(-2h_k + 1 + (z - z_k)\partial_{z_k} \right) T^a_k \mathcal{P}^{-1}_k,$$
$$J^{+a}_{(1)} = \sum_{k=1}^{n} \frac{1}{z - z_k} \left(-2\bar{h}_k + 1 + (\bar{z} - \bar{z}_k)\partial_{\bar{z}_k} \right) T^a_k \mathcal{P}^{-1}_k, \tag{A.13}$$

where

$$h_k = \frac{1}{2}(\Delta_k + s_k), \qquad \bar{h}_k = \frac{1}{2}(\Delta_k - s_k), \tag{A.14}$$

and $\mathcal{P}^{-1}_k$ implements the inverse shift on the kth operator to the one defined in (17). Treating $z, \bar{z}$ as independent complex variables, we can define the operators

$$\delta^a \equiv \oint \frac{d\bar{z}}{2\pi i} J^{-a}_{(1)}(0, \bar{z}) = \lim_{\Delta \to 0} \Delta \oint \frac{d\bar{z}}{2\pi i} O^{-a}_{\Delta}(0, \bar{z}),$$
$$\bar{\delta}^a \equiv \oint \frac{dz}{2\pi i} J^{+a}_{(1)}(z, 0) = \lim_{\Delta \to 0} \Delta \oint \frac{dz}{2\pi i} O^{+a}_{\Delta}(z, 0), \tag{A.15}$$

which have the following action on gluons

$$\delta_a O^{\pm b}_{\Delta_k}(z_k, \bar{z}_k) = -if^b_{\ ac} (2h_k - 1 + z_k \partial_{z_k}) O^{\pm c}_{\Delta_k - 1}(z_k, \bar{z}_k),$$
$$\bar{\delta}_a O^{\pm b}_{\Delta_k}(z_k, \bar{z}_k) = -if^b_{\ ac} (2\bar{h}_k - 1 + \bar{z}_k \partial_{\bar{z}_k}) O^{\pm c}_{\Delta_k - 1}(z_k, \bar{z}_k). \tag{A.16}$$

Equations (A.16) define a global symmetry associated with the subleading soft gluon theorem and constrain the gluon OPE coefficients as in (27).

$J_{(1)}$ receives corrections in the presence of gravitons. These can be deduced from the vertex (69) in which case (A.12) becomes

$$
J_{(1)}^{-a} = \sum_{k=1}^{n} \frac{1}{\bar{z} - \bar{z}_k} \left(-\frac{s_k}{\omega_k} + \partial_{\omega_k} + \frac{z - z_k}{\omega_k} \partial_{z_k} \right) T_k^a + \frac{\kappa}{2} \frac{z - z_k}{\bar{z} - \bar{z}_k} \mathcal{F}_k^{-a} - \frac{\kappa}{2} \frac{z - z_k}{\bar{z} - \bar{z}_k} \mathcal{G}_k^{-a},
$$

$$
J_{(1)}^{+a} = \sum_{k=1}^{n} \frac{1}{z - z_k} \left(\frac{s_k}{\omega_k} + \partial_{\omega_k} + \frac{\bar{z} - \bar{z}_k}{\omega_k} \partial_{\bar{z}_k} \right) T_k^a + \frac{\kappa}{2} \frac{\bar{z} - \bar{z}_k}{z - z_k} \mathcal{F}_k^{+a} - \frac{\kappa}{2} \frac{\bar{z} - \bar{z}_k}{z - z_k} \mathcal{G}_k^{+a},
$$

$$
\text{(A.17)}
$$

where

$$
\mathcal{F}_k^{\pm a} | p_k, s_k = \mp 1, a_k \rangle = \delta^{a a_k} | p_k, s_k = \mp 2 \rangle, \qquad \mathcal{F}_k^{\pm a} | p_k, s_k = \pm 1, a_k \rangle = 0,
$$

$$
\mathcal{G}_k^{\pm a} | p_k, s_k = \pm 2 \rangle = \delta^{a a_k} | p_k, s_k = \pm 1, a_k \rangle, \qquad \mathcal{G}_k^{\pm a} | p_k, s_k = \mp 2 \rangle = 0.
$$

$$
\text{(A.18)}
$$

This implies that (46) obeys

$$
\lim_{\Delta_1 \to 0} \Delta_1 O_{\Delta_1}^{+a}(z_1, \bar{z}_1) O_{\Delta_2}^{-b}(z_2, \bar{z}_2) = \frac{i f^{ab}{}_c}{z_{12}} \Delta_2 O_{\Delta_2 - 1}^{-c}(z_2, \bar{z}_2) + \frac{\kappa}{2} \frac{\bar{z}_{12}}{z_{12}} \delta^{ab} G_{\Delta_2}^{-}(z_2, \bar{z}_2),
$$

$$
\text{(A.19)}
$$

which fixes the normalization of the graviton OPE coefficient.

A.4. Subsubleading soft graviton symmetry

In this section we derive the symmetry actions (36) and (43) (for outgoing particles) from the subsubleading soft graviton theorem.

Tree-level gravity amplitudes were shown in Ref. 1 to obey the following soft relation

$$
\mathcal{A}_{n+1}(p_1, ..., p_n; q) = \left(S_{(0)} + S_{(1)} + S_{(2)} \right) \mathcal{A}_n(p_1, ..., p_n) + \mathcal{O}(q^2), \qquad \text{(A.20)}
$$

where $S_{(0)}, S_{(1)}$ and $S_{(2)}$ are the leading, subleading and subsubleading soft factors respectively. In this appendix we focus on the subsubleading soft factor,

$$
S_{(2)} = -\frac{\kappa}{4} \sum_{k=1}^{n} \frac{\varepsilon_{\mu\nu}(q_\rho \mathcal{J}_k^{\rho\mu})(q_\sigma \mathcal{J}_k^{\sigma\nu})}{q \cdot p_k}, \qquad \text{(A.21)}
$$

where $\varepsilon_{\mu\nu}$ and q are the polarization and momentum of the soft graviton and $\mathcal{J}_i, p_i$ are the total angular momenta and momenta of the hard particles. Using the parametrizations (2) of momenta and the angular momentum operators in Ref. 35, (A.21) can be shown to reduce to[25,49]

$$
S_{(2)}^{-} = -\frac{\kappa}{4} \sum_{k=1}^{n} \frac{\omega}{\omega_k} \frac{1}{(z - z_k)(\bar{z} - \bar{z}_k)} \left[(z - z_k)(s_k - \omega_k \partial_{\omega_k}) - (z - z_k)^2 \partial_{z_k} \right]^2,
$$

$$
S_{(2)}^{+} = -\frac{\kappa}{4} \sum_{k=1}^{n} \frac{\omega}{\omega_k} \frac{1}{(z - z_k)(\bar{z} - \bar{z}_k)} \left[(\bar{z} - \bar{z}_k)(-s_k - \omega_k \partial_{\omega_k}) - (\bar{z} - \bar{z}_k)^2 \partial_{\bar{z}_k} \right]^2
$$

$$
\text{(A.22)}
$$

for negative and positive-helicity soft gravitons respectively. In a conformal basis, (A.22) become

$$
\widetilde{S}^-_{(2)} = -\frac{\kappa}{4}\sum_{k=1}^n \frac{z-z_k}{\bar{z}-\bar{z}_k}\left[2h_k(2h_k-1) - 2(z-z_k)2h_k\partial_{z_k} + (z-z_k)^2\partial_{z_k}^2\right]\mathcal{P}_k^{-1},
$$

$$
\widetilde{S}^+_{(2)} = -\frac{\kappa}{4}\sum_{k=1}^n \frac{\bar{z}-\bar{z}_k}{z-z_k}\left[2\bar{h}_k(2\bar{h}_k-1) - 2(\bar{z}-\bar{z}_k)2\bar{h}_k\partial_{\bar{z}_k} + (\bar{z}-\bar{z}_k)^2\partial_{\bar{z}_k}^2\right]\mathcal{P}_k^{-1},
$$

$$
\tag{A.23}
$$

with h_k, $\bar{h}_k$ and $\mathcal{P}^{-1}$ defined in Appendix A.3. Treating z, $\bar{z}$ as independent complex variables, we define the soft operators

$$
\delta \equiv \oint \frac{d\bar{z}}{2\pi i}\partial_z\widetilde{S}^-_{(2)}(0,\bar{z}) = \lim_{\Delta\to-1}(\Delta+1)\oint\frac{d\bar{z}}{2\pi i}\partial_z G^-_\Delta(0,\bar{z}),
$$

$$
\bar{\delta} \equiv \oint \frac{dz}{2\pi i}\partial_{\bar{z}}\widetilde{S}^+_{(2)}(z,0) = \lim_{\Delta\to-1}(\Delta+1)\oint\frac{dz}{2\pi i}\partial_{\bar{z}} G^+_\Delta(z,0),
$$

$$
\tag{A.24}
$$

which act on celestial operators as follows

$$
\delta\mathcal{O}_{\Delta_k,s_k}(z_k,\bar{z}_k) = -\frac{\kappa}{4}\left[2h_k(2h_k-1) + 8h_k z_k\partial_{z_k} + 3z_k^2\partial_{z_k}^2\right]\mathcal{O}_{\Delta_k-1,s_k}(z_k,\bar{z}_k),
$$

$$
\bar{\delta}\mathcal{O}_{\Delta_k,s_k}(z_k,\bar{z}_k) = -\frac{\kappa}{4}\left[2\bar{h}_k(2\bar{h}_k-1) + 8\bar{h}_k\bar{z}_k\partial_{\bar{z}_k} + 3\bar{z}_k^2\partial_{\bar{z}_k}^2\right]\mathcal{O}_{\Delta_k-1,s_k}(z_k,\bar{z}_k).
$$

$$
\tag{A.25}
$$

Equations (A.25) define the action of the global symmetries associated with the subsubleading soft graviton theorem (36) and (43). They constrain the form of the graviton and graviton–gluon OPEs (33) and (42) as discussed in Secs. 3.2 and 3.3.

A.5. Solving the recursion relations

Consider a symmetric function of complex variables $C(\Delta_1,\Delta_2) = C(\Delta_2,\Delta_1)$ which obeys the recursion relation

$$
\Delta_1 C(\Delta_1,\Delta_2) = (\Delta_1+\Delta_2)C(\Delta_1+1,\Delta_2).
\tag{A.26}
$$

Provided $C(\Delta_1,\Delta_2)\Gamma(\Delta_1+\Delta_2)$ is holomorphic for $\mathrm{Re}(\Delta_1) > 0$ and bounded for $\mathrm{Re}(\Delta_1) \in [1,2)$, (A.26) has the unique solution

$$
C(\Delta_1,\Delta_2) = C(1,1)B(\Delta_1,\Delta_2).
\tag{A.27}
$$

This can be proven in the following way. Define a function $f(x) = C(x,y_0)\Gamma(x+y_0)$. Then (A.26) becomes

$$
xf(x) = f(x+1).
\tag{A.28}
$$

By Wieland's theorem,[50] (A.28) has the unique solution

$$
f(x) = f(1)\Gamma(x).
\tag{A.29}
$$

Eliminating $f(x)$, $f(1)$ in terms of $C(x,y_0)$, $C(1,y_0)$ we find

$$
C(x,y_0) = \frac{C(1,y_0)\Gamma(1+y_0)\Gamma(x)}{\Gamma(x+y_0)}.
\tag{A.30}
$$

For $y_0 = 1$, (A.30) implies

$$C(x,1)\Gamma(1+x) = \Gamma(x)C(1,1). \tag{A.31}$$

Now replacing x with y_0 and using symmetry in the arguments of $C(x,y)$ we deduce that

$$C(x,y_0) = C(1,1)\frac{\Gamma(y_0)\Gamma(x)}{\Gamma(x+y_0)} = C(1,1)B(x,y_0). \tag{A.32}$$

For the purposes of determining the OPE coefficients, $C(1,1)$ is often fixed by the leading soft theorems. The holomorphicity condition is obeyed by celestial amplitudes whose momentum space behavior is known from soft theorems to be no more singular than a simple pole in frequency. Boundedness in the strip is expected to be inherited from momentum-space amplitudes with sufficiently good UV behavior. Related properties were pointed out in Ref. 25. The argument can be easily generalized to functions which are not symmetric under $\Delta_1 \leftrightarrow \Delta_2$.

References

1. F. Cachazo and A. Strominger, "Evidence for a new soft graviton theorem," arXiv:1404.4091 [hep-th].
2. D. Kapec, V. Lysov, S. Pasterski and A. Strominger, "Semiclassical Virasoro symmetry of the quantum gravity S-matrix," *JHEP* **1408**, 058 (2014), arXiv:1406.3312 [hep-th].
3. A. Strominger, "Lectures on the infrared structure of gravity and gauge theory," arXiv:1703.05448 [hep-th].
4. C. Cheung, A. de la Fuente and R. Sundrum, "4D scattering amplitudes and asymptotic symmetries from 2D CFT," *JHEP* **1701**, 112 (2017), arXiv:1609.00732 [hep-th].
5. S. Pasterski and S. H. Shao, "Conformal basis for flat space amplitudes," *Phys. Rev. D* **96**, 065022 (2017), arXiv:1705.01027 [hep-th].
6. S. Banerjee, "Null infinity and unitary representation of the Poincaré group," *JHEP* **1901**, 205 (2019), arXiv:1801.10171 [hep-th].
7. L. Donnay, A. Puhm and A. Strominger, "Conformally soft photons and gravitons," *JHEP* **1901**, 184 (2019), arXiv:1810.05219 [hep-th].
8. E. Himwich and A. Strominger, "Celestial current algebra from Low's subleading soft theorem," *Phys. Rev. D* **100**, 065001 (2019), arXiv:1901.01622 [hep-th].
9. S. Banerjee, P. Pandey and P. Paul, "Conformal properties of soft-operators — 1: Use of null-states," arXiv:1902.02309 [hep-th].
10. A. Fotopoulos and T. R. Taylor, "Primary fields in celestial CFT," arXiv:1906.10149 [hep-th].
11. Y. T. A. Law and M. Zlotnikov, "Poincaré constraints on celestial amplitudes," arXiv:1910.04356 [hep-th].
12. S. Pasterski, S. H. Shao and A. Strominger, "Flat space amplitudes and conformal symmetry of the celestial sphere," *Phys. Rev. D* **96**, 065026 (2017), arXiv:1701.00049 [hep-th].
13. C. Cardona and Y. t. Huang, "S-matrix singularities and CFT correlation functions," *JHEP* **1708**, 133 (2017), arXiv:1702.03283 [hep-th].
14. S. Pasterski, S. H. Shao and A. Strominger, "Gluon amplitudes as 2D conformal correlators," *Phys. Rev. D* **96**, 085006 (2017), arXiv:1706.03917 [hep-th].

15. H. T. Lam and S. H. Shao, "Conformal basis, optical theorem, and the bulk point singularity," *Phys. Rev. D* **98**, 025020 (2018), arXiv:1711.06138 [hep-th].

16. N. Banerjee, S. Banerjee, S. Atul Bhatkar and S. Jain, "Conformal structure of massless scalar amplitudes beyond tree level," *JHEP* **1804**, 039 (2018), arXiv:1711.06690 [hep-th].

17. A. Schreiber, A. Volovich and M. Zlotnikov, "Tree-level gluon amplitudes on the celestial sphere," *Phys. Lett. B* **781**, 349 (2018), arXiv:1711.08435 [hep-th].

18. S. Stieberger and T. R. Taylor, "Strings on celestial sphere," *Nucl. Phys. B* **935**, 388 (2018), arXiv:1806.05688 [hep-th].

19. S. Stieberger and T. R. Taylor, "Symmetries of celestial amplitudes," *Phys. Lett. B* **793**, 141 (2019), arXiv:1812.01080 [hep-th].

20. W. Fan, A. Fotopoulos and T. R. Taylor, "Soft limits of Yang–Mills amplitudes and conformal amplitudes," arXiv:1903.01676 [hep-th].

21. M. Pate, A. M. Raclariu and A. Strominger, "Conformally soft theorem in gauge theory," arXiv:1904.10831 [hep-th].

22. D. Nandan, A. Schreiber, A. Volovich and M. Zlotnikov, "Celestial amplitudes: Conformal partial waves and soft limits," arXiv:1904.10940 [hep-th].

23. T. Adamo, L. Mason and A. Sharma, "Celestial amplitudes and conformal soft theorems," arXiv:1905.09224 [hep-th].

24. A. Puhm, "Conformally soft theorem in gravity," arXiv:1905.09799 [hep-th].

25. A. Guevara, "Notes on conformal soft theorems and recursion relations in gravity," arXiv:1906.07810 [hep-th].

26. F. Cachazo, S. He and E. Y. Yuan, "Scattering of massless particles in arbitrary dimensions," *Phys. Rev. Lett.* **113**, 171601 (2014), arXiv:1307.2199 [hep-th].

27. F. Cachazo, S. He and E. Y. Yuan, "Scattering of massless particles: Scalars, gluons and gravitons," *JHEP* **1407**, 033 (2014), arXiv:1309.0885 [hep-th].

28. N. Arkani-Hamed and J. Trnka, "The amplituhedron," *JHEP* **1410**, 030 (2014), arXiv:1312.2007 [hep-th].

29. L. J. Dixon, J. M. Drummond, C. Duhr, M. von Hippel and J. Pennington, "Bootstrapping six-gluon scattering in planar $N = 4$ super-Yang–Mills theory," in *Loops and Legs in Quantum Field Theory*, Weimar, Germany, 27 April–02 May, 2014, *Proc. Sci.* **211**, 077 (2014), arXiv:1407.4724 [hep-th].

30. L. Rodina, "Scattering amplitudes from soft theorems and infrared behavior," *Phys. Rev. Lett.* **122**, 071601 (2019), arXiv:1807.09738 [hep-th].

31. T. He, P. Mitra and A. Strominger, "2D Kac–Moody symmetry of 4D Yang–Mills theory," *JHEP* **1610**, 137 (2016), arXiv:1503.02663 [hep-th].

32. S. Stieberger and T. R. Taylor, "Subleading terms in the collinear limit of Yang–Mills amplitudes," *Phys. Lett. B* **750**, 587 (2015), arXiv:1508.01116 [hep-th].

33. J. Distler, R. Flauger and B. Horn, "Double-soft graviton amplitudes and the extended BMS charge algebra," *JHEP* **1908**, 021 (2019), arXiv:1808.09965 [hep-th].

34. V. Lysov, S. Pasterski and A. Strominger, "Low's subleading soft theorem as a symmetry of QED," *Phys. Rev. Lett.* **113**, 111601 (2014), arXiv:1407.3814 [hep-th].

35. D. Kapec, P. Mitra, A. M. Raclariu and A. Strominger, "2D stress tensor for 4D gravity," *Phys. Rev. Lett.* **119**, 121601 (2017), arXiv:1609.00282 [hep-th].

36. D. Kapec and P. Mitra, "A d-dimensional stress tensor for Mink$_{d+2}$ gravity," *JHEP* **1805**, 186 (2018), arXiv:1711.04371 [hep-th].

37. H. Elvang, C. R. T. Jones and S. G. Naculich, "Soft photon and graviton theorems in effective field theory," *Phys. Rev. Lett.* **118**, 231601 (2017), arXiv:1611.07534 [hep-th].

38. A. Laddha and P. Mitra, "Asymptotic symmetries and subleading soft photon theorem in effective field theories," *JHEP* **1805**, 132 (2018), arXiv:1709.03850 [hep-th].

39. H. Kawai, D. C. Lewellen and S. H. H. Tye, "A relation between tree amplitudes of closed and open strings," *Nucl. Phys. B* **269**, 1 (1986).

40. Z. Bern, J. J. M. Carrasco and H. Johansson, "New relations for gauge-theory amplitudes," *Phys. Rev. D* **78**, 085011 (2008), arXiv:0805.3993 [hep-ph].

41. Z. Bern, J. J. M. Carrasco and H. Johansson, "Perturbative quantum gravity as a double copy of gauge theory," *Phys. Rev. Lett.* **105**, 061602 (2010), arXiv:1004.0476 [hep-th].

42. S. Stieberger and T. R. Taylor, "Graviton as a pair of collinear gauge bosons," *Phys. Lett. B* **739**, 457 (2014), arXiv:1409.4771 [hep-th].

43. Z. Bern, L. J. Dixon, M. Perelstein and J. S. Rozowsky, "Multileg one loop gravity amplitudes from gauge theory," *Nucl. Phys. B* **546**, 423 (1999), arXiv:hep-th/9811140.

44. C. D. White, "Factorization properties of soft graviton amplitudes," *JHEP* **1105**, 060 (2011), arXiv:1103.2981 [hep-th].

45. R. Akhoury, R. Saotome and G. Sterman, "Collinear and soft divergences in perturbative quantum gravity," *Phys. Rev. D* **84**, 104040 (2011), arXiv:1109.0270 [hep-th].

46. S. Stieberger and T. R. Taylor, "New relations for Einstein–Yang–Mills amplitudes," *Nucl. Phys. B* **913**, 151 (2016), arXiv:1606.09616 [hep-th].

47. G. Altarelli and G. Parisi, "Asymptotic freedom in parton language", *Nucl. Phys. B* **126**, 298–318 (1977).

48. G. 't Hooft and M. J. G. Veltman, "One loop divergencies in the theory of gravitation," *Ann. Inst. H. Poincare Phys. Theor. A* **20**, 69 (1974).

49. E. Conde and P. Mao, "BMS supertranslations and not so soft gravitons," *JHEP* **1705**, 060 (2017), arXiv:1612.08294 [hep-th].

50. L. Reich, "Functional equations in the complex domain".

Chapter 21

How to Split the Electron in Half

Gordon W. Semenoff

*Department of Physics and Astronomy, University of British Columbia, 6224
Agricultural Road, Vancouver, British Columbia, Canada V6T 1Z1*

This essay is a tribute to Professor Roman Jackiw on the occasion of his eightieth
year. It discusses some ideas about fermion zero modes and fractional charges
and quantum entanglement.

1. Prologue

This is an essay in tribute to Professor Roman Jackiw on the occasion of his 80th
year. The beauty and the originality of Roman's scientific work inspired many
and it continues to form the bedrock of whole lines of investigation. I am one of
the inspired. I offer this essay about some of my modest thoughts about one of
the many interesting ideas of the Jackiw–Rebbi era [1], the connections between
topology, fermion zero modes and fractional charges. This was one of the first
relationships between the topology of states of quantized fields and their physical
properties [2].[a] Of course, like all great ideas, it inspired more in its aftermath.
In this case, it is part of the nexus of ideas which has developed into the subject
of topological insulators, one of the most active and interesting areas of modern
theoretical physics.

What I will discuss is a particular example of a topological insulator, an ex-
ceedingly simple one in fact, where the peculiarities of the system result in quantum states with unusual properties. I will not review the topological aspects of
the models, as there are already a number of beautifully explained overviews [4;
5]. I will rather concentrate on the electronic properties of some simple models.

2. Tight-binding model

Let us consider a hypothetical one-dimensional system consisting of an open chain
of sites. We shall assume that this chain can be occupied by electrons to a maximum
density of two electrons per site. To a first approximation, the quantum states of
the electrons are single one-particle bound states localized at each of the sites of the

---

[a]For a review, see Ref. [3].

chain. We will assume that the sites are charged so that the system with a density of one electron per site is charge neutral. The electrons have spin 1/2 and the maximum density of the chain of two electrons per site is achieved by the localized electrons taking up each of the two spin states. The charge neutral state of a chain is depicted in Fig. 1.

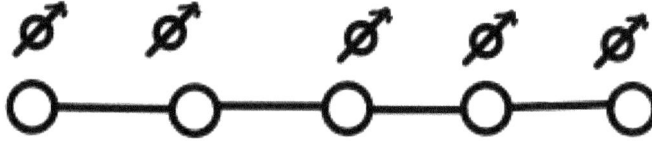

Fig. 1. We will consider a one-dimensional chain of atoms with an odd number of sites. The electron density is one electron per site. The electrons each have two spin sites. We will assume that the coupling between the atomic and electron spins is negligible so that the system has a symmetry under the rotation of electronic spins.

The tight-binding model begins with the assumption that the electrons reside in orbitals which are localized at the lattice sites as we have described. The dynamics is simply taking into account the fact that an electron in one orbital has some probability of tunnelling to a neighbouring orbital. If we put a single electron at a site on the chain, its wave-function would diffuse until, typically, it occupied the entire system.

The tight-binding Hamiltonian for the many-electron system is gotten by first introducing creation and annihilation operators for the tight-binding states. The electron bound to the n'th site and with spin polarization σ is created by $a_{\sigma,n}^\dagger$ and annihilated by $a_{\sigma,n}$. These operators have the anti-commutation relations

$$\left\{a_{\sigma_1,n_1}, a_{\sigma_2,n_2}^\dagger\right\} = \delta_{n_1 n_2}\delta_{\sigma_1\sigma_2} \tag{1}$$

$$\left\{a_{\sigma_1,n_1}, a_{\sigma_2,n_2}\right\} = 0 , \quad \left\{a_{\sigma_1,n_1}^\dagger, a_{\sigma_2,n_2}^\dagger\right\} = 0 \tag{2}$$

The Hamiltonian is

$$H = \sum_{n=1}^{N-1} \sum_{\sigma=\uparrow,\downarrow} \left(t_n a_{\sigma,n+1}^\dagger a_{\sigma,n} + t_n^* a_{\sigma,n}^\dagger a_{\sigma,n+1}\right) \tag{3}$$

where t_n is the probability amplitude for an electron to tunnel from position n to position $n + 1$. The amplitude for tunnelling from $n + 1$ to n is t_n^*. For now, we have assumed that these amplitudes are independent of the electron spin states (we will change this assumption later). Also, for the moment, apart from the assumption that all of the hopping amplitudes are non-zero, we will make no further assumptions. We note that, by redefining the phases of the creation and annihilation operators, we can make all of the parameters t_n real and positive. We shall hereafter assume that we have done that.

We shall not be overly specific about how such a chain could be realized in nature. An important example is a conducting polymer such as polyacetylene. which consists of a chain of carbon atoms bound together by strong covalent bonds.

In that case, we would assume that the electrons which are buried deep in atomic shells are so tightly bound that they do not contribute to the electronic properties of the material. A carbon atom has four valence electrons which are less tightly bound. Two of the valence electrons form the strong covalent bonds with the two carbon neighbours of each atom. A third electron forms a bond with a hydrogen atom which is bound to each carbon atom. The fourth electron on each atom is more loosely bound and it plays the role of our dynamical electron. We will assume that this single valence electron is entirely responsible for the electronic properties of the system. In particular, a model of trans-polyacetylene described in this tight-binding approximation is called the Su–Schrieffer–Heeger (SSH) model [6; 7; 8]. In that model, the strength of the hopping parameter varies with position, $t_{2n} = t$, $t_{2n+1} = t'$ and $|t| \neq |t'|$. (We review a complete solution of the SSH model in Appendix A.) Another option for engineering such a chain is to use cold atoms in an optical lattice. There are other ideas, such as linear arrays of Josephson junctions or quantum dots that can be modelled by the SSH Hamiltonian.

In the Heisenberg picture, the tight-binding model equation of motion for the creation and annihilation operators is found by taking the commutator

$$i\hbar \frac{d}{dt} a_{\sigma,n} = [a_{\sigma,n}, H] \tag{4}$$

which yields the differential-difference equation

$$i\hbar \frac{d}{dt} a_{\sigma,n} = t_{n-1} a_{\sigma,n-1} + t_n a_{\sigma,n+1} \tag{5}$$

The eigenvalue equation for the levels of the single-particle Hamiltonian can be written as a matrix equation

$$E\phi_{E,n} = t_{n-1}\phi_{E,n-1} + t_n\phi_{E,n+1} \tag{6}$$

which we could write as $h\phi_E = E\phi_E$ where ϕ_E is a vector with an amplitudes $\phi_{E,n}$, $n = 1, 2, ..., N$, or more explicitly

$$\phi_E = \begin{bmatrix} \phi_{E,1} \\ \phi_{E,2} \\ . \\ . \\ \phi_{E,N} \end{bmatrix} \tag{7}$$

and

$$h = \begin{bmatrix} 0 & t_1 & 0 & 0 & 0 & \cdots & \cdots \\ t_1 & 0 & t_2 & 0 & 0 & \cdots & \cdots \\ 0 & t_2 & 0 & t_3 & 0 & \cdots & \cdots \\ 0 & 0 & t_3 & 0 & t_4 & \cdots & \cdots \\ \cdots & \cdots & \cdots & \cdots & \cdots & \cdots & \cdots \\ \cdots & \cdots & \cdots & \cdots & \cdots & \cdots & \cdots \\ \cdots & \cdots & \cdots & \cdots & \cdots & 0 & t_{N-1} \\ \cdots & \cdots & \cdots & \cdots & \cdots & t_{N-1} & 0 \end{bmatrix} \tag{8}$$

is the single-particle Hamiltonian. The eigenvalues of this Hermitian matrix are the energy levels of the system. For the most part, we shall not need the details of the spectrum of the Hamiltonian beyond a few observations. The question most important to us, about whether there are zero modes will be discussed in the next section.

3. Fermion zero modes

First of all, we note the rather remarkable fact that, without any further specification of the hopping amplitudes, t_n, the spectrum of the Hamiltonian has particle–hole symmetry. To see this, we note that there is a matrix

$$
\Gamma \;=\;
\begin{bmatrix}
1 & 0 & 0 & 0 & 0 & \dots & \dots \\
0 & -1 & 0 & 0 & 0 & \dots & \dots \\
0 & 0 & 1 & 0 & 0 & \dots & \dots \\
0 & 0 & 0 & -1 & 0 & \dots & \dots \\
\dots & \dots & \dots & \dots & \dots & \dots & \dots \\
\dots & \dots & \dots & \dots & \dots & \dots & \dots \\
\dots & \dots & \dots & \dots & \dots & 0 & \\
\dots & \dots & \dots & \dots & \dots & 0 & \pm1
\end{bmatrix}
\tag{9}
$$

where, whether the last entry is $+1$ or -1 depends on whether N is odd or even, respectively. This matrix has the property that $\Gamma = \Gamma^\dagger$, $\Gamma^2 = I$ and

$$
\Gamma h \Gamma = -h
\tag{10}
$$

Its existence implies that if we managed to find an eigenvector of the Hamiltonian with eigenvalue E,

$$
h\Phi_E = E\phi_E
\tag{11}
$$

then we can find another eigenvector of the Hamiltonian with eigenvalue of opposite sign, $-E$, as

$$
h(\Gamma\Phi_E) = -E(\Gamma\phi_E)
\tag{12}
$$

This particle–hole conjugation, when operating on a wave-function $\phi_{E,n}$ which solves Eq. (6), is the transformation

$$
\phi_{-E,n} \;=\; (\Gamma\phi)_{E,n} \;=\; (-1)^n \phi_{E,n}
\tag{13}
$$

which changes the sign of the wave-function on the odd sub-lattice. This leads to a discrete symmetry of the many-fermion system that is described by the Hamiltonian H in Eq. (3). The transformation of the the creation and annihilation operators is

$$
a_{\sigma,n} \to (-1)^n a^\dagger_{\sigma,n} \;,\quad a^\dagger_{\sigma,n} \to (-1)^n a_{\sigma,n}
\tag{14}
$$

The Hamiltonian of the many-fermion system, H, is invariant under this transformation.

The spectrum of the single-particle Hamiltonian h thus has particle–hole symmetry, in that for each eigenvector ϕ_E with positive eigenvalue E there is an eigenvector $\Gamma\phi_E$ with negative eigenvalue $-E$. Moreover, the number of eigenvalues (per spin state) is equal to the dimension of the quantum Hilbert space, which is N, the number of sites in the chain.

If N is even, the pairing of positive and negative energy modes implies a one-to-one matching of the positive and negative eigenvalues. If they exist, zero eigenvalues must also be paired, and they must be even in number. (We will show shortly that they do not exist at all for the even-sited chain.)

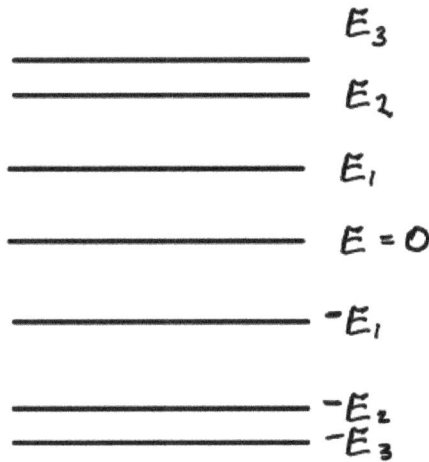

Fig. 2. *The world's simplest index theorem.* When the number of energy levels is odd and when the spectrum is particle–hole symmetric so that there is a one-to-one mapping of positive energy levels onto negative energy levels, there must be an odd number of zero energy levels. Generically, one would expect that perturbations of the Hamiltonian which preserve the particle–hole symmetry could lift the degeneracy of the zero modes, but this could only happen in positive—negative energy pairs, so an odd number of zero modes must remain. We will show by explicitly examining the zero mode equations for the tight-binding problem that the odd-sited chain always has exactly one zero mode and the even-sited chain can have no zero modes at all.

However, if N is odd, this one-to-one matching is not possible and the particle–hole symmetry of the spectrum has an interesting consequence. The positive and negative energy levels are paired and there must be an odd one out. That odd one out must be unpaired and it must reside precisely at the centre of the spectrum, at $E = 0$. This state is what we shall call the "zero mode". This circumstance is depicted in Fig. 2. This argument actually establishes that there must be an odd number of zero modes. In the following we will show that there can be only one zero mode for the odd-sited chain. (Of course the electron state that results from this one zero mode will still have the two-fold spin degeneracy.)

At this point, we do not know where the zero-mode wave-function of the odd-sited chain has support. We also do not know to what extent the zero mode is isolated from the non-zero modes. These details depend on the specific values of the hopping amplitudes, t_n.

There are explicit examples, such as the SSH model [6; 7; 8], depicted in Fig. 3, where the non-zero modes have a gap that allows the zero mode, which appears at the centre of the gap, to be well isolated from the rest of the spectrum. In that case, for the odd-sited SSH chain, there is a single zero mode with support that is concentrated at one or the other edges of the chain.

Fig. 3. The Su–Schrieffer–Heeger chain which is depicted here has alternating strong and weak bonds. The electron pairs participating in the strong bonds are sufficiently bound to make a gap in their spectrum. When there are an odd number of sites, one edge of the chain has a strong bond and the opposite edge of the chain has a weak bond. The odd electron out resides at the edge of the chain which has a weak bond and it occupies the zero mode in the electron energy spectrum. When the electron has spin, this mode can either be empty, singly occupied or doubly occupied. These states of the many-fermion Hamiltonian are thus four-fold degenerate, although the states have differing charges and spins.

The other extreme is the simple translation-invariant chain where all of the hopping amplitudes are equal, $t_n = t$ for all n. In that case, the zero-mode wave-function vanishes on the even sites and it is uniformly distributed on the odd sites. However, it is not well isolated from the rest of the spectrum since the lowest non-zero modes have energies are $\pm 2t \sin \frac{\pi}{N+1}$ which can be very small if N is large.

Note that we have shown that a zero mode must exist in the odd-sited system. Let us examine the equation that the zero mode must obey, the eigenvalue equation (6) with $E = 0$,

$$t_{2n-1}\phi_{0,2n-1} + t_{2n}\phi_{0,2n+1} = 0 \ , \quad t_{2n}\phi_{0,2n} + t_{2n+1}\phi_{0,2n+2} = 0 \qquad (15)$$

These equations have the special property that they only couple the wave-functions on even sites with even sites and odd sites with odd sites.

We can re-organize them as

$$t_1\phi_{0,2} = 0 \ , \quad t_{N-1}\phi_{0,N-1} = 0 \qquad (16)$$

$$\phi_{0,2n+2} = -\frac{t_{2n}}{t_{2n+1}}\phi_{0,2n} \ , \quad n = 1, 2, ..., (N-3)/2 \qquad (17)$$

$$\phi_{0,2n+1} = -\frac{t_{2n-1}}{t_{2n}}\phi_{0,2n-1} \ , \quad n = 1, 2, ..., (N-3)/2 \qquad (18)$$

The equations $t_1\phi_{0,2} = 0$ and $t_{N-1}\phi_{0,N-1} = 0$ in Eq. (16) are easy to see from the first and last row of the matrix equation $h\psi_0 = 0$ with h given in Eq. (8). The implication of these equations for the odd-sited chain is that the solution of Eq. (17) is $\phi_{0,2n} = 0$, that is, the zero-mode wave-functions vanish on the even sites. Then Eq. (18) has a unique solution

$$\phi_{0,2n+1} = \left(-\frac{t_{2n-1}}{t_{2n}}\right)\left(-\frac{t_{2n-3}}{t_{2n-2}}\right)\left(-\frac{t_{2n-5}}{t_{2n-4}}\right)\cdots\left(-\frac{t_3}{t_4}\right)\left(-\frac{t_1}{t_2}\right)\phi_{0,1} \qquad (19)$$

which tells us that there is precisely one non-degenerate zero mode and that its wave-function has support only on the even sub-lattice. Moreover, we know its wave-function (19) explicitly.

In passing, we notice that, if the number of lattice sites were even, rather than odd, the equations $\phi_{0,2} = 0$ and $\phi_{0,N-1} = 0$ put the wave-function to zero on both the even and the odd sub-lattices, and there are no solutions for zero modes at all. Thus, the even-sited chain can have no zero modes. This statement might not be very important for a long, almost-translation-invariant chain where there can typically be non-zero modes with very small energies, exponentially small in the length of the chain. However, for a short or meso-scopic chain, the fact that there can be no zero modes on an even-sited chain and precisely one zero mode on an odd-sited chain, no matter what the hopping parameters, could be very useful information.

In addition, what is very interesting about the odd-sited chain is that, if one could engineer the t_n's appropriately, one could determine the support of the zero mode. For example, if $|t_{\text{odd}}| < |t_{\text{even}}|$, the zero mode has support near one end of the chain, with ϕ_1 being the largest component. On the other hand, if $|t_{\text{odd}}/t_{\text{even}}| \sim 1$ the support of the zero-mode wave-function is evenly distributed over the chain. As an example, consider the chain with five sites which we can easily solve explicitly. In that case

$$\phi_0 = \frac{1}{\sqrt{1 + |t_1/t_2|^2(1 + |t_3/t_4|^2)}}\begin{bmatrix} 1 \\ 0 \\ -t_1/t_2 \\ 0 \\ (-t_3/t_4)(-t_1/t_2) \end{bmatrix} \qquad (20)$$

By appropriately engineering the ratios t_1/t_2 and t_3/t_4 we could place the zero mode at one edge:

$$t_1 < t_2, \; t_3 < t_4$$

the other edge:

$$t_1 > t_2, \; t_3 > t_4$$

both edges:

$$t_1 \ll t_2, (t_1/t_2)(t_3/t_4) \sim 1$$

or the centre of the chain:

$$t_1 \gg t_2, (t_1/t_2)(t_3/t_4) \sim 1$$

Placing the support of the wave-function at both edges of the chain is interesting as it is located non-locally, reminiscent of Kitaev's Majorana fermions on a finite chain [9]. In this case, if we put

$$t_1 = t_4 \ , \ t_2 = t_3$$

with

$$t_1 \ll t_2$$

we find the non-zero energy levels are

$$\pm t_1 \ , \quad \pm \sqrt{t_1^2 + 2t_2^2}$$

The gap between the zero mode and the first non-zero mode is t_1, the smaller of the two dimensional scales in the problem. However, one could imagine that, in this small system it might be possible to engineer the system so that t_1 is larger than the characteristic energies of processes which would affect the zero mode.

In the following, we will concentrate on the case where the zero mode is located at one end of the chain. The Su–Schrieffer–Heeger chain where $t_{2n} = t$ and $t_{2n+1} = t'$ with $t \neq t'$, is a beautiful example of such a situation which has the added nice feature that the zero mode is isolated in a gap whose magnitude is $|t' - t|$. We can clearly see that if $t_{2k}/t_{2k+1} = t/t' < 1$ for all $k = 1, ..., (N-1)/2$, the wave-function is localized near $n = 1$ and it decays exponentially as $\exp\left(-n \ln \frac{t'}{t}\right)$ as we move away from the edge.

Before we continue, let us mention that the continuum analog of the odd-sited chain is the continuum problem governed by the Dirac Hamiltonian

$$E \begin{bmatrix} u(x) \\ v(x) \end{bmatrix} = \begin{bmatrix} 0 & \frac{d}{dx} + m \\ -\frac{d}{dx} + m & 0 \end{bmatrix} \begin{bmatrix} u(x) \\ v(x) \end{bmatrix}$$

where $0 \leq x \leq L$. The analog of the odd-sited spin chain is this Dirac equation with the two boundary conditions $v(0) = 0$ and $v(L) = 0$. This boundary condition is sufficient for the current density, $j(x) = -iu^*(x)v(x) + iv^*(x)u(x)$ to vanish at the boundaries of the system. The zero-mode solution is

$$\phi_0(x) = \begin{bmatrix} \sqrt{\frac{m}{e^{mL}-1}} \exp(mx) \\ 0 \end{bmatrix}$$

where c is a normalization constant. The existence of this zero mode does not depend on the sign of m, since $u(x)$ has open boundary conditions, this is always a solution.

On the other hand, the even-sited chain analog is of this Dirac equation with boundary conditions $u(0) = 0$ and $v(L) = 0$. In this case, there is no zero energy mode at all. To find a solution, we could begin with the ansatz

$$u(x) = \frac{1}{c} \sin kx$$

which satisfies the first boundary condition. Then the Dirac equation tells us that

$$v(x) = \sqrt{\frac{m}{e^{mL} - 1}} \frac{1}{E} (m \sin kx - k \cos kx)$$

and that

$$E = \pm\sqrt{k^2 + m^2}$$

The boundary conditions are satisfied if k obeys the transcendental equation

$$\tan kL = \frac{k}{m}$$

This equation always has an infinite number of solutions which are the allowed wave-vectors. To find a bound state, we must look for a solution with imaginary k, that is, a solution of

$$\tanh \kappa L = \frac{\kappa}{m}$$

This equation always has a solution when $m > 0$ and it does not have a solution when $m < 0$. When $m > 0$ and $mL \gg 1$, $\kappa \sim m - 2me^{-2mL}$ and $E \sim \pm 2me^{-mL}$. When $m < 0$, there are no states at all within the energy gap. The continuum problem with $m < 0$ is the analog of the even-sited SSH chain with strong bonds at each edge and the problem with $m > 0$ is the analog of the even-sited SSH chain with weak bonds at each edge.

4. Zero modes and the quantum numbers of many-fermion states

We are interested in the influence of the zero energy modes on the many-body quantum states of electrons on the odd-sited chain. Let us begin by studying the lowest energy state of the system when it is charge neutral, that is, when it contains N electrons, one for each of its N sites. We note that there are also N energy levels, that is, there are N eigenvalues of the single-particle Hamiltonian, h.

The occupation of the energy levels by electrons is governed by the Pauli principle. The electrons have two spin states and therefore two electrons are allowed to occupy each energy level. Let us assume that N is odd. Then $N - 1$ is an even integer. The single-particle energy levels consist of $(N-1)/2$ positive energy states, $(N - 1)/2$ negative energy states and one zero energy state. A neutral state of the system will have N electrons distributed amongst these states. The lowest energy state with N electrons has $N - 1$ of those electrons forming $(N - 1)/2$ spin up–spin down pairs occupying the $(N-1)/2$ negative energy states. Then the one additional electron must occupy the zero energy state. This is the state of N electrons which has the lowest possible energy. It is charge neutral. This is due to the fact that the electric charge of the N electrons is canceled by the electric charge of the N sites. However, the state is two-fold degenerate, the zero energy state could have been occupied by a spin-up or a spin-down electron. This leads to two distinct many-electron charge neutral ground states with the same energy. They carry a doublet representation of the electron spin.

This ground state has more degeneracy yet. We could remove the electron which is in the zero energy mode. This results in a state with charge $-e$, where e is the electron charge. Since we de-populated a zero mode, the energy of the many-electron system is unchanged. What is more, this state is a singlet of the electron spin. The result is a state which has charge but no spin. Similarly, we could add an electron to the original neutral state so that there are then two electrons in the zero-mode state. These electrons would have paired spins, they would form a spin singlet and we would have a spinless state with charge e. In this way we see that there are four degenerate ground states of the many-electron system, two of them neutral states which form a spin doublet and the other two are spin singlet states which have charges e and $-e$. These are the unusual spin-charge quantum numbers which result from the presence of a single fermion zero mode in the spectrum.

5. Entangled states

We begin with an even-sited chain, where the strong bonds occur at the two edges of the chain, as depicted in Fig. 4. In this case, the electron spectrum on the chain is entirely gapped. It has no energy states in the gap between $+|t-t'|$ and $-|t-t'|$ at all. Then we split the chain into two odd-sited chains by adiabatically decreasing one of the t_n's to zero. The electronic state which is concentrated on that link contains a pair of electrons in a spin singlet. The spin remains a singlet throughout the process. What is more, the chain is charge neutral and the total charge remains at zero throughout.

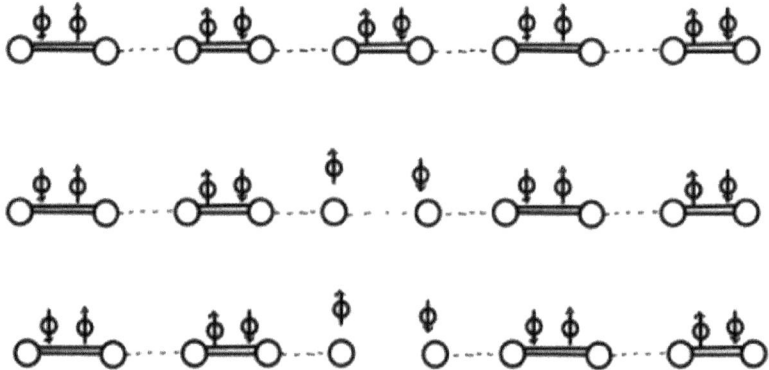

Fig. 4.   Consider an even-sited Su–Schrieffer–Heeger chain where we adiabatically weaken a bond in such a way as to split it into two odd-sited Su–Schreiffer–Heeger chains. If it was a strong bond, as is depicted in the figure, the zero modes of the two odd-sited chains which are produced reside on the edges of the odd-sited chains that are adjacent to the broken bond. The resulting electronic state is a highly entangled state of the zero mode sectors on the two odd-sited chains.

To describe this system, we introduce two sets of creation and annihilation operators with the (non-vanishing) anti-commutation relations

$$\left\{a_{R,\sigma,0}, a^\dagger_{R,\sigma',0}\right\} = \delta_{\sigma\sigma'} , \quad \left\{a_{L,\sigma,0}, a^\dagger_{L,\sigma',0}\right\} = \delta_{\sigma\sigma'} \tag{21}$$

one for each zero mode which arises on each of the odd-sited chains that are produced, which we can label as R and L for right and left. To understand what state is produced by the process, we could imagine a neutral state of the two odd-sited chains with the edges which support the zero modes brought into proximity. They can interact simply by virtue of the fact that an electron can tunnel between them. We can describe this by turning on an interaction Hamiltonian

$$H_{\text{int}} = \sum_\sigma \tau \left\{a^\dagger_{L,\sigma,0} a_{R,\sigma,0} + a^\dagger_{R,\sigma,0} a_{L,\sigma,0}\right\} \tag{22}$$

The charge neutral, spin-singlet ground state of this Hamiltonian is

$$|\text{gs}\rangle = \frac{1}{2}\left(a^\dagger_{L,\uparrow,0} - a^\dagger_{R,\uparrow,0}\right)\left(a^\dagger_{L,\downarrow,0} - a^\dagger_{R,\downarrow,0}\right)|0\rangle$$

where $|0\rangle$ is the state of the two odd-sited chains where all of the zero modes are empty. This should be the state which is produced by turning off the link adiabatically. This is a highly entangled state of the two chains. To see this we could form the reduced density matrix for the R chain by taking $|\text{gs}\rangle\langle\text{gs}|$ and tracing out the degrees of freedom of the L chain. What we end up with is the unit matrix

$$\rho = \frac{1}{4}\left\{|0\rangle\langle0| + a^\dagger_{R,\uparrow,0}|0\rangle\langle0|a_{R,\uparrow,0} + a^\dagger_{R,\downarrow,0}|0\rangle\langle0|a_{R,\downarrow,0}\right.$$
$$\left. + a^\dagger_{R,\uparrow,0}a^\dagger_{R,\downarrow,0}|0\rangle\langle0|a_{R,\downarrow,0}a_{R,\uparrow,0}\right\}$$

which has the maximum possible entanglement entropy for a four-level system, $S_{\text{entanglement}} = -\text{Tr}\rho\ln\rho = \ln 4$. An algorithm which uses the entanglement that is generated in this way to teleport the spin wave-function of an electron from the vicinity of one edge to the other edge [11].

6. Spin-dependent hopping

In the previous section, we reviewed the sense in which the fermion zero mode leads to exotic quantum numbers of the many-fermion system. Because of the spin degeneracy, the states did not really have fractional charges, they had "fractional charge per spin state" which, with the two-fold spin degeneracy, resulted in the states having whole integer charges only. It did have the interesting upshot that the quantum numbers of the states were unusual, charge zero with spin 1/2 or charge $\pm e$ with spin 0.

One might wonder whether we could find more interesting states if the Hamiltonian contained a spin-dependent term which removes the spin degeneracy. An example would be a Zeeman interaction with a magnetic field which would shift the energies of one of the spin states to higher energies and the other spin state to lower energies. This can be interesting and we will return to it. However, first, let us consider a more exotic possibility, one where the hopping amplitudes, t_n, of the tight-binding model are spin-dependent. We found in the last section that, generally, the fermion zero mode is an edge mode if, generically, either $\frac{t_{\text{odd}}}{t_{\text{even}}} < 1$ or $\frac{t_{\text{odd}}}{t_{\text{even}}} > 1$, the zero mode has support at one or the other edge of the one-dimensional chain. If the hopping amplitudes could be spin dependent in such a way that $\frac{t_{\text{odd}}}{t_{\text{even}}} < 1$ for spin-up electrons and $\frac{t_{\text{odd}}}{t_{\text{even}}} > 1$ for spin-down electrons, the spin-up and spin-down zero modes would be localized at opposite edges of the one-dimensional chain.

To see the implications for charge density, let us examine the expectation value of the charge density in low energy, overall neutral states. The charge density of a given state is the expectation value of the operator, ρ_n, which measures the number of electrons at site n multiplied by the electron charge e

$$\rho_n = e a_{\sigma,n}^{\dagger} a_{\sigma,n} \quad - \quad e \tag{23}$$

where the last term are the charges of value $-e$ residing at each site of the chain (charges of the atoms when it is an atomic chain). The operator corresponding to the total charge is the sum of this charge density over the lattice sites,

$$Q = \sum_{n=1}^{N} \rho_n = \sum_n e \left[a_{\sigma,n}^{\dagger} a_{\sigma,n} - 1 \right] \tag{24}$$

The charge density obeys the continuity equation

$$\frac{d}{dt} \rho_n = \mathbf{j}_n - \mathbf{j}_{n-1} \tag{25}$$

where the electric current $\mathbf{j}_n$ across the link between the site n and the site $n+1$ is given by

$$\mathbf{j}_n = i t_n \left[a_{\sigma,n+1}^{\dagger} a_{\sigma,n} - a_{\sigma,n}^{\dagger} a_{\sigma,n+1} \right] \tag{26}$$

If we consider the second quantization of the electron,

$$a_n = \sum_{E>0} \phi_{\sigma,E,n} e^{-iEt} a_{\sigma,E} + \sum_{E<0} \phi_{\sigma,E,n} e^{-iEt} b_{\sigma,E}^{\dagger} + \phi_{\sigma,0,n} a_{\sigma,0} \tag{27}$$

where the non-vanishing anti-commutators are

$$\left\{ a_{\sigma,E}, a_{\tau,E'}^{\dagger} \right\} = \delta_{EE'} \delta_{\sigma\tau} \tag{28}$$

$$\left\{ b_{\sigma,E}, b_{\tau,E'}^{\dagger} \right\} = \delta_{EE'} \delta_{\sigma\tau} \tag{29}$$

The wave-functions for each spin state are orthonormal

$$\sum_n \phi_{\sigma,E,n}^{*} \phi_{\sigma,E',n} = \delta_{EE'} \ , \ \forall E, E' \tag{30}$$

and they obey a completeness relation

$$\sum_E \psi_{\sigma,E,n} \psi_{\sigma,E,n'}^{*} = \delta_{nn'} \tag{31}$$

In Eq. (31), the summation is over positive, negative and zero energy states.

Let us consider the ground state where all of the negative energy states, for both spin states, are filled and one of the zero modes (here the spin-up zero mode) is filled and the other zero mode is empty. We denote this state by $|\uparrow\rangle$. It is defined as the state of the second quantized system with the properties

$$a_{\sigma,E}|\uparrow\rangle = 0 \ , \ b_{\sigma,E}|\uparrow\rangle = 0 \ , \ \sigma =\uparrow,\downarrow \tag{32}$$

$$a_{\uparrow,0}^{\dagger}|\uparrow\rangle = 0 \ , \ a_{\downarrow,0}|\uparrow\rangle = 0 \tag{33}$$

$$\langle\uparrow\,|\uparrow\rangle = 1 \tag{34}$$

We denote the other overall charge neutral state as $|\downarrow\rangle$ and it is defined by

$$a_{\sigma,E}|\downarrow\rangle = 0 \ , \ b_{\sigma,E}|\downarrow\rangle = 0 \ , \ \sigma =\uparrow,\downarrow \tag{35}$$

$$a_{\downarrow,0}^{\dagger}|\downarrow\rangle = 0 \ , \ a_{\uparrow,0}|\downarrow\rangle = 0 \tag{36}$$

$$\langle\downarrow\,|\downarrow\rangle = 1 \tag{37}$$

The expectation value of the charge density in the state $|\uparrow\rangle$ is given by

$$\langle\uparrow\,|\rho_n|\uparrow\rangle = e \sum_{E<0} \left[\phi_{\uparrow,E,n}^{*}\phi_{\uparrow,E,n} + e\phi_{\downarrow,E,n}^{*}\phi_{\downarrow,E,n}\right] + \phi_{\uparrow,0,n}^{*}\phi_{\uparrow,0,n} - e \tag{38}$$

Now, we use the fact that the negative and positive energy states are related by a simple transformation $\phi_{\sigma,-E} = \Gamma\phi_{\sigma,E}$ or $\phi_{\sigma,-E,n} = (-1)^n\phi_{\sigma,E,n}$ to write the sum over negative energy states in Eq. (38) as half of the sum over all non-zero mode states,

$$\langle\uparrow\,|\rho_n|\uparrow\rangle = \frac{e}{2} \sum_{E\neq0} \left[\phi_{\uparrow,E,n}^{*}\phi_{\uparrow,E,n} + \phi_{\downarrow,E,n}^{*}\phi_{\downarrow,E,n}\right] + e\phi_{\uparrow,0,n}^{*}\phi_{\uparrow,0,n} - e \tag{39}$$

We can then use the completeness relation (31) with $n = n'$ to get

$$\langle\uparrow\,|\rho_n|\uparrow\rangle = \frac{e}{2} \left[2 - \phi_{\uparrow,0,n}^{*}\phi_{\uparrow,0,n} - \phi_{\downarrow,0,n}^{*}\phi_{\downarrow,0,n}\right] + e\phi_{\uparrow,0,n}^{*}\phi_{\uparrow,0,n} - e \tag{40}$$

$$\tag{41}$$

which, upon canceling some terms, yields the expression

$$\boxed{\langle\uparrow\,|\rho_n|\uparrow\rangle = \frac{e}{2}\phi_{\uparrow,0,n}^{*}\phi_{\uparrow,0,n} - \frac{e}{2}\phi_{\downarrow,0,n}^{*}\phi_{\downarrow,0,n}} \tag{42}$$

In addition,

$$\langle\downarrow\,|\rho_n|\downarrow\rangle = -\langle\uparrow\,|\rho_n|\uparrow\rangle \tag{43}$$

and

$$\langle\uparrow\,|\rho_n|\downarrow\rangle = 0 = \langle\downarrow\,|\rho_n|\uparrow\rangle \tag{44}$$

These are our central results.

The charge density is concentrated equally (and with opposite signs) on the densities of the zero-mode wave-functions for each of the spin polarizations. What is interesting here is that they can be in two different locations. In the spin-dependent SSH model, which we present a complete solution of in Appendix B, the zero-mode density $\phi_{\uparrow,0,n}^* \phi_{\uparrow,0,n}$ has support that is localized near one edge of the chain and $\phi_{\downarrow,0,n}^* \phi_{\downarrow,0,n}$ has support that is localized near the other edge of the chain. This fact does not depend on the length of the chains, so in principle the separation of the locations can be very large. The wave-functions of the zero modes decay exponentially with distance from the edge, so for a long chain they would have vanishing overlap. Moreover, all of the other electronic energy levels have an energy gap, $|t-t'|$, so these zero modes can be isolated from the rest of the energy spectrum.

One might wonder in what sense this has to do with fractional charge, as the states that we have discussed are overall charge neutral. They are simultaneous eigenstates of the Hamiltonian and of the total charge operator $Q = \sum_n \rho_n$. The eigenvalue of Q is zero. The expectation value of the charge density is indeed concentrated at the edges of the chain. However, the charge density ρ_n, or the charge density in a subregion of the chain, in spite of being Hermitian operators which could in principle be diagonalized, do not commute with the Hamiltonian, so their eigenstates are not stationary but are an admixture of states with different energies. If we imagine measuring the charge residing, say, on the first site, $n = 1$ of the chain, the measurement collapses the wave-function into the space of states corresponding to an outcome of the experiment, that is, eigenstates of ρ_1, which are integer charges $e, 0, -e$. After such a measurement, one would expect that, with the help of a little dissipation, the system relaxes back to the same neutral ground state and the measurement can be repeated. The average of the outcomes of many such measurements should yield the expectation value of the charge density on the first site of the chain, that is, the value of $\langle \uparrow |\rho_1| \uparrow \rangle$ in Eq. (42) with $n = 1$.

There are other diagnostics of a charge distribution which would be sensitive to its asymmetry in the states $| \uparrow \rangle$ or $| \downarrow \rangle$. For example, in an external electric field of strength E, the chain would experience a torque of magnitude

$$\mathcal{T} = E\alpha \sin \theta \sum_n n \langle \rho_n \rangle \tag{45}$$

where θ is the angle between the field and the chain. The torque exerted on the chain in the states $| \uparrow \rangle$ and $| \downarrow \rangle$ would be in opposite directions.

In the limit where the chain is long and we consider states such that the chain resembles a semi-infinite wire, if the limit is taken in such a way that the Hamiltonian of the semi-infinite system is a self-adjoint operator, the charge of half of the system is time-independent and it and the Hamiltonian have simultaneous eigenstates. The limit of a long chain and the charge operator for one half of it can be defined by a limiting procedure where we define a smeared charge $Q_f = \sum_{n=1}^{N} f(n)\rho_n$ with a test function $f(n)$ which has support near one edge of the chain and then goes to zero in the central region. We then first take the limit $N \to \infty$ where the chain is infinitely long and then we take the limit $f(n) \to 1$. This will yield a charge operator whose eigenvalues are fractional. This procedure was carried out in the context of a slightly different model, with fractionally charged solitons in Ref. [10].

Let us consider the analog of the odd-sited spin chains that we have been discussing in continuum Dirac equations. In this case, the spin-up and the spin-down wave-functions, which we still label as $\begin{bmatrix} u_\sigma(x) \\ v_\sigma(x) \end{bmatrix}$, have the boundary conditions $v_\sigma(0) = 0$ and $v_\sigma(L) = 0$ and they satisfy different equations

$$E \begin{bmatrix} u_\uparrow(x) \\ v_\uparrow(x) \end{bmatrix} = \begin{bmatrix} 0 & \frac{d}{dx} + m \\ -\frac{d}{dx} + m & 0 \end{bmatrix} \begin{bmatrix} u_\uparrow(x) \\ v_\uparrow(x) \end{bmatrix} \tag{46}$$

$$E \begin{bmatrix} u_\downarrow(x) \\ v_\downarrow(x) \end{bmatrix} = \begin{bmatrix} 0 & \frac{d}{dx} - m \\ -\frac{d}{dx} - m & 0 \end{bmatrix} \begin{bmatrix} u_\downarrow(x) \\ v_\downarrow(x) \end{bmatrix} \tag{47}$$

The mass term appears with opposite signs for the spin-up and spin-down wave-functions. Now, the zero-mode wave-functions are

$$\phi_{\uparrow,0} = \begin{bmatrix} \sqrt{\frac{2m}{e^{2mL}-1}} \exp(mx) \\ 0 \end{bmatrix} \tag{48}$$

$$\phi_{\downarrow,0} = \begin{bmatrix} \sqrt{\frac{2m}{e^{2mL}-1}} \exp(m(L-x)) \\ 0 \end{bmatrix} \tag{49}$$

The spin-up zero mode is concentrated near $x = 0$ and the spin-down zero mode is concentrated near $x = L$.

7. Epilogue: The $\{|\uparrow\rangle, |\downarrow\rangle\}$ system as a qubit

The two neutral states $|\uparrow\rangle$ and $|\downarrow\rangle$ are the charge neutral states of the many-fermion system. A transition from these states to a state with excited particles and holes requires energy input to overcome the gap, $|t - t'|$ in the spectrum. There are two further states that are degenerate in energy with them, the state with both zero modes empty,

$$a_\downarrow |\downarrow\rangle \text{ or } a_\uparrow |\uparrow\rangle$$

and the state with both zero modes full,

$$a_\downarrow^\dagger |\uparrow\rangle \text{ or } -a_\uparrow^\dagger |\downarrow\rangle$$

A transition to one of these from either of the states $|\uparrow\rangle$ or $|\downarrow\rangle$ requires the chain to absorb or emit an electron. If such processes are suppressed, the doublet $\{|\uparrow\rangle, |\downarrow\rangle\}$ forms an isolated and interesting two-state system. A dynamical mixing of the states $|\uparrow\rangle$ and $|\downarrow\rangle$ by perturbations to the SSH Hamiltonian would be suppressed as the chain is sufficiently long that the wave-functions of the zero modes which reside at either edge have negligible overlap. The transition would then require an electron to be transferred from one end of the chain to the other, a process which would be inhibited by the fact that the bulk material in between is an insulator with an energy gap $2|t - t'|$. The Zeeman interaction

$$H_{\text{int},z} = \sum_n \frac{1}{2} \mu B_z a_n^\dagger \sigma_z a_n \tag{50}$$

would still work to implement a single-particle z-gate as once it is turned on the phases of the spin-up and spin-down wave-functions evolve with opposite signs,

$$|\uparrow\rangle \to e^{iB_z t/2}|\uparrow\rangle \quad \text{and} \quad |\downarrow\rangle \to e^{-iB_z t/2}|\downarrow\rangle \tag{51}$$

If the system were smaller, so that the zero-mode wave-functions overlap, the Zeeman interactions with magnetic fields perpendicular to the direction of the spin polarization

$$H_{\text{int,x}} = \sum_n \frac{1}{2}\mu B_x\, a_n^\dagger \sigma_x a_n \tag{52}$$

or

$$H_{\text{int,y}} = \sum_n \frac{1}{2}\mu B_y\, a_n^\dagger \sigma_y a_n \tag{53}$$

could be used to implement the single-qubit X- and Y-gates, respectively. However, at the same time, a shorter chain is more vulnerable to accidental environmentally induced transitions between $|\uparrow\rangle$ and $|\downarrow\rangle$.

Appendix A. The SSH chain

In this appendix we will present the explicit solution of the SSH model for an odd-sited chain. The SSH model is a particular one-dimensional tight-binding model where the hopping amplitudes are given by

$$t_{2n} = t, \; t_{2n+1} = t' \tag{54}$$

so that the tight-binding Hamiltonian is

$$H = \sum_{\sigma=\uparrow,\downarrow}^{N} \sum_{n \text{ even}} \left\{ t' a_{\sigma n+1}^\dagger a_{\sigma n} + t' a_{\sigma n}^\dagger a_{\sigma n+1} + t a_{\sigma n+1}^\dagger a_{\sigma n} + t a_{\sigma n}^\dagger a_{\sigma n+1} \right\} \tag{55}$$

The equations for the wave-functions are

$$n \text{ even}: \; E\phi_{E,n} = t\phi_{E,n-1} + t'\phi_{E,n+1} \tag{56}$$

$$n \text{ odd}: \; E\phi_{E,n} = t'\phi_{E,n-1} + t\phi_{E,n+1} \tag{57}$$

with the boundary conditions $\phi_{E,0} = 0$ and $\phi_{E,N+1} = 0$. If N is odd, both boundary conditions apply to the even sites and the wave-functions on even sites must have the form $\sin\frac{\pi kn}{N+1}$ where $k = 1, 2, ..., (N-1)/2$. This function has the property that it vanishes when $n = 0$ and when $n = N + 1$. With this observation, the explicit solutions for the electron wave-functions and energies are

$$n \text{ even}: \; \phi_{E,n} = \sqrt{\frac{2}{N+1}} \sin\frac{\pi kn}{N+1} \tag{58}$$

$$n \text{ odd}: \; \phi_{E,n} = \sqrt{\frac{2}{N+1}} \frac{1}{E(k)} \left[t\sin\frac{\pi k(n+1)}{N+1} + t'\sin\frac{\pi k(n-1)}{N+1} \right] \tag{59}$$

$$E(k) = \pm\sqrt{t^2 + t'^2 + 2tt'\cos\frac{2\pi k}{N+1}} \;, \quad k = 1, 2, ..., (N-1)/2 \tag{60}$$

$$\phi_{0,2n} = 0 \;, \quad \phi_{0,2n+1} = \sqrt{\frac{1 - t'/t}{1 - (t'/t)^{(N+1)/2}}} \left(-\frac{t'}{t}\right)^n \tag{61}$$

The wave-functions in Eqs. (58) and (59) contain the integer parameter k. There are $(N-1)/2$ values of k, each of which corresponds to a positive–negative pair of energy levels with the dispersion relation given in Eq. (60). This makes $N-1$ energy levels for each spin state. Then, there is the additional zero mode whose wave-function is given in Eq. (61). This adds up to the expected N energy levels.

If on the other hand, N is even, we could make the same ansatz for the even sites as we do in Eq. (58). This guarantees the boundary condition $\phi_{E,0} = 0$. However, the condition $\psi_{E,N+1} = 0$ belongs to the odd sites and Eq. (59) with $n = N+1$ gives us the transcendental equation for k

$$0 = t \sin\left(\frac{\pi k(N+2)}{N+1}\right) + t' \sin\left(\frac{\pi k N}{N+1}\right) \tag{62}$$

The solutions of this equation are the allowed wave-vectors.

Appendix B. The spin-dependent SSH model

In this appendix, we will present the solution of the spin-dependent SSH chain. Again, this is a tight-binding model where the hopping parameters differ for the different, up and down, spin polarizations,

$$\text{spin} \uparrow \quad t_{2n} = t, \ t_{2n+1} = t' \tag{63}$$
$$\text{spin} \downarrow \quad t_{2n} = t', \ t_{2n+1} = t \tag{64}$$

The tight-binding Hamiltonian is

$$H = \sum_{n \text{ even}}^{N} \left\{ t' a_{\uparrow n+1}^\dagger a_{\uparrow n} + t' a_{\uparrow n}^\dagger a_{\uparrow n+1} + t a_{\downarrow n+1}^\dagger a_{\downarrow n} + t a_{\downarrow n}^\dagger a_{\downarrow n+1} \right\} \tag{65}$$

$$+ \sum_{n \text{ odd}}^{N} \left\{ t a_{\uparrow n+1}^\dagger a_{\uparrow n} + t a_{\uparrow n}^\dagger a_{\uparrow n+1} + t' a_{\downarrow n+1}^\dagger a_{\downarrow n} + t' a_{\downarrow n}^\dagger a_{\downarrow n+1} \right\} \tag{66}$$

This model still has a partial spin symmetry in that the number operators for the total number of up spins and for the total number of down spins commute with the Hamiltonian separately. The SSH model of Appendix 7 had $U(1) \times SU(2)$ global symmetry. In the present case, this symmetry has been reduced to $U(1) \times U(1)$.

In this case, the spin-up and spin-down electrons each have a zero mode and in fact the energy eigenvalues for the spin-up and spin-down components of the wave-functions are still identical, and are in fact identical to the SSH model of Appendix 7.

What has changed is the fact that the spin-up and spin-down electrons with a given energy no longer have identical wave-functions. In fact, in this highly symmetric situation, the wave-functions are related by the transformation

$$\phi_{\downarrow,E,n} = \phi_{\uparrow,E,N+1-n} \tag{67}$$

Moreover, when $t \neq t'$, the spectrum of non-zero modes has a gap, it obeys $E \geq |t - t'|$ or $E \leq -|t - t'|$.

We shall assume that at charge neutrality the chain has a density of one electron per site. Due to the energy gap for mobile sites, the charge neutral chain is an insulator.

Here, we are assuming that the structure of the chain, that is, the even–odd nature of the hopping amplitudes t and t' is rigid so that the polyacetylene solitons are not allowed. What is exotic here is that, when $t \neq t'$, the spin-up and the spin-down zero modes have spatially separated support. Our arguments of the previous sections tell us that they live at opposite ends of the odd-sited chain. It is easy to confirm this by finding the solutions, which we shall write below.

The equations for the wave-functions of the spin-dependent SSH chain are

$$n \text{ even}: \quad E\phi_{\uparrow,E,n} = t\phi_{\uparrow,E,n-1} + t'\phi_{\uparrow,E,n+1} \tag{68}$$

$$n \text{ odd}: \quad E\phi_{\uparrow,E,n} = t'\phi_{\uparrow,E,n-1} + t\phi_{\uparrow,E,n+1} \tag{69}$$

$$n \text{ even}: \quad E\phi_{\downarrow,E,n} = t'\phi_{\downarrow,E,n-1} + t\phi_{\downarrow,E,n+1} \tag{70}$$

$$n \text{ odd}: \quad E\phi_{\downarrow,E,n} = t\phi_{\downarrow,E,n-1} + t'\phi_{\downarrow,E,n+1} \tag{71}$$

The explicit solutions for the electron wave-functions and energies are

$$n \text{ even}: \quad \phi_{\uparrow,E,n} = \sqrt{\frac{2}{N+1}} \sin \frac{\pi k n}{N+1} \tag{72}$$

$$n \text{ odd}: \quad \phi_{\uparrow,E,n} = \sqrt{\frac{2}{N+1}} \frac{1}{E(k)} \left[t \sin \frac{\pi k(n+1)}{N+1} + t' \sin \frac{\pi k(n-1)}{N+1} \right] \tag{73}$$

$$n \text{ even}: \quad \phi_{\downarrow,E,n} = \sqrt{\frac{2}{N+1}} \sin \frac{\pi k n}{N+1} \tag{74}$$

$$n \text{ odd}: \quad \phi_{\downarrow,E,n} = \sqrt{\frac{2}{N+1}} \frac{1}{E(k)} \left[t' \sin \frac{\pi k(n+1)}{N+1} + t \sin \frac{\pi k(n-1)}{N+1} \right] \tag{75}$$

$$E(k) = \pm\sqrt{(t-t')^2 + 4tt' \cos^2 \frac{\pi k}{N+1}} \ , \quad k = 1, 2, ..., (N-1)/2 \tag{76}$$

$$\phi_{\uparrow,0,2n} = 0 \ , \quad \phi_{\uparrow,0,2n+1} = \sqrt{\frac{1-t'/t}{1-(t'/t)^{(N+1)/2}}} \left(-\frac{t'}{t}\right)^n \tag{77}$$

$$\phi_{\downarrow,0,2n} = 0 \ , \quad \phi_{\downarrow,0,2n+1} = \sqrt{\frac{1-t/t'}{1-(t/t')^{(N+1)/2}}} \left(-\frac{t}{t'}\right)^n \tag{78}$$

The wave-functions in Eqs. (72)–(75) contain the integer parameter k. There are $(N-1)/2$ values of k, each of which corresponds to a positive–negative pair of energy levels with the dispersion relation given in Eq. (76). This makes $N-1$ energy levels for each spin state. Then, there is the additional zero mode for each spin, whose wave-functions are given in Eqs. (77) and (78). This adds up to the expected N energy levels for each spin state.

Acknowledgment

This work is supported by NSERC of Canada.

References

1. R. Jackiw and C. Rebbi, "Solitons with fermion number 1/2," *Phys. Rev. D* **13**, 3398 (1976).
2. R. Jackiw, Dirac Prize Lecture, arXiv:hep-th/9903255.
3. A. J. Niemi and G. W. Semenoff, "Fermion number fractionization in quantum field theory," *Phys. Rep.* **135**, 99 (1986).
4. J. K. Asbóth, L. Oroszlány, A. Pályi, "A short course on topological insulators," arXiv:1509.02295.
5. N. Batra, G. Sheet, "Understanding basic concepts of topological insulators through Su–Schrieffer–Heeger (SSH) model," arXiv:1906.08435.
6. W. P. Su, J. R. Schrieffer and A. J. Heeger, "Solitons in polyacetylene," *Phys. Rev. Lett.* **42**, 1698 (1979).
7. W. P. Su, J. R. Schrieffer and A. J. Heeger, "Soliton excitations in polyacetylene," *Phys. Rev. B* **22**, 2099 (1980).
8. R. Jackiw and J. R. Schrieffer, "Solitons with fermion number 1/2 in condensed matter and relativistic field theories," *Nucl. Phys. B* **190**, 253 (1981).
9. A. Kitaev, "Unpaired Majorana fermions in quantum wires", arXiv:cond-mat/0010440.
10. R. Jackiw, A. K. Kerman, I. R. Klebanov and G. W. Semenoff, "Fluctuations of fractional charge in soliton anti-soliton systems," *Nucl. Phys. B* **225**, 233 (1983), doi:10.1016/0550-3213(83)90051-2.
11. M. Ghrear, B. Mackovic and G. W. Semenoff, "Edge modes and teleportation in a topologically insulating quantum wire," *J. Mod. Phys.* **9**, 2090 (2018) [arXiv:1805.08924 [quant-ph]].

Chapter 22

Anomalies and Topologies in Quantum Field Theories[*]

Gerard 't Hooft

Institute for Theoretical Physics
Utrecht University

Postbox 80.089
3508 TB Utrecht, the Netherlands

e-mail: g.thooft@uu.nl
internet: http://www.staff.science.uu.nl/~hooft101/

The $\pi^0 \to \gamma\gamma$ decay raised peculiar questions in quantum field theory. This led to discussions that culminated in the realisation that topology can be an important feature, implying that quantum corrections can be less symmetric than classical equations. Physicists referred to *anomalies*. Different but related anomalies are also found elsewhere in particle physics, ranging from proton decay in Grand Unified Theories and low energy bound states in technicolor theories, to tetraquarks and beyond.

1. Introduction

Throughout my career as a theoretical physicist, but particularly at the beginning, Roman Jackiw and his work had a big influence on me. In fact when I was still an undergraduate student, my first assignment for a study concerned a delicate effect where an apparent impeccable symmetry was explicitly broken by the effects of renormalization. My advisor, Martinus Veltman, was interested in the decay of a neutral pion into two photons, and had concluded that the decay was forbidden by chiral symmetry, but a good friend of his at CERN, Geneva, John Bell, had written a paper together with Jackiw that led to a different conclusion. There was what was called an 'anomaly'. The Bell–Jackiw anomaly [1] and conceivable ideas on how to 'cure' it, was the subject of my undergraduate thesis. It turned out that there was no cure. The neutral pion indeed does decay into two photons, but totally unexperienced as I was, I thought I could 'remove' the anomaly. That was a mistake I would make once more later, until we realised what it was that really goes on here. I later also came to know Steve Adler and W. Bardeen [2, 3], who had written a more extensive study on the phenomenon showing that higher order

---

[*] Dedicated to Roman Jackiw, on the occasion of his 80th birthday.

corrections do not have any effect on the anomaly, which also means that any non-perturbative version of this kind of field theory should produce the same anomalous phenomenon.

The problem turned out to be very important for the discussion of renormalization of gauge theories, which had caught my interest as a graduate student. How colossal this enterprise would actually be, and how important for particle physics in general, would only become clear later. Not only did we need to study all effects of local gauge invariance on renormalization, but after this was all understood we also had to investigate whether Adler–Bell–Jackiw anomalies, and possibly many sorts of generalisations of that, could jeopardise our renormalization program by completely ruining gauge invariance. After trying all sorts of procedures it eventually turned out that 'dimensional renormalization' did the job: whenever you can generalise your theory to different numbers of space-time dimensions, and you can replace the number of dimensions by a continuous variable, this variable can be used in the renormalization, and no anomalies can affect the symmetry properties needed for renormalization, if the symmetry holds in any number of dimensions. The interaction that makes the pion decay into two photons only has chiral symmetry in the four-dimensional case.

After my appointment as a fellow at CERN, I had a sabbatical at Harvard University, where I frequently met Roman, and he and his wife So-Young Pi became friends. Roman became interested in magnetic monopoles and together with Claudio Rebbi he had found zero-energy solutions for chiral particles that form bound states in a monopole [4,5]. These bound states had special topological properties, and only later we would learn about the connection between this and the anomalies.

I had hit upon a problem with instanton solutions [6] in gauge theories. Due to the Bell–Jackiw anomaly, this instanton should generate interactions that are ruled out by the symmetries of the classical system. How can this be explained?

It turned out that there are Jackiw–Rebbi types of solutions for fermions in an instanton, and this phenomenon put all jig-saw pieces together. The instanton in the Standard Model of the subatomic particles was found to violate the symmetry that stabilises the baryons. Thus, a combination of the Bell–Jackiw anomaly and the Jackiw–Rebbi solutions would generate decays of baryons into lighter particles. The decay is suppressed by large exponentials, but it does occur, and would play important roles in cosmology.

Linking zero-energy solutions to symmetry breaking effects had been the source of purely mathematical discoveries, and we would learn how to link our results to what was known as 'index theorems' in mathematics, for which Michael Atiyah and Isadore Singer would receive the 2004 Abel Prize.

Much later I became involved in discussions concerning the physical interpretation of quantum mechanics. My problem was that I could not believe the stories made up by philosophers of science, who were seriously talking of 'pilot wave functions' and of 'many parallel worlds'. I still don't think these ideas describe accurately what happens. The physical world as I experienced it in my research does not look like this, and there are all sorts of indications that the real reason for what is generally perceived as 'quantum behaviour' must be something much more basic in the ways particles and fields interact. To me it looks as if the physical world can be modelled by what is called a 'cellular automaton'. When Roman saw what I had written down about this idea he insisted to introduce me to a world expert in cellular automaton designs in computer models, Edward Fredkin [7]. Fredkin has also visions about attributing properties of particles and fields to underlying mechanical laws of cellular automata. We became friends as well. Fredkin had done well in the computer world, owning several houses and a private airplane. He and his wife Joyce invited me and my wife to his island in the Caribbean. We discussed physics there, but it was difficult to overcome the gap between the physics of elementary particles and computers programmed on a space-time grid.

2. Anomalies in the Standard Model

When a classical field theory is replaced by its quantum analogue, some of its symmetries might become violated, which highly contrasts against the (older) standard wisdom that every ordinary dynamical, classical system can be viewed as the classical limit of a quantum system exhibiting exactly the same symmetries.

This startling feature was discovered after the first calculations were done for the decay rate of neutral pions into two photons. The amplitude for this rate, $\pi^0 \to \gamma\gamma$, was estimated first by J. Steinberger [8], and later understood more precisely by J. S. Bell and R. Jackiw [1] (see also a much more recent review [9]): the pion first dissociates into a virtual proton plus antiproton, which then annihilate by emitting two photons, the famous triangle diagram. The amplitude can be computed by applying current algebra. The pion field is then assumed to be the divergence of an axial vector current. The virtual protons are generated by this current in a calculable manner, and one finds reasonable agreement with the experimental observation of the decay. Remarkably, the contribution from the virtual proton stays exactly the same if the proton is considered to be composed of three quarks, each occurring in three colors, see Fig. 1. Thus, the quarks out of which the proton is built can be held responsible for the axial vector current regardless whether they first combine to make nucleons, or whether the photons are produced by the quarks directly. This remarkable feat would be used again later, see Section 3.

There was however a deep problem with this result. The axial vector current is directly associated with the conservation of chiral symmetry. If we consider the Lagrangian of either the Gell-Mann–Lévy–Lee–Gervais model [10–12] for nucleons and pions, or a Lagrangian describing the quarks from which these pions and nucleons are built, we find that it is almost invariant under isospin and electromagnetic gauge transformations, while also the chiral isospin symmetry appears to be almost perfect. The pion acts as the Goldstone boson of chiral symmetry, as this symmetry is spontaneously but not explicitly broken. There is one term in the Gell-Mann–

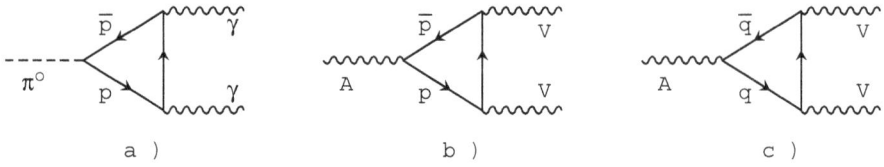

Fig. 1.   The triangle diagram. (a) The $\pi^0 \to \gamma\gamma$ decay, through a virtual $p\bar{p}$ pair. (b) Using current algebra, this amplitude is seen to be related to the amplitude for a vector–vector–axial vector (AVV) interaction, mediated by a closed triangular loop of a virtual proton. This amplitude is proportional to the charge of the proton squared, e^2. (c) The same amplitude can also be obtained by regarding the proton to be built from colored up and down quarks, having charges $+\frac{2}{3}$ and $-\frac{1}{3}$: $\pi^0 = (u\,\bar{u} - d\,\bar{d})$, so the amplitude is proportional to $\sum_{c=1}^{3} \left(\frac{2}{3}e\right)^2 - \left(\frac{1}{3}e\right)^2) = e^2$. In the diagram, $q = u\,or\,d$.

Lévy–Lee model that breaks the symmetry explicitly, which is the pion mass term, $-\frac{1}{2}m_\pi^2\,\pi(x)^2$, where $\pi(x)$ is the pion field, while in the quark model the same term takes the form $-m_q\,\bar{q}(x)\,q(x)$. In these expressions, $m_\pi^2 \propto m_q$ is a relatively small perturbation.

The divergence of the axial vector current betrays to what extent chiral isospin symmetry is explicitly broken. This means that this divergence term must be proportional to the pion mass squared, or equivalently, the masses of the up and the down quark. But these numbers are too small to account for the pion decay into two gammas. It was very puzzling, in the late 1960s, that not only the neutral pion does decay predominantly into two gammas, but also the actual calculation gives, roughly, the correct amplitude. There was something wrong with the symmetry argument. It had been emphasised as a real problem by Crewther [13–15].

Let

$$J_\mu^A = i\bar{\psi}\gamma_\mu\gamma_5\psi \qquad (1)$$

be the axial vector current. To describe the evolution of the π^0, one finds [2,3,16] for the divergence of the axial vector current J_μ^A

$$\partial_\mu J_\mu^A = 2im\,\bar{\psi}\gamma_5\psi - \frac{iNg^2}{16\pi^2}\,F_{\mu\nu}\tilde{F}_{\mu\nu}\ , \qquad (2)$$

The first term at the r.h.s. describes the symmetry breaking effect of the quark mass (or the nucleon mass) m, but the second term, where $\tilde{F}_{\mu\nu} = 1/2\ \varepsilon_{\mu\nu\alpha\beta}F_{\alpha\beta}$, is the surprise. It is due to the renormalization of the one-loop diagrams. It does not contain the fermion fields at all, just the index N counting the number of field species participating. In Fig. 1, we see that the fermions contribute in closed loops (the triangles), which explains why this contribution is proportional to the number of chiral fermion species;[a] g is a gauge coupling constant. It is easy to see that, when we expand this last term in terms of the vector potential, we get products of two creation operators of photons.

---

[a]There is a triangle for every particle species that can go around, and they all add up.

Why does chiral symmetry not protect the vector current from diverging? One can actually redefine the axial current so that it is conserved: $F_{\mu\nu}\tilde{F}_{\mu\nu}$ is the divergence of a Chern–Simons current:

$$F_{\mu\nu}\tilde{F}_{\mu\nu} = \partial_\mu(2A_\nu\tilde{F}_{\mu\nu}) \ . \tag{3}$$

However, the current $A_\nu\tilde{F}_{\mu\nu}$ is not gauge invariant, which is why one cannot require it to vanish at the boundary, and thus the space-time integral of $\partial_\mu J_\mu^A$ does not vanish: a net axial charge may be produced, which here consists of the two formally forbidden photons. Thus, the Adler–Bell–Jackiw anomaly causes a *frustration*: either axial vector gauge symmetry, or the electromagnetic vector symmetry must be explicitly broken; one can't have these symmetries coexisting locally.

In the mid-1970s, the Lagrangian for the strong interactions, Quantum Chromodynamics [17–20], became an important subject for study. Did this theory properly explain the hadronic spectrum? The fact that the pions are more than 6 times lighter than most other hadrons, pointed towards an approximate global chiral symmetry, which is spontaneously broken [10–12].

A small breaking of this symmetry was understood to be caused by the tiny but not vanishing mass values for the up- and the down quarks, the mass term in Eq. (2). In the limit where these two mass terms vanish, the vector symmetry group $U(2)$ is enhanced to become an axial symmetry of the form $U(2)_L \otimes U(2)_R$. This has four more generators than the manifestly unbroken vector symmetry $U(2)_{\text{vector}}$, so that one expects 4 Nambu–Goldstone bosons. There are three pion fields, π^+, π^-, and π^0. Where is the fourth? The missing fourth Goldstone boson would represent a subgroup $U(1)^{\text{axial}}$ of the global gauge symmetry $U(2) \otimes U(2)$.[b]

We actually do have a particle among the hadrons that has the desired quantum numbers, the η meson. Its mass is 547.9 MeV$/c^2$, much more than the masses of the pions, around 135 MeV$/c^2$. So we could also ask: *Why is this thing so heavy?*

The answer seemed to be staring us in the face: it's the Adler–Bell–Jackiw anomaly, again. But now the question arose how to quantify this. How can an anomaly in a technical calculation give rise to a new, calculable, effective interaction term? We saw in Eq. (2) that the anomalous term $F\tilde{F}$ does not contain the chiral fermion field explicitly. Yet we do need a separate explicit interaction that gives mass to the η. That's not just the Chern–Simons term (2). It was in fact proposed that this answer cannot be correct, since the $F\tilde{F}$ term can be written as a divergence [13–15]. Even if we note that, in Eq. (3), the field of which this is a divergence isn't gauge invariant, how do the Feynman rules change if you have a gauge non-invariant boundary contribution?

The solution came when we studied a four-dimensional, topologically stable field configuration, the BPST instanton [6].[c] The fermion field equations have zero-action solutions in this instanton configuration [22]. This would imply that, due to the action of these chiral fermions, the instanton contribution to any functional

---

[b]The Abelian vector part, $U(1)_{\text{vector}}$ represents baryon number conservation.
[c]The word 'instanton' was proposed by the author in Ref. [21].

integral should vanish. Now if we add fermion sources at infinity, we find that these
may lift the action of the fermion bound states away from zero. Apparently, the
Feynman functional integral only does not vanish when we add sources and sinks
for chiral fermions at infinity. It now makes sense to interpret this by saying that
the instanton field configuration actually generates chiral fermion mass terms or
interaction terms, which explicitly flip the chiral charges of these fermions, just
like symmetry breaking terms do. This way, the Chern–Simons term acts as an
interaction term for chiral fermions.

Thus we have to add instanton configurations to the functional integral. As had
been found already by BPST, the instanton generates a contribution to the integral

$$\int d^4 x F_{\mu\nu} \tilde{F}_{\mu\nu} = \frac{32\pi^2}{g^2} \; . \tag{4}$$

Each instanton therefore contributes to the total chiral charge change:

$$Q_2^A - Q_1^A = \int d^4 x \frac{g^2}{16\pi^2} F_{\mu\nu} \tilde{F}_{\mu\nu} = 2N \; . \tag{5}$$

This means that, every flavor contributes a flip of its own contribution to the total
chiral charge from -1 to $+1$. That is, $\Delta Q^A = 2$ for each flavor, totalling to
$\Delta Q^A = 2N$ if there are N flavors.

The amplitude for this transition, in a Feynman-like diagram, is non-vanishing
only if we add the appropriate sources and sinks of chiral charge at infinity. The
physical process, and the mathematical reason as to why zero-action solutions are
involved, was neatly phrased by Jackiw and Rebbi in Refs. [23] and [24].

Thus, the instanton generates interactions of a type that is not directly impli-
cated by the Lagrangian of the theory [16]. The new interaction in essence only
violates chiral symmetries, but if the primary gauge interaction treats left and right
helicity sectors of the fermions differently, the resulting instanton interaction can
violate symmetries in a more spectacular manner. This is indeed what happens
in the electroweak force section of Standard Model. According to the Lagrangian,
baryon number is a strictly conserved quantum number of all well-known elementary particle states. Yet, since the weak interaction handles the right helicity sectors differently from the left helicities, one now finds that the electroweak instanton
causes violation of baryon number conservation. Since the electroweak gauge group
$SU(2) \otimes U(1)$ acts on three families of quarks and leptons, while mixing helicities
with gauge charges, one finds that a single instanton gives rise to only $\Delta B = \pm 3$
interactions (where B stands for baryon number).

The total amplitude for this baryon-number-breaking interaction is exponen-
tially suppressed, typically $\mathcal{O}(e^{-\pi/\alpha})$, where α is the fine structure constant,
$\approx 1/137$ at SM energies, but larger at unification scale energies.[d] This interaction must be of vital importance for cosmological theories: where did the apparent
baryon number excess, as opposed to antibaryons, originate? Unfortunately, the

---

[d]It is actually more accurate to write $e^{-\mathcal{O}(\pi/\alpha)}$.

amplitude just mentioned seems to be too small to explain the observed situation today, but of course today's models of the Big Bang and all associated phenomena may well be far too premature to fit all data available. There are still many aspects of the electroweak and possibly other forces that are not yet understood, but efforts to improve our understanding will continue, and we won't stop until everything comes together.

3. Other anomalies of similar kinds

Having stumbled upon zero-energy solutions of fermions in three-dimensional topological objects (magnetic monopoles), and zero-action solutions on instantons in four space-time dimensions, one could ask what the effects are of related mathematical structures in one or two dimensions. In one dimension, topologically stable structures take the form of domain boundaries (sheets), and the zero-energy solutions are massless chiral fermions living on these sheets. In two dimensions, we have the Abrikosov vortex lines, where fermions may be attached to the bulk region of such string-like objects.

In lattice approaches to QCD, it was found that space-time can be represented as a domain wall in a five-dimensional theory, where fermions are attached, as states, to the domain wall. This is an elegant way to introduce massless fermions on the lattice, without explicitly breaking the chiral symmetry.

A related application of instantons was found in technicolor theories [25]. These are theories in which particles considered to be elementary today, are represented as bound states, much like baryons and mesons in QCD. A problem in such theories is the question how one can obtain very light states — think of electrons and neutrinos — to act as bound states. One might argue that this is impossible, considering the ultra-strong forces involved. But QCD itself may be showing possible openings here: the pions are bound states, and they are light, as they are protected by chiral symmetry. One would like to use similar protection mechanisms in technicolor theories to explain why electrons and neutrinos are light; even the electroweak gauge bosons and the Higgs particle do not carry much mass in the light of the high energy scale where the technicolor fields are roaming.

In technicolor theory also, we attribute the light masses to some form of chiral symmetry. How could we attribute the observed algebra of known, light particles to chiral forces?

Here, we proposed to use the observation mentioned in the introduction, that the Adler–Bell–Jackiw anomaly generates the $\pi^0 \to \gamma\gamma$ decay amplitude, regardless whether we use the baryons P and N or the quarks u, d, and s as our basic fields. Keeping this in mind, it was now proposed to use *spectator gauge fields* taking the shape of instantons with different possible topological charges, and compare the anomalies in the framework of the baryonic bound states, with what contributions would be generated by the elementary 'techni-quarks'. The theory would not be consistent if these calculations would not generate the same anomalies, as the $\pi^0 \to$

$\gamma\gamma$ decay would not be consistent if the result would be different for the nucleons on one hand and the quarks on the other. As we can calculate the anomalies exactly, we can impose this 'anomaly matching condition', in order to establish what kinds of light fermionic or bosonic bound states might be formed [26].

In spite of extensive experimental searches, technicolor effects were not found up to the highest energies available in accelerators, so the theory of the anomaly matching constraints could not be checked against observations.

4. Tetraquarks and pentaquarks

In more recent times, QCD interactions could be investigated with more accuracy. We now have many more hadronic states, and could inspect how accurately the properties are explained by the theory. A new puzzle appeared [27]: the masses of mesons and baryons depend on the number of strange quarks present, since strangeness carries some 100 MeV more mass than the up and down quarks. For some of the heavier hadron states, the strangeness dependence of the masses appears to be different from this expectation. It was suspected that some of the mesons are 'tetraquarks', containing four quarks rather than two, and some baryonic states contain five quarks rather than three ('pentaquarks'). But even then, it looks as if strong transitions take place between strange and non-strange constituent quarks.

Now enter the instanton configuration of the QCD color fields. Ignoring temporarily the heaviest flavor quarks, from which only small correction effects are expected, the lighter flavor quarks all have zero-action configurations in an instanton. This means that, at the space-time location of one instanton, effectively one u quark, one d and one s quark flip their helicities (charm, top and bottom quarks also flip their helicities but their mass terms compensate for that). This results in an effective 6-quark interaction (three flavors in with helicities $+1$, and three out with helicities inverted). This effective interaction can cause a state with two quarks to transform into a state with three different quarks and one antiquark, and back. Thus baryons consisting of three quarks can transform into states containing four quarks and an antiquark. Similarly, mesons with the familiar quark composition can transform into a state of two quarks and two antiquarks. Since strangeness is strongly involved, the strangeness contents can vary anomalously by this interaction (by violating the Zweig rule). It was found that this effect can explain quite neatly the mass spectra observed [28].

Topologically stable field configurations in any dimensions, in any kind of theory, are still a fruitful source of inspiring physics. In particular when imagining quantum theories of gravity, we feel like opening Pandora's box. The world has not yet seen the end of it.

References

1. J. S. Bell and R. Jackiw, "A *PCAC* puzzle: $\pi^0 \to \gamma\gamma$ in the σ model", *Nuovo Cimento* **A60**, 47 (1969).
2. S. L. Adler, "Axial vector vertex in spinor electrodynamics", *Phys. Rev.* **117**, 2426 (1969).
3. S. L. Adler and W. A. Bardeen, "Absence of higher order corrections in the anomalous axial vector divergence equation", *Phys. Rev.* **182**, 1517 (1969).
4. R. Jackiw and C. Rebbi, "Solitons with fermion number 1/2", *Phys. Rev.* **D13**, 3398 (1976).
5. R. Jackiw and C. Rebbi, "Spin from isospin in a gauge theory", *Phys. Rev. Lett.* **36**, 1116 (1976).
6. A. A. Belavin, A. M. Polyakov, A. S. Schwartz and Yu. S. Tyupkin, "Pseudoparticle solutions of the Yang–Mills equations", *Phys. Lett.* **59B**, 85 (1975).
7. E. Fredkin, "Digital information mechanics," Privately circulated paper (1982).
8. J. Steinberger, *Phys. Rev.* **76**, 1180 (1949).
9. A. M. Bernstein and B. R. Holstein, "Neutral pion lifetime measurements and the QCD chiral anomaly", *Rev. Mod. Phys.* **85**, 49 (2013), arXiv:1112.4809 [hep-ph].
10. M. Gell-Mann and M. Lévy, "The axial vector current in beta decay", *Il Nuovo Cimento (1955-1965)* **16**, 705 (1960).
11. B. W. Lee, "Renormalization of the sigma model", *Nucl. Phys.* **B9**, 649 (1969).
12. J.-L. Gervais and B. W. Lee, "Renormalization of the sigma-model (II) fermion fields and regularization", *Nucl. Phys.* **B12**, 627 (1969).
13. R. Crewther, *Phys. Lett.* **70B**, 349 (1977).
14. R. Crewther, *Riv. Nuovo Cim.* **2**, 63 (1979).
15. R. Crewther, *Acta Phys. Austriaca Suppl.* **XIX**, 47 (1978).
16. G. 't Hooft, "How instantons solve the $U(1)$ problem", *Phys. Rep.* **142**, 357 (1986).
17. D. J. Gross and F. Wilczek, "Ultraviolet behavior of non-Abelian gauge theories", *Phys. Rev. Lett.* **30**, 1343 (1973).
18. D. J. Gross, J. A. Harvey, E. Martinec and R. Rohm, "Heterotic string theory (I). The free heterotic string". *Nucl. Phys.* **B256**, 253 (1985).
19. H. D. Politzer, "Reliable pertubative results for strong interactions?" *Phys. Rev. Lett.* **30**, 1346 (1973).
20. H. D. Politzer, "Asymptotic freedom: An approach to strong interactions", *Phys. Rep.* **14C**, 129 (1974).
21. G. 't Hooft, "Symmetry breaking through Bell–Jackiw anomalies", *Phys. Rev. Lett.* **37**, 8 (1976).
22. G. 't Hooft, "Computation of the quantum effects due to a four-dimensional pseudoparticle", *Phys. Rev.* **D14**, 3432 (1976).
23. R. Jackiw and C. Rebbi, "Vacuum periodicity in a Yang–Mills quantum theory", *Phys. Rev. Lett.* **37**, 172 (1976).
24. R. Jackiw and C. Rebbi, "Spinor analysis of Yang–Mills theory", *Phys. Rev.* **D16**, 1052 (1977).
25. E. Eichten and K. D. Lane, "Dynamical breaking of weak interaction symmetries", *Phys. Lett.* **90B**, 125 (1980).
26. G. 't Hooft, "Naturalness, chiral symmetry, and spontaneous chiral symmetry breaking", in *Cargèse Summer Inst.*, Aug. 26–Sep. 8, 1979, *Recent Developments in Gauge Theories*, eds. G. 't Hooft *et al.* (Plenum Press, New York, 1980), pp. 135–157. Reprinted in *Dynamical Symmetry Breaking, A Collection of Reprints*, eds. E. Farhi *et al.* (World Scientific, Singapore, 1982), pp. 345–367.

27. L. Maiani *et al.*, "Diquark–antidiquarks with hidden or open charm and the nature of $X(3872)$", *Phys. Rev. Lett.* **93**, 212002 (2004).
28. G. 't Hooft, G. Isidori, L. Maiani, A. D. Polosa and V. Riquer, "A theory of scalar mesons", *Phys. Lett.* **B662**, 424–430 (2008), arXiv:0801.2288.

© 2020 World Scientific Publishing Company
https://doi.org/10.1142/9789811210679_0023

Chapter 23

Superinsulators, a Toy Realization of QCD in Condensed Matter

M. Cristina Diamantini

NiPS Laboratory, INFN and Dipartimento di Fisica e Geologia, University of Perugia,
via A. Pascoli, I-06100 Perugia, Italy,
cristina.diamantini@pg.infn.it

Carlo A. Trugenberger

SwissScientific Technologies SA,
rue du Rhone 59, CH-1204 Geneva, Switzerland,
c.trugenberger@swissscientific.com

Superinsulators are dual superconductors, dissipationless magnetic monopole condensates with infinite resistance. The long-distance field theory of such states of matter is QED with dynamical matter coupled via a compact BF topological interaction. We will quantize the 2D model in the functional Schrödinger picture and show how strong entanglement of charges leads to a phase which is a single-color, asymptotically free version of QCD in which the infinite resistance is caused by the linear confinement of charges. This phase has been experimentally detected in TiN, NbTiN and InO superconducting thin films, including signatures of asymptotically free behaviour and of the dual, electric Meissner effect. This makes superinsulators a "toy realization" of QCD with Cooper pairs playing the role of quarks.

Foreword (by Carlo A. Trugenberger)

I first met Roman when I knocked on the door of his office at MIT to ask him if I could come to work with him as a postdoc and he gave me ten minutes at the blackboard to convince him. When I eventually took up my post a year later he showed me around the Centre for Theoretical Physics and told me I could either have lunch at Legal Seafood just outside MIT or have a sandwich with the "boys" in the common room. When I showed up there, I realized that the "boys" were actually the senior faculty members, including such luminaries as Jeffrey Goldstone and Manhattan project legend Viktor Weisskopf. It is thus that I understood that Roman's world was divided between the "boys" and the rest and I realized the importance of entering his circle of "young boys". Once you had his consideration as such, he gave you his full attention and treated you as "scientific family". Roman was a wonderful teacher to me and, over the years, has become a good friend to me and my wife Cristina Diamantini, with whom I am writing this contribution. It is fair to say that most physics I know today I have learnt from him, including the meaning of "frustrated system", which was impressed upon me when I had to bring Roman to shop on Newbury street and he ordered me to stop the car and wait for him right in the middle of the street while a cop was urging me insistently to move on.

Roman's contributions to physics are manyfold, essential and of great importance to both particle physics and condensed matter physics. Barring the axial anomaly and charge fractionalization, which made him famous and earned him the Dirac medal, he was paramount in the development of high-temperature field theory and the study of topological properties of gauge theories. Above all, he introduced in physics the famed topological Chern–Simons term, which was to dominate a large part of gauge theory studies for the next 30 years, till today. This is the subject on which I had the privilege to collaborate with him in the late 80s and it is also the topic of the present contribution.

Chern–Simons theories have emerged as the long-distance effective field theories for a host of strongly correlated condensed matter systems, most notably the fractional quantum Hall liquids. When the Chern–Simons term couples two different gauge fields one speaks of a mixed, or doubled Chern–Simons term. Cristina Diamantini, Pasquale Sodano and I have introduced such mixed Chern–Simons terms as effective field theories for condensed matter systems, which has led to the prediction of a new state of matter, superinsulators, whose realization as a phase of 2D superconducting films is now established. Superinsulation, the dual of superconductivity, is the topic of the present essay, whose title could as well be "The Physics of Mixed Topologically Massive Gauge Theories". I hope that Roman will enjoy reading how his ideas have had profound consequences and lead directly to a toy version of QCD which is concretely realized in condensed matter systems. Happy birthday and thank you Roman!

1. Introduction

In (2+1) dimensions, gauge theories can be augmented by a topological term, the famed Chern–Simons (CS) term.[1-3] When this couples two different Abelian gauge fields, a vector and a pseudovector, one speaks of a mixed, or doubled CS term. In this paper we will consider an Abelian gauge model involving such a mixed CS term and the two corresponding Maxwell actions. This model is the (2+1)-dimensional version of topological BF models[4,5] in any number of dimensions.

Such mixed CS and BF models were introduced as long-distance effective theories of condensed matter systems in Ref. 6, where it was shown that they model the superconductor to insulator transition (SIT)[7-11] in Josephson junction arrays and thin superconducting films. Specifically, three phases were found when both

gauge symmetries are compact, with gauge group U(1), and contain thus topological excitations describing vortices and point charges. First, the U(1) × U(1)
phase with dilute topological excitations describes[12] the intermediate Bose metal
phase,[13,14*] realizing a U(1) ⋊ $\mathbb{Z}_2^T$ bosonic topological insulator.[16–18] When one
of the two gauge symmetries is broken to $\mathbb{Z}$, instead, we have the superconductor
and the superinsulator phases,[6,19–27] respectively. The latter is the subject of the
present paper.

Superinsulators are a condensed matter realization of compact QED,[26,27] the
simplest example of a strongly coupled gauge theory with a massive photon and
linear confinement of charges.[28,29] While the pure gauge model, with only closed
string excitations[30,31] is non-renormalizable, it is known that coupling fermions, instead, does lead to a non-trivial fixed point of the renormalization group flow.[32] The
situation is, indeed controversial in the compact Abelian Higgs model, one result
pointing to the existence of a fixed point,[33] another showing its absence.[34] Here we
show that deep non-relativistic QED coupled to dynamical matter via a compact
mixed Chern–Simons term has a Berezinskii–Kosterlitz–Thouless (BKT)[35–37] fixed
point separating an integer topological phase[16–18] from a confined phase. The integer topological phase[16–18] corresponds to a functional first Landau level and consists
of an intertwined incompressible fluid of charges and vortices with a gap set by the
CS mass. The confined phase is a highly entangled vortex condensate in which
charges get bound to the ends of electric strings, and the theory is asymptotically
free, the BKT transition representing the strong-coupling infrared (IR) confinement
phase. This is the superinsulation phase, realizing 't Hooft's old idea of quark confinement as dual superconductivity.[38] Remarkably, this superinsulation phase has
been experimentally observed in TiN, NbTiN and InO thin films,[20–25] including the
explicit realization of confinement, asymptotic freedom and of the electric Meissner
state.[39]

2. The model

We consider a (2+1)-dimensional, non-relativistic model of dynamical matter coupled to electromagnetic gauge fields A_μ via a mixed CS term,

$$S = \int dt\, d^2x\, \frac{-v}{2e_0^2} F_0 F^0 + \frac{-1}{2e_0^2 v} F_i F^i + \frac{q}{2\pi} A_\mu \epsilon^{\mu\alpha\nu} \partial_\alpha b_\nu + \frac{-v}{2g_0^2} f_0 f^0 + \frac{-1}{2g_0^2 v} f_i f^i\,, \quad (1)$$

where e_0^2 is the gauge coupling constant, with dimension [mass] in 2+1 dimensions
(we use natural units $c = 1$, $\hbar = 1$), and v is the speed of light, a dimensionless
number smaller than one in our units. Matter is formulated itself in terms of a
fictitious pseudovector gauge field b_μ so that $j^\mu = (q/2\pi) f^\mu$, with $f^\mu = \epsilon^{\mu\alpha\nu} \partial_\alpha b_\nu$
the dual field strength, represents the conserved charge current. Correspondingly,
$\phi^\mu = (1/2\pi) F^\mu$, $F^\mu = \epsilon^{\mu\alpha\nu} \partial_\alpha A_\nu$, is the vortex current. When the gauge symmetries
are taken as compact, with radius 2π and $2\pi/q$ for U(1)$_b$ and U(1)$_A$, respectively,

---

*For a review, see Ref. 15.

$q \in \mathbb{Z}$ plays the role of the charge quantum. The coupling g_0^2, also with dimension [mass], sets the energy scale of matter fluctutations.

In applications to condensed matter systems, the relevant limit is the one in which the speed of light $v = 1/\sqrt{\varepsilon\mu} \ll 1$ due to a very high dielectric permittivity ε (while the magnetic susceptibility $\mu = O(1)$). We will thus consider the model (1) in the limit $v \to 0$, in which only electric fields survive. This limit has been called the "strong coupling limit" in Ref. 40; it is however, rather the deep non-relativistic limit (DNRL) and we will henceforth call it like that. The action is thus

$$S = \int dt d^2x \, \frac{-1}{2e^2} F_i F^i + \frac{q}{2\pi} A_\mu \epsilon^{\mu\alpha\nu} \partial_\alpha b_\nu + \frac{-1}{2g^2} f_i f^i \,, \tag{2}$$

where we have reabsorbed the factor v in a redefinition of the coupling constants, $e^2 = e_0^2 v$, $g^2 = g_0^2 v$.

3. Functional Landau levels

We shall quantize the model (2) in the functional Schrödinger picture. As usual for a gauge theory, the gauge components A^0 and b^0 are not dynamical fields, since they never appear with time derivatives. They are Lagrange multipliers, whose associated Gauss law constraints implement gauge invariance. They can be set to zero, $A^0 = 0$ and $b^0 = 0$, after imposing the corresponding Gauss law constraints. This is called the Weyl gauge.

The two canonical momenta conjugate to the canonical variables A^i and b^i are:

$$\mathcal{P}_A^i = \frac{\delta\mathcal{L}}{\delta(\partial_0 A^i)} = \frac{1}{e^2} F^{0i} + \frac{q}{4\pi} \epsilon^{ij} b^j \,,$$

$$\mathcal{P}_b^i = \frac{\delta\mathcal{L}}{\delta(\partial_0 b^i)} = \frac{1}{g^2} f^{0i} + \frac{q}{4\pi} \epsilon^{ij} A^j \,. \tag{3}$$

They are realized as functional derivatives,

$$\mathcal{P}_A^i = -i\frac{\delta}{\delta A^i} \,, \qquad \mathcal{P}_b^i = -i\frac{\delta}{\delta b^i} \,. \tag{4}$$

The Hamiltonian density, when written in canonical variables takes the form

$$\mathcal{H} = \frac{e^2}{2} \left(\Pi_A^i\right)^2 + \frac{g^2}{2} \left(\Pi_b^i\right)^2 \,, \tag{5}$$

where

$$\Pi_A^i = \mathcal{P}_A^i - \frac{q}{4\pi} \epsilon^{ij} b^j \,,$$

$$\Pi_b^i = \mathcal{P}_b^i - \frac{q}{4\pi} \epsilon^{ij} A^j \,, \tag{6}$$

are the kinetic momenta. Due to the Chern–Simons term, the kinetic momenta do not commute,

$$\left[\Pi_a^i(\mathbf{x}), \Pi_b^j(\mathbf{y})\right] = -i \frac{q}{2\pi} \epsilon^{ij} \delta^2(\mathbf{x} - \mathbf{y}) \,. \tag{7}$$

which is tantamount to the presence of a non-trivial functional gauge connection

$$\mathcal{A}_A^i(A,b) = \frac{q}{4\pi}\epsilon^{ij}b_j \ ,$$

$$\mathcal{A}_b^i(A,b) = \frac{q}{4\pi}\epsilon^{ij}A_j \ , \tag{8}$$

in the theory. Canonical momenta are not functional gauge invariant quantities anymore, since they can be traded with the connection by a functional gauge transformation of the wave functionals,

$$\Psi\left[A^i,b^i\right] \to e^{i\frac{q}{4\pi}\Lambda(A,b)}\,\Psi\left[A^i,b^i\right] \ ,$$

$$\epsilon^{ij}b^j \to \epsilon^{ij}b^j + \frac{\delta}{\delta A^i}\Lambda(A,b) \ ,$$

$$\epsilon^{ij}A^j \to \epsilon^{ij}A^j + \frac{\delta}{\delta b^i}\Lambda(A,b) \ ,$$

$$\partial_i\frac{\delta}{\delta A^i}\Lambda\left(A^i,b^i\right) = 0 \ , \quad \partial_i\frac{\delta}{\delta b^i}\Lambda\left(A^i,b^i\right) = 0 \ , \tag{9}$$

where the last conditions are required to respect traditional gauge invariance, encoded in the Gauss law constraints, see below. Only the electric fields Π_A^i and the charge currents Π_b^i are well-defined gauge invariant quantities.

The functional connection (8) is not pure gauge. The quantity

$$\mathcal{B}^{ij}(\mathbf{x}-\mathbf{y}) = \frac{\delta}{\delta A^i(\mathbf{x})}\mathcal{A}_b^j(A(\mathbf{y}),b(\mathbf{y})) - \frac{\delta}{\delta b^j(\mathbf{y})}\mathcal{A}_A^i(A(\mathbf{x}),b(\mathbf{x})) = -\frac{q}{2\pi}\epsilon^{ij}\delta^2(\mathbf{x}-\mathbf{y}) \ , \tag{10}$$

plays the role of a functional uniform magnetic field and appears as the commutator of kinetic momenta, exactly as in the traditional Landau problem of electrons in an external magnetic field. This shows that the Chern–Simons term plays the role of a functional magnetic field, i.e. of a non-trivial curvature in configuration space.

Exactly as in the standard problem of Landau levels we can define lowering and raising operators

$$\mathcal{A}^i = \sqrt{\frac{\pi}{qeg}}\left(e\Pi_A^i - ige^{ij}\Pi_b^j\right) \ ,$$

$$\mathcal{A}^{i\dagger} = \sqrt{\frac{\pi}{qeg}}\left(el l_A^i + ige^{ij}l l_b^j\right) \ , \tag{11}$$

with commutation relation

$$\left[\mathcal{A}^i(\mathbf{x}),\mathcal{A}^{j\dagger}(\mathbf{y})\right] = \delta^{ij}\,\delta^2(\mathbf{x}-\mathbf{y}) \ . \tag{12}$$

In terms of these, the Hamiltonian takes the familiar form

$$H = m\sum_i\int d^2\mathbf{x}\left(\mathcal{A}^{i\dagger}(\mathbf{x})\mathcal{A}^i(\mathbf{x}) + \frac{1}{2}\,\delta^{ii}\,\delta^2(\mathbf{0})\right) \ , \tag{13}$$

where the second term represents the infinite ground state energy that has to be subtracted and $m = egq/2\pi$ is the topological energy gap. Finally, the Gauss law operators, implementing standard gauge invariance, are the constraints associated with the Lagrange multipliers A_0 and b_0,

$$G_A \equiv \partial_i \mathcal{P}_A^i + \frac{q}{4\pi} \partial_i \epsilon^{ij} b^j \ ,$$

$$G_b \equiv \partial_i \mathcal{P}_b^i + \frac{q}{4\pi} \partial_i \epsilon^{ij} A^j \ . \tag{14}$$

At the quantum level these constraints must be imposed as conditions on physical states:

$$G_A \Psi[A^i, b^i] = 0 \ , \qquad G_b \Psi[A^i, b^i] = 0 \ . \tag{15}$$

The ground state wave functional Ψ_0 is thus given by the symmetric gauge functional first Landau level, defined by $\mathcal{A}^i(\mathbf{x})\Psi_0[A^i, b^i] = 0$, subject to the gauge constraints (15).

Localized excited states of unit norm and energy m are created by the operators

$$\mathcal{A}_f^\dagger = \int d^2\mathbf{x} \ f(\mathbf{x} - \mathbf{x}_0) \ \hat{n}^i \mathcal{A}^{i\dagger} \ ,$$

$$\int d^2\mathbf{x} \ f^2(\mathbf{x} - \mathbf{x}_0) = 1 \ , \tag{16}$$

with commutation relations

$$\left[\mathcal{A}_f, \mathcal{A}_f^\dagger \right] = 1 \ ,$$

$$\left[H, \mathcal{A}_f^\dagger \right] = m \mathcal{A}_f^\dagger \ . \tag{17}$$

These represent extended superpositions of matter and gauge fields. Their energy does not depend on the form factor $f(\mathbf{x}-\mathbf{x}_0)$, as long as it satisfies the normalization condition (16).

In the original variables of (1), the topological energy gap is $m = q e_0 g_0 v/2\pi$. In applications to condensed matter physics the relevant length scales are given by $1/e_0^2 = d/\alpha$ and $1/g_0^2 = \alpha\lambda^2/\pi^2 d$, where α is the fine structure constant, λ is the London penetration depth of the superconducting material and d is the film thickness, of the order of the coherence length. Therefore, the topological energy gap reduces to $m = (1/k)(v/d)$, where k is the Ginzburg–Landau parameter of the material and we have considered $q = 2$ for Cooper pairs. We shall consider the 2D non-relativistic limit $d \to 0$, $v \to 0$ so that v/d is the highest frequency in the problem. Therefore, the higher Landau levels decouple and all relevant physics takes place in the lowest Landau level.

Following Refs. 41 and 42, we write the ground state functional as the product of a phase and a contribution that depends only on the transverse components of the two dynamical variables, A_T^i and b_T^i:

$$\Psi_0[A^i, b^i] = e^{i\chi\left(A^i, b^i\right)} \ \Phi(A_T^i, b_T^i) \ ,$$

$$\chi[A^i, b^i] = \frac{q}{4\pi} \int d^2\mathbf{x} \left(b \frac{\partial_i}{\Delta} A^i + A \frac{\partial_i}{\Delta} b^i \right) \ ,$$

$$\Phi[A_T^i, b_T^i] = \exp\left[\frac{-q}{4\pi} \int d^2\mathbf{x} \left(\frac{g}{e}(A_T^i)^2 + \frac{e}{g}(b_T^i)^2 \right) \right] \ , \tag{18}$$

where $A = \epsilon^{ij}\partial_i A^j$, $b = \epsilon^{ij}\partial_i b^j$, $\Delta = \partial_i \partial_i$ and $A_T^i = P^{ij} A^j$, $b_T^i = P^{ij} b^j$, with the projector P^{ij} onto the transverse part of the gauge fields given by $P^{ij} = \left(\delta^{ij} - \frac{\partial^i \partial^j}{\Delta}\right)$. Using the Hodge decomposition for the spatial components of the two gauge fields A^i and b^i:

$$A^i = \partial_i \xi + \epsilon^{ij}\partial_j \phi \,,$$
$$b^i = \partial_i \lambda + \epsilon^{ij}\partial_j \psi \,, \tag{19}$$

we can rewrite $\Psi_0[A^i, b^i]$ as:

$$\Psi_0[A^i, b^i] = \exp\left[\frac{iq}{4\pi}\int d^2\mathbf{x}\,(\psi\Delta\xi + \phi\Delta\lambda)\right]\exp\left[\frac{-q}{4\pi}\int d^2\mathbf{x}\,\left(\kappa(\partial_i\phi)^2 + \frac{1}{\kappa}(\partial_i\psi)^2\right)\right]\,, \tag{20}$$

where $\kappa = g/e$ represents the dimensionless coupling constant of the theory. As always in Chern–Simons gauge theories, gauge invariance is realized with a 1-cocycle,[41] which manifests itself in the phase in (18) and (20). This can be expressed also as

$$e^{i\chi(A^i, b^i)} = \exp\left[\frac{i}{2}\int d^2\mathbf{x}\,(q\lambda\phi^0 + \xi j^0)\right]\,. \tag{21}$$

Two possibilities have to be considered. In the simplest case both gauge symmetries are non-compact, with gauge group $\mathbb{R}$. In this case, neither charges nor vortices are quantized and ground state quantum correlation functions of their densities are given by

$$\langle j^0(\mathbf{x})j^0(\mathbf{y})\rangle_c = \frac{1}{4\pi^2}\frac{1}{Z_\psi}\int \mathcal{D}\psi\,\Delta\psi(\mathbf{x})\Delta\psi(\mathbf{y})\,\exp\left[\frac{-q}{2\pi\kappa}\int d^2\mathbf{x}\,(\partial_i\psi)^2\right]\,,$$
$$Z_\psi = \int \mathcal{D}\psi\,\exp\left[\frac{-q}{2\pi\kappa}\int d^2\mathbf{x}\,(\partial_i\psi)^2\right]\,, \tag{22}$$

where the subscript "c" denotes connected correlation functions and with an analogous expression for vortices in terms of the field ϕ and with $\kappa \to 1/\kappa$. This gives

$$\langle j^0(\mathbf{x})j^0(\mathbf{y})\rangle_c = \frac{\pi\kappa}{q}\Delta\delta^2(\mathbf{x}-\mathbf{y})\,, \quad \langle \phi^0(\mathbf{x})\phi^0(\mathbf{y})\rangle_c = \frac{\pi}{q\kappa}\Delta\delta^2(\mathbf{x}-\mathbf{y})\,, \tag{23}$$

which represent short-range, screened correlations. The vanishing screening length implied by these expressions is a consequence of the deep non-relativistic limit $v \to 0$ in (2). In the general case (1) the real part of the ground state wave functional (18) is modified to

$$\Phi[A_T^i, b_T^i] = \exp\left[-\int d^2\mathbf{x}\,\left(\frac{1}{2e^2}A_T^i\sqrt{m^2 - v^2\Delta}\,A_T^i + \frac{1}{2y^2}b_T^i\sqrt{m^2 - v^2\Delta}\,b_T^i\right)\right]\,, \tag{24}$$

which shows that both charge and vortex densities are correlated on a typical length $\lambda_{\rm corr} = v/m = \lambda$.

In the Hamiltonian formalism, the operator that inserts external point charges $\pm Q$ at $\mathbf{x}_1$ and $\mathbf{x}_2$ is the exponential line integral,[43]

$$\Psi_0^\rho[A^i, b^i] = \exp\left[iQ\int_{\mathbf{x}_1}^{\mathbf{x}_2} A^i dx^i\right]\Psi_0[A^i, b^i]\,,$$
$$\rho = Q\,\delta^2(\mathbf{x}-\mathbf{x}_1) - Q\,\delta^2(\mathbf{x}-\mathbf{x}_2)\,. \tag{25}$$

Using the functional gauge invariance (9) (without the last, zero divergence condition due to the presence of an external charge distribution), this can be reabsorbed by a simple shift

$$\epsilon^{ij}b^j \rightarrow \epsilon^{ij}b^j + \frac{4\pi Q}{q}\frac{\partial_i}{\Delta}\rho \,,$$

(26)

which amounts to a corresponding shift

$$\Delta E_0 = \int d^2\mathbf{x}\, \rho\frac{1}{\Delta}\rho \,,$$

(27)

in the ground state energy, showing that, in this phase, external probe charges interact logarithmically. Note that this is not in contradiction with short-range charge density correlation functions. It is only *external* probes that interact logarithmically, *dynamical* charges in the model are screened by the topological interactions.

4. Confinement and asymptotic freedom

Things change when the two gauge symmetries are compact U(1) instead of $\mathbb{R}$. In this case, the fields ξ and λ in (19) are angles and the identity $\epsilon^{ij}\partial_i\partial_j\theta = 2\pi\delta^2(\mathbf{x})$, in polar coordinates r, θ, implies the existence of quantized vortices and point charges. As we now show, in this U(1) × U(1) case, the fields ξ and λ become new dynamical fields embedding the dynamics of these additional degrees of freedom.

We start by noting that, in case of compact gauge symmetries, the cocycle is changed to

$$e^{i\chi\left(A^i, b^i\right)} = \exp\left[\frac{i}{2}\int d^2\mathbf{x}\left(q\lambda\phi^0 + \xi j^0\right)\right]$$

(28)

$$\times \exp\left[i\frac{q}{2}\sum_i\int d^2\mathbf{x}\xi(\mathbf{x})N(\mathbf{x},\mathbf{x}_i N_i) + \frac{i}{2}\sum_i\int d^2\mathbf{x}\lambda(\mathbf{x})\Phi(\mathbf{x},\mathbf{x}_i\Phi_i)\right] \,,$$

where

$$N\left(\mathbf{x},\mathbf{x}_i, N_i\right) = N_i\,\delta^2\left(\mathbf{x} - \mathbf{x}_i\right) \,,$$
$$\Phi\left(\mathbf{x},\mathbf{x}_i, \Phi_i\right) = \Phi_i\,\delta^2\left(\mathbf{x} - \mathbf{x}_i\right) \,,$$

(29)

represent the additional point particle and vortex degrees of freedom. The integers N_i and Φ_i encode the particle and vortex numbers, while $\mathbf{x}_i$ denotes their locations. We shall now consider gauge sector observables in entangled mixed states in which the charge degrees of freedom $N(\mathbf{x},\mathbf{x}_i, N_i)$ and $\psi(\mathbf{x})$ are considered as the non-observed environment over which we trace. The expectation values of gauge sector operators $O(\phi, \xi)$ in this mixed state are given by

$$\langle O\rangle = \frac{1}{Z}\int \mathcal{D}\phi\mathcal{D}\xi\mathcal{D}\psi \sum_N \frac{z^N}{N!}\sum_{\mathbf{x}_1\ldots\mathbf{x}_N}\sum_{N_1\ldots N_N=\pm1} O(\phi, \xi)$$

$$\times \exp\left[\frac{iq}{2}\sum_i\int d^2\mathbf{x}\,\xi(\mathbf{x})N\left(\mathbf{x},\mathbf{x}_i N_i\right) + i\int d^2\mathbf{x}\,\frac{q}{4\pi}\xi\Delta\psi\right]$$

$$\times \exp\left[\frac{-q}{2\pi}\int d^2\mathbf{x}\left(\kappa(\partial_i\phi)^2 + \frac{1}{4\kappa}(\partial_i\psi)^2\right)\right] \,.$$

(30)

where Z is the normalization factor and ψ denotes the difference $\psi = \psi_{\text{bra}} - \psi_{\text{ket}}$ between bra and ket states. We have also used the dilute charge approximation in which only interferences between point charge states differing by one unit are taken into account. The quantum fugacity parameter z governs the entanglement. For small z we have a highly entangled state of charge degrees of freedom, for $z \to \infty$ charges are liberated as independent degrees of freedom.

At this stage both the integration over the transverse field ψ and the summation over charge interference configurations can be done explicitly,[28,29] with the result

$$\langle O \rangle = \frac{1}{Z} \int \mathcal{D}\phi \mathcal{D}\xi \; O(\phi, \xi) \; \exp\left[-\int d^2\mathbf{x} \; \frac{q\kappa}{2\pi} (\partial_i \phi)^2 + \frac{\kappa}{8\pi q} (\partial_i \xi)^2 - 2z \cos \xi\right] , \tag{31}$$

where we have renormalized the angle ξ to lie in the range $[0, 2\pi]$ for comparison with results in the literature. Gauge field observables in the entangled mixed state are thus determined by the classical partition function of the 2D sine–Gordon model, or equivalently the 2D XY model.[44] While the original ϕ field plays the role of the spin wave field, the new dynamical sine–Gordon field ξ describes the vortex dynamics. We can now use the classical results on the 2D sine–Gordon (or XY) renormalization group flow[44] also here.

The 2D XY model undergoes the famed Berezinskii–Kosterlitz–Thouless (BKT)[35–37] phase transition. The value $\kappa = q/2$ separates a weak-coupling phase for $\kappa < q/2$ from a strong coupling phase for $\kappa > q/2$. For the case $q = 2$ of Cooper pairs this value corresponds to the self-dual point $\kappa = 1$. In both phases the coupling constant κ flows to large values in the IR limit. In the weak coupling phase, the fugacity $z \to 0$ in the IR limit while in the strong coupling phase $z \to \infty$ in the IR limit. The renormalization flow depends on a constant C, which represents a particular combination of the initial conditions for the flow. For $C < 0$, there is a direct transition between two dual high entanglement regimes. In the case $C > 0$, of interest here, there is no IR fixed-point in the strong-coupling phase, while the BKT critical point $\kappa_{\text{crit}} = q/(2 + C)$ is a confining IR fixed point ($z = 0$). As in QCD, the coupling constant κ flows to small values in the ultraviolet (UV) regime. Contrary to QCD, it is the confining IR fixed point which is perturbative, while the asmyptotically free UV regime is non-perturbative.

Let us now prove that this high entanglement regime is linearly confining. To this end we first note that it is a condensate phase for vortices. As is evident from (31), near the IR fixed point, the phase ξ of the gauge field has correlations screened on the length $\lambda_{\text{vor}} = \sqrt{\kappa/2\pi qz}$, which diverge at the fixed point. These diverging correlations imply diverging fluctuations of the vortex number near the fixed point, the characteristics of a condensate. As has been first ponted out in Refs. 28 and 29, this vortex condensate phase in a 2D compact gauge theory is characterized by the presence of instantons. Let us compute thus the effect of instantons on the charge-anticharge potential. To this end we shall consider again the insertion of two external probe charges of different sign by the exponential line integral factor (25), focusing specifically on the effect of the new dynamical field ξ. Since this is a phase,

at first sight it would seem that this effect reduces simply to two phase factors at
the end of the path. Instantons, however, can cause the phase to jump on the path,
leading to large effects. Following Ref. 43 we write the energy shift caused by single
instantons/anti-instantons on the path connecting the two charges as

$$\Delta E_{\text{inst}} = E_{\text{inst}} \int_{\text{path}} d\mathbf{x} \left(1 - \text{Re} \left[K^{-1}(\mathbf{x}_1, \mathbf{x}) K(\mathbf{x}_2, \mathbf{x})\right]\right) , \tag{32}$$

where E_{inst} is the contribution of instantons in absence of external charges and
$K(\mathbf{x}_2, \mathbf{x})$ denotes the exponential of the instanton function valued at $\mathbf{x}_2$ for an
instanton located at some $\mathbf{x}$ on the path. Since the instanton solution corresponds
to a phase jump at $\mathbf{x}$ we obtain

$$\Delta E_{\text{inst}} = E_{\text{inst}} \int_{\text{path}} d\mathbf{x} \left(1 - \cos[Q\Delta\lambda_{\text{inst}}(\mathbf{x})]\right) . \tag{33}$$

In compact U(1) gauge theory, the instantons[28,29] correspond to unit magnetic
monopoles in 3D Euclidean space,[45] which implies a magnetic flux $2\pi/q$ on the 3D
unit sphere. The corresponding phase jumps at fixed time in Minkowski space-time
are thus $\Delta\lambda_{\text{inst}} = \pi/q$. This gives the final result

$$\Delta E_{\text{inst}} = E_{\text{inst}} \left(1 - \cos\left(\frac{\pi Q}{q}\right)\right) R , \tag{34}$$

where R is the separation of the charge-anticharge pair. This is Polyakov's classical[28,29] result that instantons in the vortex condensation phase cause linear confinement for charges $Q = q$ satisfying the Dirac quantization condition, while double
charges $Q = 2q$ are non-confined.

These results show that 2D QED with compact Chern–Simons dynamical matter
is a linearly confining, asymptotically free theory due to the strong entanglement
of charge with a vortex condensate. As such it is a single-colour "toy model"
for QCD. The BKT weak-coupling fixed-point represents a new state of matter,
superinsulation, which is a condensed matter realization of the QCD vacuum. It is
remarkable that this is explicitly realized in condensed matter as the superinsulating
phase of thin superconducting films.[20–25,39]

References

1. R. Jackiw and S. Templeton, "How super-renormalizable interactions cure their infrared divergences", *Physical Review D* **23**, 2291–2304 (1981).
2. S. Deser, R. Jackiw and S. Templeton, "Three-dimensional massive gauge theories",
   Physical Review Letters **48**, 975–978 (1982).
3. S. Deser, R. Jackiw and S. Templeton, "Topologically massive gauge theories", *Annals
   of Physics* **140**, 372–411 (1982).
4. D. Birmingham, "Topological field theory", *Physics Reports* **209**, 129–340 (1991).
5. M. Bergeron, G. Semenoff and R. Szabo, "Canonical BF-type topological field theory
   and fractional statistics of strings", *Nuclear Physics B* **437**, 695–721 (1995).
6. M. Diamantini, P. Sodano and C. Trugenberger, "Gauge theories of Josephson junction
   arrays", *Nuclear Physics B* **474**, 641–677 (1996).

7. K. B. Efetov, "Phase transition in granulated superconductors", *Sov. Phys. JETP* **51**, 1015–1022 (1980).
8. D. Haviland, Y. Liu and A. Goldman, "Onset of superconductivity in the two-dimensional limit", *Physical Review Letters* **62**, 2180–2183 (1989).
9. A. Hebard and M. Paalanen, "Magnetic-field-tuned superconductor-insulator transition in two-dimensional films", *Physical Review Letters* **65**, 927–930 (1990).
10. M. Fisher, G. Grinstein and S. Girvin, "Presence of quantum diffusion in two dimensions: Universal resistance at the superconductor-insulator transition", *Physical Review Letters* **64**, 587–590 (1990).
11. R. Fazio and G. Schön, "Charge and vortex dynamics in arrays of tunnel junctions", *Physical Review B* **43**, 5307–5320 (1991).
12. M. C. Diamantini, A. Yu. Mironov, S.V. Postolova, X. Liu, Z. Hao, D. M. Silevitch, Ya. Kopelevich, P. Kim, C. A. Trugenberger and V. M. Vinokur, "Bosonic topological intermediate state in the superconductor-insulator transition" (2019), arXiv:1906.07969.
13. D. Das and S. Doniach, "Existence of a Bose metal at $T = 0$", *Physical Review B* **60**, 1261–1275 (1999).
14. D. Das and S. Doniach, "Bose metal: Gauge-field fluctuations and scaling for field-tuned quantum phase transitions", *Physical Review B* **64**, 134511 (2001).
15. A. Kapitulnik, S. Kivelson and B. Spivak, "Anomalous metals: Failed superconductors", *Reviews of Modern Physics* **91**, 011002 (2019).
16. Y. Lu and A. Vishwanath, "Theory and classification of interacting integer topological phases in two dimensions: A Chern–Simons approach", *Physical Review B* **86**, 125119 (2012).
17. C. Wang and T. Senthil, "Boson topological insulators: A window into highly entangled quantum phases", *Physical Review B* **87**, 235122 (2013).
18. X. Chen, Z. Gu, Z. Liu and X. Wen, "Symmetry protected topological orders and the group cohomology of their symmetry group", *Physical Review B* **87**, 155114 (2013).
19. A. Krämer and S. Doniach, "Superinsulator phase of two-dimensional superconductors", *Physical Review Letters* **81**, 3523–3526 (1998).
20. V. Vinokur, T. Baturina, M. Fistul, A. Mironov, M. Baklanov and C. Strunk, "Superinsulator and quantum synchronization", *Nature* **452**, 613–615 (2008).
21. T. Baturina and V. Vinokur, "Superinsulator–superconductor duality in two dimensions", *Annals of Physics* **331**, 236–257 (2013).
22. T. Baturina, A. Mironov, V. Vinokur, M. Baklanov and C. Strunk, "Localized superconductivity in the quantum-critical region of the disorder-driven superconductor-insulator transition in tin thin films", *Physical Review Letters* **99**, 257003 (2007).
23. A. Mironov, D. Silevitch, T. Proslier, S. Postolova, M. Burdastyh, A. Gutakovskii, T. Rosenbaum, V. Vinokur and T. Baturina, "Charge berezinskii-kosterlitz-thouless transition in superconducting nbtin films", *Scientific Reports* **8**, 4082 (2018).
24. G. Sambandamurthy, L. Engel, A. Johansson, E. Peled and D. Shahar, "Experimental evidence for a collective insulating state in two-dimensional superconductors", *Physical Review Letters* **94**, 017003 (2005).
25. M. Ovadia, D. Kalok, I. Tamir, S. Mitra, B. Sacépé and D. Shahar, "Evidence for a finite-temperature insulator," *Scientific Reports* **5**, 13503 (2015).
26. M. Diamantini, C. Trugenberger and V. Vinokur, "Confinement and asymptotic freedom with Cooper pairs", *Communications Physics* **1**, 77 (2018).
27. M. Diamantini, L. Gammaitoni, C. Trugenberger and V. Vinokur, "Vogel–Fulcher–Tamman criticality of 3D superinsulators", *Scientific Reports* **8**, 15718 (2018).
28. A. Polyakov, "Compact gauge fields and the infrared catastrophe", *Physics Letters B* **59**, 82–84 (1975).

29. A. M. Polyakov, *Gauge Fields and Strings* (Harwood Academic Publisher, Switzerland, 1987).
30. M. Caselle, M. Panero and D. Vadacchino, "Width of the flux tube in compact U(1) gauge theory in three dimensions", *Journal of High Energy Physics* **2016**, 180 (2016).
31. M. Panero, "A numerical study of confinement in compact QED", *Journal of High Energy Physics* **2005**, 066–066 (2005).
32. F. Nogueira and H. Kleinert, "Compact quantum electrodynamics in2+1 dimensions and spinon deconfinement: A renormalization group analysis", *Physical Review B* **77**, 045107 (2008).
33. H. Kleinert, F. Nogueira and A. Sudbø, "Kosterlitz–Thouless-like deconfinement mechanism in the (2+1)-dimensional Abelian Higgs model", *Nuclear Physics B* **666**, 361–395 (2003).
34. N. Nagaosa and P. Lee, "Confinement and bose condensation in gauge theory of high-T_c superconductors", *Physical Review B* **61**, 9166–9175 (2000).
35. V. L. Berezinskii, "Destruction of long-range order in one-dimensional and twodimensional systems having a continuous symmetry group I. Classical systems", *Sov.Phys. JETP* **32**, 493–500 (1970).
36. J. Kosterlitz and D. Thouless, "Long range order and metastability in two dimensional solids and superfluids. (Application of dislocation theory)", *Journal of Physics C: Solid State Physics* **5**, L124–L126 (2001).
37. J. Kosterlitz and D. Thouless, "Ordering, metastability and phase transitions in two-dimensional systems", *Journal of Physics C: Solid State Physics* **6**, 1181–1203 (2002).
38. G. 't Hooft, "On the phase transition towards permanent quark confinement", *Nuclear Physics B* **138**, 1–25 (1978).
39. M. C. Diamantini, L. Gammaitoni, C. Strunk, S. V. Postolova, A. Yu. Mironov, C. A. Trugenberger and V. M. Vinokur, "Electrostatics of a superinsulator" (2019), arXiv:1906.12265.
40. G. Grignani, G. Semenoff, P. Sodano and O. Tirkkonen, "G/G models as the strong coupling limit of topologically massive gauge theory", *Nuclear Physics B* **489**, 360–384 (1997).
41. S. Treiman, R. Jackiw, B. Zumino and E. Witten, *Current Algebra and Anomalies* (World Scientific, 2011).
42. G. Dunne, R. Jackiw and C. Trugenberger, "Chern–Simons theory in the Schrödinger representation", *Annals of Physics* **194**, 197–223 (1989).
43. J. Gervais and B. Sakita, "Gauge degrees of freedom, external charges, and quark confinement criterion in thea0=0canonical formalism", *Physical Review D* **18**, 453–462 (1978).
44. C. Itzykson and J. Drouffe, *Statistical Field Theory* (Cambridge University Press, 2012).
45. P. Goddard and D. Olive, "Magnetic monopoles in gauge field theories", *Reports on Progress in Physics* **41**, 1357–1437 (2001).

Chapter 24

Three Easy Pieces
(in Tribute to Roman Jackiw)*

Frank Wilczek

Center for Theoretical Physics, MIT, Cambridge, MA 02139, USA;
T. D. Lee Institute and Wilczek Quantum Center,
Shanghai Jiao Tong University, Shanghai, China;
Arizona State University, Tempe, AZ, USA;

Stockholm University, Stockholm, Sweden

Roman Jackiw has made highly original and influential contributions to several areas of physics that have grown and blossomed, notably including the quantum physics of domain walls, magnetic monopoles, and fractional quantum numbers. Here I offer three small pieces that take off from those themes. I discuss the emergence of topological surface structure in materials, the emergence of a shape-space magnetic monopole in a simple mechanical system, and the emergence of fractional angular momentum in an even simpler quantum mechanical (molecular) system.

Introduction

Roman Jackiw has had an uncanny knack for identifying "curiosities" that have grown into fertile, vibrant areas of physics research [1]. His seminal contributions to the theory of anomalies, the interplay of topology with quantum theory, and fractional quantum numbers are a rich legacy which has become central to both fundamental physics and modern quantum engineering. It has influenced my own work both directly and indirectly. The direct influence should be obvious. Among others, my work with Goldstone on fractional quantum numbers started as an attempt to understand, and then generalize, the remarkable discoveries of Jackiw and Rebbi [2] and of Su, Schrieffer, and Heeger [3]; anyons and braiding arose from coming to terms with fractional angular momentum; axions arose as the possible solution to a big conceptual problem raised by the topology of gauge fields in QCD. The indirect influence is perhaps less visible but no less profound. I became aware of Dirac's recommendation to "play with equations" in my student days, but it came alive for me only a bit later, in large part because I saw how well it had worked for Roman. That helped me feel liberated to indulge in such play.

---

*Contribution to *Roman Jackiw: 80th Birthday Festschrift*, ed. A. Niemi, T. Tomboulis, K. K. Phua.

Here I will present three brief case studies in playing with equations, which have close ties to Roman's work. Roman, I hope this little tribute brightens your day.

1. Domain walls and boundary surfaces

In this part I want to make a few simple but important observations which highlight the close connection of Jackiw and Rebbi's pioneering work to earlier problems and later developments, and may suggest other significant directions. To make the discussion self-contained for both condensed matter and high energy physicists, I will start from scratch.

Consider a one-dimensional chain of identical molecules with spacing a. We want to model the situation where there is one conduction electron per molecule. Let us adopt the drastic simplifying assumption that the molecules simply provide a weak periodic background potential and ignore interactions among the electrons, aside from their mutual influence through fermionic quantum statistics. To begin, we will also assume that the potential is spin-independent and invariant under spatial reflection.

According to standard band theory we should expect to have a metal, since the electrons exhaust only half the available states. Indeed, the carrying capacity of the lowest band, allowing for the spin degree of freedom, is twice its extent in (quasi-)momentum space, and so the actual density corresponds to half filling:

$$N/L = 1/a = \frac{1}{2} \cdot 2 \cdot \frac{\frac{2\pi}{a}}{2\pi} \tag{1}$$

In the electrons' ground state, they occupy all the momentum states for $-k_F \leq k \leq k_F$ with $k_F = \pi/2a$. The low-energy excitations around this ground state involve changes in the occupation of modes near the boundaries of that region. There we can approximate the change in their energy with momentum — that is, the deviation in momentum from $\pm k_F$ — as a linear relationship, bringing in the Fermi velocity

$$v_F = \frac{dE}{dk}(k_F) \tag{2}$$

Thus we derive, from the two points of the Fermi surface, left- and right-moving excitations obeying

$$(\partial_t - \partial_x)\psi_R = 0$$
$$(\partial_t + \partial_x)\psi_L = 0 \tag{3}$$

where we adopt the unit of velocity $v_F \to 1$. This can be summarized in relativistic form, as a Dirac equation

$$i\gamma \cdot \partial \psi = 0 \tag{4}$$

with

$$\psi \equiv \begin{pmatrix} \psi_L \\ \psi_R \end{pmatrix} ; \ \gamma_0 \equiv \sigma_2 ; \ \gamma_1 \equiv i\sigma_1 \tag{5}$$

This empowers us to visualize a Dirac cone.

The modes from the two sides of the cone are separated, as we saw earlier, by $\Delta k = \frac{\pi}{a}$. Insofar as that wavenumber is not represented in the background potential, those modes remain independent. On the other hand, were the potential to contain that wave number it could connect the left- and right-movers, and open a gap in the Dirac cone. This mixing pushes down the energies of occupied states, so it is a favorable effect, energetically. We should inquire whether it can be triggered dynamically, and occur spontaneously.

To generate the required potential, we need to let the molecules on neighboring lattice sites displace in alternate directions. Displacements of the molecules correspond to phonons, and in that language we are asking whether the "optical" phonons at $k = \frac{\pi}{a}$ condense. If we denote the amplitude of the displacement (phonon) $k = \frac{\pi}{a}$ field by $\phi(x,t)$, then we have a coupling

$$\mathcal{L} = \bar{\psi}(i\gamma \cdot \partial - g\phi)\psi \tag{6}$$

so that condensation in ϕ opens a gap and gives us a fermion mass $m = g\langle\phi\rangle$.

This entire discussion can be phrased in the language of relativistic quantum field theory. In that language, we are studying how mass generation through coupling to a scalar condensate back-reacts on the energetics of a scalar field which triggers it. In particle physics this effect is called the Coleman–Weinberg mechanism [4]. It plays an important role in the theory of the Standard Model, and in many speculations that go beyond the Standard Model.

We can exploit that mapping to bring in the machinery of Feynman graphs, easing our computational challenge. Our energy gain corresponds to changes in the one fermion loop vacuum amplitude brought in by changes in the fermion mass.

To assess the possible instability, we can focus on infinitesimal displacements. The most singular part takes the form

$$\mathcal{E} \sim \int_0^\Lambda \frac{d^2k}{(2\pi)^2} \frac{\mathrm{Tr}(\gamma \cdot k)^2}{(k^2 + m^2)^2}$$
$$\propto (g\phi)^2 \ln(\Lambda/g\phi) \tag{7}$$

Note that it is appropriate to supply an ultraviolet cutoff, since our simple "relativistic" description of the electron modes is only valid for low-energy excitations, involving small deviations from the Fermi points.

Of course, displacement of the molecules also involves ordinary, Hooke's law elastic energy. But that energy, being proportional to ϕ^2, is dominated by Eq. (7) for small ϕ. Thus the lattice will always deform. This is Peierls' instability [5]. It is strikingly reminiscent of the Cooper instability which drives superconductivity.

The most favorable displacement amplitude $\langle\phi\rangle \equiv \pm v$ has two possible values. They correspond to two distinct homogeneous ground states. They differ from one another by displacement through a — that is, by the transformation which used to be, but is no longer, a symmetry. In these two configurations our electrons acquire, according to Eq. (6), effective mass $\pm gv$. The negative value looks a bit strange, at first sight, but the sign of the mass can be absorbed, formally, into a redefinition of ψ. Indeed, the transformation $\psi' = \gamma_0\gamma_1\psi$ changes the sign of the mass term, while leaving the kinetic term invariant.

A simple defect allows one to interpolate between the two distinct homogeneous states. If we let the simple defect configuration relax, energetically, it will settle down to a minimum energy configuration — a domain wall — that is stable for topological reasons. In this configuration the effective electron mass $m(x)$ interpolates between $-gv$ (say) on the far left and gv on the far right. (The *relative* sign, of course, cannot be undone by re-definition.) The effective mass will vanish somewhere in between, and we might anticipate that something interesting could happen as a result. In fact, as discovered by Jackiw and Rebbi, there is a remarkably robust and characteristic consequence: the existence of zero energy (midgap) states, centered on the wall.

Fixing the center of the wall to be at $x = 0$, we have an effective mass $m(x)$ with $m(-x) = -m(x)$. The electron equation for zero energy reads

$$0 = (-i\gamma_1\partial_x + m(x))\psi(x)$$
$$= (\sigma_1\partial_x + m(x))\psi(x) \tag{8}$$

Projecting this on spinor structures with $\sigma_1 s_\pm = \pm s_\pm$, we find that for $\psi(x) \equiv \psi_+(x)s_+ + \psi_-(x)s_-$

$$(\partial_x \pm m(x))\psi_\pm(x) = 0 \tag{9}$$

with solutions

$$\psi_+^0(x) = e^{-\int_0^x du\, m(u)} \tag{10}$$
$$\psi_-^0(x) = e^{\int_0^x du\, m(u)} \tag{11}$$

The first of these, i.e. Eq. (10) shrinks exponentially to zero in both directions away from the wall, and defines a normalizable state. It is the zero energy, midgap state advertised just above.

The existence of this zero energy mode has several interesting consequences. It should be plausible, given the symmetry of the problem, that the zero energy mode draws half its strength from below and gap half from above. Thus if we fail to occupy the zero mode, we have a deficit of half a unit of electron charge, relative to the ground state; while if we do occupy the zero mode, we have an excess of half a unit of electric charge. One can also show this formally, either directly from the definition of the charge operator, or by calculating the flow of charge as one builds up the domain wall (see below). Thus the domain wall induces a special kind of vacuum polarization, where a fractional charge $\pm\frac{1}{2}$ accumulates [2].

When we take into account the spin degree of freedom, the accounting takes a different form. Now there are two zero modes: one for spin up, one for spin down. By occupying, or not, each of the two zero modes, we have:

- *zero occupancy*: charge $-e$, spin 0
- *single occupancy*: charge zero, spin $\frac{1}{2}$ doublet
- *double occupancy*: charge $+e$, spin 0

Though the charge spectrum is normal, the relation between charge and spin is unusual. We can express the general situation through the equation

$$(-1)^{2S}(-1)^{Q/e} = (-1)^{W} \tag{12}$$

where W is a topological quantity, indicating the number of domain walls. Since the quantities that appear in this equation are intrinsically discrete the relationship it expresses will be exact, unless our approximations have been very bad. Moreover, it is the only non-trivial relation among $2S, Q/e$, and W consistent with the property that two domain walls can annihilate into states with normal quantum numbers. We expect this property to be valid, because the lattice of molecules with two minimal defects differs from a correctly ordered (i.e., ground state) lattice only within a bounded region, and is topologically trivial.

The logic of Peirels' instability is not restricted to half filling [3]. One can, for example, consider conduction bands that fill $1/k$ of the available states, and trigger an instability toward charge density waves at $2\pi/k$, with k an integer. Then we will have domain walls that can annihilate in k-tuples. If such annihilation is accompanied by emission or absorption of l electrons, we can deduce a generalization of Eq. (12), in the form

$$e^{i2\pi(Q/e - lW/k)} = 1 \tag{13}$$

(Here, for simplicity, I have not kept track of the spin quantum number, which need not be conserved separately.) Equivalently, we can write

$$Q/e = lW/k + \text{integer} \tag{14}$$

Earlier we had $k = 2, l = 1$, and half-integer charge. In that case, we could understand the fractional charge based on the existence of a zero energy solution with equal particle–hole character. But in more general situations, say for example $k = 3, l = 1$, where we have third-integer charge, it cannot be understood in that way.

An appropriate, minimal model for these more general situations allows the field ϕ in Eq. (6) to become complex. The overall phase of ϕ adjusted by re-defining ψ, according to $\psi' = e^{i\lambda}\psi$, but relative phases in $\phi(x,t)$ are physically meaningful. In particular, we can have situations where there are domain walls which interpolate between values

$$\phi(x) \rightarrow \quad v; \qquad x \rightarrow -\infty$$
$$\phi(x) \rightarrow e^{2\pi i/k}\, v; \quad x \rightarrow +\infty \tag{15}$$

Such domain walls can annihilate in k-tuples. We expect, based on the preceding discussion, that fractional charge may accumulate on such walls.

An efficient way to calculate the charge is first to imagine building up configurations with the wall asymptotics of Eq. (15) gradually, starting from trivial
asymptotics [6]. As long as the magnitude of the local gap exceed the field gradients, i.e.

$$\frac{|\partial\phi|}{g|\phi|} \ll 1 \tag{16}$$

there will be no particle production, and we can calculate the current flow (to lowest
order in gradients) by means of a simple vacuum polarization graph, similar in spirit
to how one calculates the correction to the phonon energy, but now with insertion
of the electron number current in place of a phonon field. An elegantly simple result
emerges, in the form

$$\langle j^\mu \rangle = \frac{1}{2\pi} \epsilon^{\mu\nu} \partial_\nu \operatorname{Arg}\phi \tag{17}$$

and for the integrated charge

$$Q/e = \int_{-\infty}^{\infty} dx\, j^0$$

$$= \frac{1}{2\pi} \int_{-\infty}^{\infty} dx\, \partial_x \operatorname{Arg}\phi$$

$$= \frac{1}{2\pi} \left(\operatorname{Arg}\phi(\infty) - \operatorname{Arg}\phi(-\infty) \right) \tag{18}$$

Now the realistic, minimum energy wall configuration may not satisfy Eq. (16)
everywhere, though of course it does so asymptotically (where $\partial\phi \to 0$). So we
must imagine a second step, where we build up the steeper gradients. During that
process electrons can be radiated to, or absorbed from, infinity. But since the
electrons are normal out there, any such radiation will change the electron number
by an integer. Also, from a complementary perspective, the angle function $\operatorname{Arg}\phi$
becomes ill-defined at $\phi = 0$, and we need to allow for extra 2π jumps, as we
integrate Eq. (17) through such points. For both these reasons, we should generally
interpret Eq. (18) as a relation modulo integers, i.e. as a formula for the fractional
part of the charge. As such, it precisely embodies Eq. (14).

We can "explain" the existence of the zero energy solution we found earlier, based
on these more global considerations, as follows. The zero energy solution occurred
in the model with g and ϕ real. Within that framework we cannot achieve the non-
trivial domain wall asymptotics, where $\langle\phi\rangle$ changes sign, without encountering a
zero of ϕ, where the requirement of Eq. (16) cannot be met. We can get around that
difficulty by adding a small infinitesimal imaginary piece to ϕ, and then removing
it at the end. Since there is a gap, the limit is harmless. But depending on the sign
of the added piece, we will get $Q/e = \pm\frac{1}{2}$. So there must be degenerate states with
those quantum numbers, and therefore a zero energy mode of the electron field,
whose occupancy (or not) distinguishes those charges.

We can also contemplate incommensurate density waves. This allows domain walls where the change in Arg, and thus the accumulated charge, takes any value, rational or irrational.

The possibility of zero energy states at the termination of a 1-dimensional lattice was an early discovery of Shockley [7]. It gave a microscopically-based model of Bardeen's theory of "surface states", which in turn played an important role in the discovery of solid-state transistors. Shockley's model is closely related to the models with which we began our discussion. We can put his discovery into the same framework, and generalize it, by considering how we might model boundaries of chains (or, of course, surfaces) as extreme versions of domain walls, as follows.

Taking Eq. (6) as our basic model, we can have some value $g\phi_+$ (not necessarily real) for the effective mass in bulk, in for $x > 0$, while taking $g\phi_- \to \infty$ for $x < 0$, to make it difficult for the electrons to penetrate there. (If desired we can have a small bridging region, and let ϕ interpolate continuously between those values.) Thus, we realize the boundary as a specific kind of domain wall, to which our general analysis can be applied.

Topological insulators [8,9], in their simplest form, fit neatly into this framework, as follows. Assuming T symmetry, we let g and ϕ be real. Then the relevant issue is the relative sign of $m(x)$ for between $x < 0$ and $x > 0$. If the sign changes, we have a zero energy surface state; if not, not. If we keep spin as a passive (bookkeeping) quantum number throughout, then we will get even numbers of zero energy states, as in Eq. (1); but if there are significant spin-dependent forces, then spin is not a good label, and in general we will find an odd number. One simple version of how this can occur is to let the "vacuum" mass at $x \to -\infty$ be spin-dependent, asymptotically in the form $\sigma_3 m(x)$, with $m(x) \to \infty$ rapidly as $x \to -\infty$ and $m(x)$ approaches a constant — possibly zero — as $x \to +\infty$. These constructions highlight that there need not be a bulk signature of the topological insulator state. Of course, we could write things alternatively so as to put the exotic structure in the bulk, which might (or might not) correspond to what emerges, in realistic cases, from a conventional band theory calculation.

If we drop the requirement of T symmetry, we bring in possibilities for surface fractional charge, according to the same basic mechanisms, exploiting the domain wall $\to$ boundary principle. Note that the T breaking can be a purely surface effect, but of course it need not be. In the presence of an appropriate discrete symmetry the surface charge can be quantized, but in general it will be irrational. In higher-dimensional situations we would predict, taking the models at face value, a surface charge density. When we include Coulomb energy, there will be a strong incentive for the material to enforce neutrality on large scales, for example by fractionally populating conduction band states near the surface.

2. A mechanical monopole

The ability of falling cats and of divers to re-orient themselves by changing shape, despite having zero angular momentum, is an instructive puzzle whose mathematical solution includes an elegant emergent gauge structure. Let me briefly recall how this works [10].

In describing how much a deformable body has rotated, we immediately confront a basic issue, that no rigid rotation can connect different (non-congruent) shapes. To make comparison possible, we can set up "reference shapes" with definite positions, and compare the position of our body, with its current shape, to the position of the associated reference shape. This procedure introduces two issues:

- It introduces an element of convention, because one is free to place the reference shapes differently. If we write the relationship between body shapes and reference shapes as

$$r^{(j)}(t) = \mathcal{R}(t)s^{(j)}(r^{(j)}(t)) \tag{19}$$

  then a re-definition

$$\tilde{s}^{(j)} = \mathcal{U}(s^{(j)}) \, s^{(j)} \tag{20}$$

  induces the re-definition

$$\tilde{\mathcal{R}}(t) = \mathcal{R}(t)\mathcal{U}^{-1}(s(r(t))) \tag{21}$$

  where we have, mercifully, dropped the superscripts. The net rotation between shapes at times t_f, t_i is described alternatively as

$$\mathcal{R}(t_f)\,\mathcal{R}^{-1}(t_i) \tag{22}$$

  or

$$\tilde{\mathcal{R}}(t_f)\,\tilde{\mathcal{R}}^{-1}(t_i) = \mathcal{R}(t_f)\mathcal{U}^{-1}(s(r(t_f)))\mathcal{U}(s(r(t_i))\,\mathcal{R}^{-1}(t_i) \tag{23}$$

  If the initial and final shapes are the same then the $\mathcal{U}$ factors cancels, and the ambiguity disappears.
- It fails when the shape degenerates to a point or a line. In those cases there are congruences which leave the shape invariant, so the transformation to a reference shape is ambiguous.

The first point indicates that we have introduced a gauge structure, while the second indicates the possibility of singularities.

Having that background in mind, we are prepared to discuss the general relation between shape changes at zero angular momentum and re-orientation in space. With the transformation Eq. (19) the zero angular momentum condition

$$0 = \sum_j m^{(j)} r^{(j)} \times \dot{r}^{(j)} \tag{24}$$

becomes

$$0 \; = \; \sum_j m^{(j)} s^{(j)} \times \dot{s}^{(j)} + \sum_j m^{(j)} s^{(j)} \times \mathcal{R}^{-1} \dot{\mathcal{R}} s^{(j)} \tag{25}$$

Here the first term defines an effective angular momentum L^{shape}, while $\mathcal{R}^{-1}\dot{\mathcal{R}}$ defines an effective angular velocity, in tensor form.

With a bit of algebra we can cast Eq. (25) into the form

$$V(\mathcal{R}^{-1}\dot{\mathcal{R}}) \; = \; \hat{I}^{-1} L^{\text{shape}} \tag{26}$$

where

$$I_{pq} \; = \; \sum_j m^{(j)} s_p^{(j)} s_q^{(j)} \tag{27}$$

$$\hat{I}_{pq} \; \equiv \; (I_{ll}\delta_{pq} - I_{pq}) \tag{28}$$

defines an effective inertia tensor and

$$V(M_{jk})_l \; \equiv \; \frac{1}{2}\epsilon_{jkl} M_{jk} \tag{29}$$

defines the vector equivalent of the antisymmetric matrix M. Equation (26) relates effective angular momentum and velocity through an effective inertia tensor. The division between effective angular momentum and velocity is gauge (i.e. reference shape) dependent, but the effective inertia tensor is not.

This construction provides a gauge connection in shape space, according to

$$\mathcal{R}^{-1}\dot{\mathcal{R}}\, dt \; \equiv \; A_j\, dx^j \tag{30}$$

The rotation resulting from a given trajectory in shape space is given by the ordered line integral ("Wilson line") of this connection. Note that $\frac{d}{dt}$ cancels, so that the accumulated rotation depends only on the geometry of the trajectory in shape space, not on how fast it is traversed. Alternatively, we can say that it is time-reparameterization invariant.

Now let us turn to evaluating the connection for a system of three point masses. For our standard shapes, we can put the first mass at the origin, the second at distance $\sqrt{m_2}r \equiv \kappa \cos\theta$ along the x axis, and the third in the x-y plane at distance $\sqrt{m_3 s} \equiv \kappa \sin\theta$ and azimuthal angle ϕ. θ ranges between 0 and $\frac{\pi}{2}$. Both θ and ϕ are ill-defined when $\kappa = 0$, when all the masses coincide. At that singular shape the preceding procedures to obtain V fail, so we should only consider shape trajectories which avoid it. When the distribution of masses is along a line, i.e. at $\phi = 0$, we have the ambiguity mentioned previously. We can relieve it by demanding that the motion is restricted to a plane, which for simplicity let us do.

Even then, since ϕ is ill-defined for $\sin\theta = 0$, we will need to use a different parameterization to cover that case. A natural choice is to interchange the role of the second and third masses. Calling the new parameters $\tilde{\theta}, \tilde{\phi}$ (with $\tilde{\kappa} = \kappa$), we have

$$\tilde{\theta} = \frac{\pi}{2} - \theta$$

$$\sin\tilde{\theta} = \cos\theta$$

$$\tilde{\phi} = 2\pi - \phi \tag{31}$$

whenever both make sense.

Given these definitions, a simple calculation yields

$$L^{\text{shape}} = \begin{pmatrix} 0 \\ 0 \\ \kappa^2 \sin^2 \theta \, \dot{\phi} \end{pmatrix} \tag{32}$$

while $\hat{I}$ has the form

$$\hat{I} = \begin{pmatrix} * & * & 0 \\ * & * & 0 \\ 0 & 0 & \kappa^2 \end{pmatrix} \tag{33}$$

so that

$$\hat{I}^{-1} = \begin{pmatrix} * & * & 0 \\ * & * & 0 \\ 0 & 0 & \kappa^{-2} \end{pmatrix} \tag{34}$$

and finally

$$V(\mathcal{R}^{-1}\dot{\mathcal{R}}) = \begin{pmatrix} 0 \\ 0 \\ \sin^2 \theta \, \dot{\phi} \end{pmatrix} \tag{35}$$

(One might be concerned that our choice of shapes does not enforce that the center of mass stays fixed. However, since under a change of origin the rigid motion $(\mathcal{R}, \vec{a})$ — applying first the rotation $\mathcal{R}$ and then the translation $\vec{a}$ — changes into $(\mathcal{R}, \mathcal{R}\vec{b}-\vec{b}+\vec{a})$, our determination of $\mathcal{R}$ is unaffected. We could include the necessary compensating translations as a gauge potential, too, taking us into the group of rigid motions rather than just rotations. But because the effect of enforcing the center of mass constraint vanishes for any closed path in shape space, the added piece introduces no curvature, and for simplicity I will discuss it no further here.)

Since the motion is planar, we can regard interpret this result for the connection as an $SO(2)$ vector potential

$$A_\phi(\rho, \theta, \phi) = \sin^2 \theta \tag{36}$$

on our shape space, with the accumulated rotation angle (relative to the standard shape) for a given trajectory in shape space given by

$$\Delta \alpha = \int A_\phi d\phi = \int_{\text{initial}}^{\text{final}} \sin^2 \theta \, d\phi \tag{37}$$

integrated over the trajectory.

The only non-vanishing component of the field strength associated with this potential is

$$F_{\theta\phi} = 2 \sin \theta \cos \theta \tag{38}$$

The potential and field strength are independent of κ, but develop a singularity at $\kappa = 0$, where θ is ill-defined. The lack of κ dependence reflects that we may re-scale (dilate) the shape at any time without affecting the overall rotation dynamics. Recall that $\kappa = 0$ corresponds to the degenerate shape of three coincident points, where singularity might be anticipated.

Now let us fix κ and evaluate the total flux emanating from the singularity. We find

$$\oint F_{\theta\phi} \, d\theta \, d\phi \;=\; \int_0^{\frac{\pi}{2}} 2\sin\theta \, \cos\theta \, d\theta \int_0^{2\pi} d\phi \;=\; 2\pi \tag{39}$$

Here the limits on the θ integral are obvious from its definition, but the limits on the ϕ integral call for comment. Indeed, one might be tempted to restrict the integral to $0 \le \phi \le \pi$, since the shapes with ϕ and $2\pi - \phi$ are congruent. That congruence is, however, not implemented by an $SO(2)$ rotation, but requires reflection in the x axis. To work with continuous $SO(2)$ rotations, we must use a patching construction, which effectively distinguishes ϕ from $2\pi - \phi$. (These reflective congruences can be implemented continuously within $SO(3)$, but not in a way that applies continuously to all three-dimensional realizations of the base shapes. Note that while three-dimension inversion, as opposed to plane-dependent reflection, can be implemented globally, it does not make the relevant planar configurations equivalent.) Equation (39) indicates that our connection has, for each fixed value of κ, the magnetic flux corresponding to the minimal Dirac charge. Thus, we can identify the singularity at $\kappa = 0$ as a magnetic monopole in shape space.

We can also bring the monopole structure out more topologically, by examining the patching construction required to knit together the $\tilde{\theta}, \tilde{\phi}$ and the θ, ϕ coordinates. We have

$$A_\phi = \sin^2\theta$$
$$A_{\tilde{\phi}} = -\sin^2\tilde{\theta} = -\cos^2\theta \tag{40}$$

The two potentials are related by a globally non-trivial gauge transformation, according to

$$A_\phi \;=\; \partial_\phi \psi + A_{\tilde{\phi}} \tag{41}$$

Finally, let us note that in terms of the alternative angle $\lambda \equiv \frac{\theta}{2}$ then we have conventional spherical coordinates λ, ϕ on our constant κ spheres. On those spheres, with the canonical metric, the magnetic flux is uniform.

3. A fractionated molecule

Upon considering the quantum version of this problem we encounter a related but different emergent gauge field, which illustrates another interesting phenomenon.

To keep things simple, in our system of three planar bodies let us take one to be heavy and fixed at the origin, and the other two to be at fixed distances from the origin. Then we are reduced to two dynamical degrees of freedom, i.e. two angles ϕ_1, ϕ_2. Assigning them moments of inertia I_1, I_2, we have the rotational kinetic Hamiltonian

$$H_{\text{rot.}} = \frac{p_1^2}{2I_1} + \frac{p_2^2}{2I_2} \tag{42}$$

with

$$p_1 \equiv -i\frac{\partial}{\partial \phi_1}$$
$$p_2 \equiv -i\frac{\partial}{\partial \phi_2} \tag{43}$$

We can re-write Eq. (42) as

$$H_{\text{rot.}} = \frac{1}{2(I_1 + I_2)}(p_1 + p_2)^2$$
$$+ \frac{I_1 + I_2}{2I_1 I_2}\left(\frac{p_1 + p_2}{2}\frac{I_2 - I_1}{I_1 + I_2} + \frac{p_1 - p_2}{2}\right)^2 \tag{44}$$

Here $p_1 + p_2$ is the total rotation generator. In our previous notation, it rotates the angle α. Now let us separate α, as is appropriate for a rotationally invariant system, and impose the quantization condition

$$p_1 + p_2 = m \tag{45}$$

Then in the second part of expression (44) we have the bracketed term

$$\left(\frac{m}{2}\frac{I_2 - I_1}{I_1 + I_2} + \frac{p_1 - p_2}{2}\right)^2 \rightarrow \left(p_2 - m\frac{I_2}{I_1 + I_2}\right)^2 \tag{46}$$

But p_2 generates changes in the relative angle in shape space, i.e., our previous ϕ. Thus, we see that non-trivial angular momentum for α generally yields fractional (kinetic) angular momentum [11] for the shape-space angle ϕ. The fractional part is

$$\delta l = \frac{mI_2}{I_1 + I_2} \tag{47}$$

Note that this approaches an integer for $I_1/I_2 \to 0$ or $\to \infty$, as it should.

The bracketed term in Eq. (44) might look strange at first sight, but if we substitute $p_1 = I_1\omega_1$, $p_2 = I_2\omega_2$ we find that it is proportional to $(\omega_2 - \omega_1)^2$. Thus, it represents the contribution of relative angular velocity. It is only the passage from Lagrangian to Hamiltonian that introduces complications. But of course quantum theory requires the Hamiltonian.

In an alternate description, corresponding to the other patch in our preceding monopole construction, we interchange the roles of bodies 1 and 2. In that way, we replace Eq. (46) by

$$\left(\frac{m}{2}\frac{I_2 - I_1}{I_1 + I_2} + \frac{p_1 - p_2}{2}\right)^2 \rightarrow \left(p_1 + m\frac{I_1}{I_1 + I_2}\right)^2 \tag{48}$$

and hence Eq. (47) by

$$\delta l = -\frac{mI_I}{I_1 + I_2} \tag{49}$$

Consistency requires that the difference between Eq. (47) and Eq. (49) must be an integer, and indeed it is the integer m. Thus we see an interesting connection between the monopole in shape space, which we originally uncovered classically, and the quantum theory of the same mechanical system.

Another consistency check is to consider the free limit, where Eq. (42) is the entire Hamiltonian. Then of course by substituting integer values for p_1, p_2 we get the spectrum, either directly from Eq. (42) or after some algebra from Eq. (44), with identical results. The significance of the separation procedure is that overall rotational symmetry is more generic than shape independence, and by exploiting it we always reduce the dimensionality of the potentially non-trivial dynamics.

If we restore the dynamics that allows I_1, I_2 to vary, then the preceding construction gives an emergent dynamical compact $U(1)$ gauge field in shape space, which however is not governed by the classic Maxwell–Yang–Mills action. In that interpretation, the total angular momentum supplies an effective charge.

Acknowledgement

The first section partly adapts material from [12]. The second section adapts material from a forthcoming paper with X. Peng, J. Dai, and A. Niemi [13]. This work is supported by the U.S. Department of Energy under grant Contract Number DE-SC0012567, by the European Research Council under grant 742104, and by the Swedish Research Council under Contract No. 335-2014-7424.

References

1. R. Jackiw, *Diverse Topics in Theoretical and Mathematical Physics* (World Scientific, 1994).
2. R. Jackiw and C. Rebbi, *Phys. Rev.* **D13**, 2298 (1976).
3. W. P. Su, J. R. Schrieffer and A. J. Heeger, *Phys. Rev. Lett.* **42**, 1698 (1979).
4. S. Coleman and E. Weinberg, *Phys. Rev.* **D7**, 1888 (1973).
5. R. Peirels, *Quantum Theory of Solids* (Oxford, 1955).
6. J. Goldstone and F. Wilczek, *Phys. Rev. Lett.* **47**, 968 (1981).
7. W. Shockley, *Phys. Rev.* **56**, 317 (1939).
8. B. A. Bernevig, *Topological Insulators and Topological Superconductors* (Princeton, 2013).
9. X. L. Li and S. C. Zhang, *Rev. Mod. Phys.* **83**, 1057 (2011).
10. A. Shapere and F. Wilczek, *Am. J. Phys.* **57**, 514 (1989).
11. F. Wilczek, *Phys. Rev. Lett.* **48** 1144 (1982).
12. F. Wilczek, *Physica Scripta* **T168**, (2016).
13. X. Peng, J. Dai, A. Niemi and F. Wilczek, *Shape Change and External Motion*, to appear.

Chapter 25

A Note on Boundary Conditions in Euclidean Gravity

Edward Witten

School of Natural Sciences, Institute for Advanced Study
Einstein Drive, Princeton, NJ 08540, USA

We review what is known about boundary conditions in General Relativity on a spacetime of Euclidean signature. The obvious Dirichlet boundary condition, in which one specifies the boundary geometry, is actually not elliptic and in general does not lead to a well-defined perturbation theory. It is better-behaved if the extrinsic curvature of the boundary is suitably constrained, for instance if it is positive- or negative-definite. A different boundary condition, in which one specifies the conformal geometry of the boundary and the trace of the extrinsic curvature, is elliptic and always leads formally to a satisfactory perturbation theory. These facts might have interesting implications for semiclassical approaches to quantum gravity. April, 2018

1. Introduction

The goal of this note is to make accessible some basic properties of boundary conditions in Euclidean gravity [1–4]. The facts described here are not new. The motivation for presenting this material is that it may have applications to semiclassical quantization of gravity. There is an extensive literature on this subject, a small selection being Refs. [5–12].

In Euclidean signature, let X be a $D = d + 1$-dimensional Riemannian manifold with boundary a d-manifold M. The most obvious boundary condition in Euclidean quantum gravity is to specify the geometry of M (that is, its Riemannian metric) and integrate over all metrics on X that are consistent with this boundary geometry. Another equally simple boundary condition is to specify the conformal geometry of M and the trace K of its extrinsic curvature or second fundamental form (which is roughly the normal derivative of the metric along M). We will call these the Dirichlet and conformal boundary conditions, respectively. The main point that we aim to explain is that in general the Dirichlet boundary condition does not lead to a sensible perturbation expansion, while the conformal boundary condition always does lead formally (that is, modulo the usual ultraviolet divergences) to a sensible perturbation expansion about any given classical solution. Technically, the conformal boundary condition is elliptic, ensuring its good behavior, and the

Dirichlet boundary condition is not.

The conformal boundary condition is natural in the context of using the conformal structure of an initial value surface (rather than the full Riemannian geometry of that surface) and the trace of the extrinsic curvature as a maximal set of commuting variables. This idea has a long history [13].

Although the Dirichlet boundary condition is ill-behaved in general, one can show, by using the ellipticity of a certain alternate boundary condition that is similar to the conformal one, that the Dirichlet boundary condition is well-behaved at least in some respects if the extrinsic curvature of the boundary satisfies a certain condition, for instance if it is positive- or negative-definite. (The precise statement here is a little subtle, as we discuss momentarily.)

In Sec. 2, we review the concept of an elliptic boundary condition, and then in Sec. 3, we explore this concept in the context of gravity. In gauge theory, there is a very natural elliptic boundary condition in which the boundary value of the gauge field is specified. Gravity is different, basically because of the second-order nature of the Hamiltonian constraint equation. As already remarked, one can get an elliptic boundary condition by specifying the conformal class of the boundary metric and the trace of the second fundamental form, but not by specifying the boundary metric.

To avoid confusion, we should stress that ellipticity of a boundary condition in General Relativity does not guarantee either existence or uniqueness of a solution of the Einstein equations with specified boundary data. Nor does it have anything to do with positivity of the operator that governs gravitational perturbations. Ellipticity does guarantee the properties that are needed to construct perturbation theory. It ensures that the gauge-fixed Einstein equations (on a compact manifold) have only a finite number of zero-modes, and that modulo these zero-modes, the linearized Einstein equations have a propagator with the usual properties.[a] Mathematically, a differential operator L with a finite-dimensional kernel and cokernel[b] is said to be Fredholm; and a propagator defined after removing a finite-dimensional space of zero-modes is called a parametrix.

Quantum perturbation theory is usually constructed by expanding around a classical solution. Generically ellipticity is needed to ensure the properties that make perturbation theory possible.[c] However, for the specific case of General Relativity,

---

[a] Technically, ellipticity is the right condition on a boundary condition to guarantee these properties if gauge-fixing is carried out in the standard way, so that the gauge-fixed action for metric fluctuations has a nondegenerate second-order kinetic energy. In some other approaches to gauge-fixing, one would encounter a slight generalization of the notion of ellipticity. See the remark at the end of Sec. 2.2.

[b] The kernel of an operator L is the space of solutions of $Lu = 0$. To define the cokernel, consider the more general equation $Lu = f$ with a source f. The cokernel is the space of all f's modulo those for which the equation can be solved. If L is self-adjoint — as are the linearized Einstein equations with some boundary conditions — the condition of finite dimensionality of the cokernel is redundant as there is a natural isomorphism between the kernel and the cokernel.

[c] Perturbation theory also requires the absence of certain one-loop anomalies, which involve topological considerations. Further issues arise if a theory has gauge symmetries that do not allow any

one can show that if the extrinsic curvature of the boundary is positive- or negative-definite (and under somewhat more general conditions), the linearized Einstein equations with Dirichlet boundary conditions are Fredholm even though not elliptic. It seems plausible that they admit a parametrix, though this does not seem to be rigorously known. So this may be a case in which perturbation theory is possible with a boundary condition that is not elliptic.

A lecture by the author on issues possibly related to what is described here can be found in Ref. [14].

It is a pleasure to dedicate this article to Roman Jackiw on the occasion of his birthday. Roman has introduced many important ideas in physics. The triangle anomaly was a milestone and a turning point in the understanding of the strong interactions. His work on zero-modes of fermions in the field of a soliton or instanton has been very important in both particle physics and condensed matter physics. Also important for both relativistic physics and condensed matter physics has been Roman's work on topologically massive gauge theories in three spacetime dimensions. All these things have in particular been important for my own work.

Finally, since the article that I am contributing to this volume is largely concerned with gravity, I think I should mention Roman's work with Curt Callan and Sidney Coleman in which they constructed a new 'improved' energy-momentum tensor with a softer trace, more natural for the coupling of a quantum field theory to gravity.

2. Background

2.1. *Elliptic boundary conditions*

First let us recall the definition of an elliptic differential operator. The symbol of a differential operator is defined, roughly speaking, by replacing derivatives $-i\partial/\partial x^\mu$ by momenta p_μ. To be more exact, if L is a differential operator of order n on a manifold X, then its "leading symbol" is defined by making the substitution $-i\partial/\partial x^\mu \to p_\mu$ in the terms of order n and dropping terms of order less than n. Thus the leading symbol of L is, for each point $x \in X$, a polynomial $\sigma_x(p)$ in p that is homogeneous of degree n. For example, the leading symbol of the Laplacian

$$\Delta = -g^{\mu\nu} D_\mu D_\nu \tag{1}$$

is $\sigma_x(p) = g^{\mu\nu}(x)p_\mu p_\nu$. Likewise, the leading symbol of the Dirac operator

$$i\slashed{D} = i\Gamma^\mu D_\mu \tag{2}$$

(where Γ^μ are gamma matrices) is $\sigma_x(p) = \slashed{p} = \Gamma \cdot p$.

If L has order n, then the leading symbol $\sigma_x(p)$ is naturally understood as a function — or more generally, as in the Dirac case, a matrix-valued function — on the cotangent bundle T^*X, homogeneous of degree n on the fibers. L is called

---

useful regularization. This is relatively uncommon.

"elliptic" if for all $x \in X$ and all real nonzero p, $\sigma_x(p)$ is invertible. For example, the leading symbols $\sigma_x(p)$ for the Laplacian and the Dirac operator have this property so these are elliptic operators.

An elliptic operator L on a compact manifold is Fredholm. If L maps the space of sections of some vector bundle E to itself (so that the eigenvalue problem $L\psi = \lambda\psi$ makes sense and one can define the spectrum of L), then L has a discrete spectrum. In particular, this is so if L is self-adjoint, a common case. The eigenvalues of an elliptic operator tend to infinity[d] in a simple way that can be described semiclassically in terms of $\sigma_x(p)$. In particular, the space of zero-modes is at most finite-dimensional. L can be inverted, on a subspace transverse to the zero-modes, by a Green's function $G(x, x')$ that is regular for $x \neq x'$, and whose singularities for $x \to x'$ are controlled in the standard way by an operator product expansion. (In particular, the leading singularity for $x \to x'$ depends only on $\sigma_x(p)$.) Such a Green's function defined after removing zero-modes is called a parametrix; physically, it is used as a propagator in constructing perturbation theory. Ellipticity, along with some topological considerations involving anomalies, also makes it possible to define a "determinant" of L (or in appropriate circumstances, a Pfaffian) with standard properties. In short, ellipticity guarantees the properties that are needed in constructing perturbation theory.[e]

Ellipticity is an "open" condition on differential operators, in the sense that if L is an elliptic of order n, then any small perturbation of L (by terms of order n or less) does not affect ellipticity. This is true because invertibility of $\sigma_x(p)$ is similarly an open condition, invariant under small perturbations.

Now suppose that X has a boundary M of dimension $d = D-1$, and that we are given some boundary condition for L along M that reflects a boundary condition on underlying quantum fields. To be able to do perturbation theory in this situation, the boundary condition must satisfy a condition that ensures that L will still have a propagator and determinant with appropriate properties. This will always be true if the boundary condition is "elliptic." Otherwise it is typically not true.

Ellipticity of a boundary condition involves a condition that must be checked at each boundary point. In checking ellipticity at a given boundary point $x \in M$, we only care about short distance behavior, so we can approximate X by a half-space $\mathbb{R}^D_+ \subset \mathbb{R}^D$, with boundary $M \cong \mathbb{R}^{D-1}$, and we can drop from L all terms of order

---

[d] Here "infinity" means $+\infty$ if $\sigma_x(p)$ is hermitian and positive-definite — as for the Laplacian — or $\pm\infty$ if $\sigma_x(p)$ is hermitian but not positive-definite, as for the Dirac operator. If $\sigma_x(p)$ is invertible but not hermitian, then L still has a discrete spectrum but its eigenvalues are not necessarily real, and can tend to infinity in any direction in the complex plane in which an eigenvalue of $\sigma_x(p)$ can tend to infinity.

[e] The definitions just given are adequate for most purposes, but in general one has to take into account different scaling weights of different fields in defining what one means by the leading symbol $\sigma_x(p)$. The definition of $\sigma_x(p)$ is modified, but a satisfactory theory still requires that $\sigma_x(p)$ should be invertible for nonzero real p. This situation can arise in gauge-fixing of Yang–Mills theory or gravity if one does not integrate out the auxiliary field. See the concluding remarks of Sec. 2.2.

less than n. Moreover, ellipticity at a given boundary point x only depends on $\sigma_x(p)$ for that value of x, so we can treat L as an operator with constant coefficients. In other words, if $\sigma_x(p) = \sum \sigma^{\mu_1 \mu_2 \cdots \mu_n} p_{\mu_1} p_{\mu_2} \cdots p_{\mu_n}$, then in testing ellipticity, we can replace L with

$$(-\mathrm{i})^n \sum \sigma^{\mu_1 \mu_2 \cdots \mu_n} \frac{\partial}{\partial x^{\mu_1}} \cdots \frac{\partial}{\partial x^{\mu_n}}. \tag{3}$$

For example, if L is the Laplacian of Eq. (1), we can approximate it by the flat space Laplacian $\Delta_0 = -\sum_{\mu=1}^{D} \frac{\partial^2}{\partial x_\mu^2}$, on a half-space, say the half-space $x^D \geq 0$.

Let us write $\vec{x} = (x^1, x^2, \cdots, x^{D-1})$ for boundary coordinates and $\vec{p} = (p_1, p_2, \ldots, p_{D-1})$ for the momentum along the boundary. Also, let us write $x_\perp$ for the coordinate x_D in the normal direction to the boundary, and $p_\perp$ for the momentum in the $x_\perp$ direction. We work on the half-space $x_\perp \geq 0$. In the approximation of treating L as an operator with constant coefficients, the equation $L\Phi = 0$ has plane-wave solutions $\exp(\mathrm{i}\vec{p} \cdot \vec{x} + \mathrm{i}x_\perp p_\perp(\vec{p}))$. Here $p_\perp(\vec{p})$ is found by solving the equation $\sigma_x(\vec{p}, p_\perp) = 0$. Let us restrict to the case that $\vec{p}$ is real and nonzero, so that the solution behaves as an oscillatory plane wave along the boundary. In this case, ellipticity of L away from the boundary means that the equation $\sigma_x(\vec{p}, p_\perp) = 0$ has no solutions for real $p_\perp$.

Let us assume for the moment that $\sigma_x(p)$ is hermitian for real p (as in Yang–Mills theory and gravity with the usual gauge-fixing). Then, for given real nonzero $\vec{p}$, the solutions of $\sigma_x(\vec{p}, p_\perp) = 0$ occur in complex conjugate pairs, half with positive imaginary part of $p_\perp$ and half with negative imaginary part. The space of solutions is thus, for some nonnegative integer s, a $2s$-dimensional vector space $V_{2s}(\vec{p})$. The importance of the sign of $\mathrm{Im}\, p_\perp$ is simply that a plane-wave solution $\exp(\mathrm{i}\vec{p} \cdot \vec{x} + \mathrm{i}x_\perp p_\perp(\vec{p}))$ is exponentially decaying or exponentially growing as $x_\perp$ increases, depending on the sign of $\mathrm{Im}\, p_\perp$.

An elliptic boundary condition is one that selects, for every real nonzero $\vec{p}$, a middle-dimensional subspace $W_s(\vec{p}) \subset V_{2s}(\vec{p})$ of allowed solutions, with the property that for sufficiently large $|\vec{p}|$, none of the solutions in $W_s(\vec{p})$ is exponentially decaying with increasing $x_\perp$. The intuitive idea is that if a boundary condition allows solutions with arbitrarily large $|\vec{p}|$ that are exponentially decaying as $x_\perp$ increases, then the operator L with this boundary condition has too many near zero-modes that are localized at short distances along the boundary, and cannot have a discrete spectrum (even when X is compact) or a satisfactory propagator.

For a boundary condition to be local as well as elliptic means that $W_s(\vec{p})$ must be defined by vanishing of all appropriate function of the fields and their derivatives

Let us verify that the usual Dirichlet and Neumann boundary conditions on the Laplacian are elliptic. The plane-wave solutions of the Laplace equation are $\exp(\mathrm{i}\vec{p} \cdot \vec{x} \pm |\vec{p}|x_\perp)$. Dirichlet boundary conditions $\phi| = 0$ (where $\phi|$ denotes the restriction of ϕ to $x_\perp = 0$) are satisfied by the linear combination $\exp(\mathrm{i}\vec{p} \cdot \vec{x})(e^{|\vec{p}|x_\perp} - e^{-|\vec{p}|x_\perp})$, which is exponentially growing with $x_\perp$. Neumann boundary conditions $\frac{\partial \phi}{\partial x_\perp}\big| = 0$ are satisfied by the linear combination $\exp(\mathrm{i}\vec{p} \cdot \vec{x})(e^{|\vec{p}|x_\perp} + e^{-|\vec{p}|x_\perp})$, which also is exponentially growing. So both of these boundary conditions are elliptic. In the case of the Dirac operator, the most commonly studied boundary conditions are $\Gamma_D \psi| = \pm \psi$ (with some choice of the sign). We leave it to the reader to verify that these are elliptic boundary conditions, by showing that a solution of the Dirac equation on the half-space that has plane wave behavior along the boundary and satisfies either of these boundary conditions is exponentially growing with increasing $x_\perp$.

If L is self-adjoint in the absence of a boundary, then in the presence of a boundary, one usually wants to pick a boundary condition that is self-adjoint as well as elliptic, in other words a boundary condition that ensures that L remains self-adjoint even in the presence of a boundary. This is a stronger condition. For example, in the case of the Laplacian acting on a complex-valued field, the mixed boundary condition $\left(\frac{\partial \phi}{\partial x_\perp} - c\phi \right)\Big| = 0$ (which is sometimes called a Robin boundary condition) is elliptic for any constant c, but it is only self-adjoint if c is real.[f]

If $\sigma_x(p)$ is not hermitian, then solutions of $\sigma_x(\vec{p}, p_\perp) = 0$ do not necessarily come in complex conjugate pairs. Still, an elliptic boundary condition is one that selects in the space $V(\vec{p})$ of plane wave solutions with given $\vec{p}$ a middle-dimensional subspace $W(\vec{p})$ with the property that for sufficiently large $|\vec{p}|$, no solution in $W(\vec{p})$ is an exponentially decreasing function of $x_\perp$. In general, the operator L may not admit any local elliptic boundary condition. The most obvious obstruction is that $V(\vec{p})$ might be odd-dimensional, and there also are further obstructions of topological nature. In quantum field theory, the most important example with $\sigma_x(p)$ not hermitian is the chiral Dirac operator for even D; it does not admit any local elliptic boundary condition.

We note that the condition of ellipticity — $W(\vec{p})$ does not contain any exponentially decaying solutions, for any nonzero real $\vec{p}$ — has the property that if it is true for any one boundary condition, then it is true for any sufficiently nearby boundary condition. In this sense, ellipticity is an "open" condition on boundary conditions.

2.2. Yang–Mills theory on a closed manifold

Let A be a Yang–Mills gauge field with gauge group G and field strength $F = dA + A \wedge A$. The usual action is[g]

$$ I = -\frac{1}{4} \int_X d^D x \sqrt{g} \, \text{Tr} \, F_{\mu\nu} F^{\mu\nu} \tag{4} $$

and the field equations are

$$ D^\mu F_{\mu\nu} = 0. \tag{5} $$

---

[f]With Robin boundary conditions, provided $c < 0$, there are solutions $\exp(i\vec{p} \cdot \vec{x} - |\vec{p}|x_\perp)$ with $|\vec{p}| = -c$ that decay exponentially away from the boundary. But for sufficiently large $|\vec{p}|$, there are no such solutions, so this boundary condition is elliptic.

[g]Here g is the metric tensor of X. We consider A and F to be real and antihermitian, so the trace on the Lie algebra is a negative-definite quadratic form. This accounts for the minus sign in Eq. (4).

For future reference, we recall the Bianchi identity:

$$D^\mu(D^\nu F_{\mu\nu}) = 0. \tag{6}$$

Now let A_0 be a classical solution and set $A = A_0 + \mathsf{a}$. However, it is clumsy to write A_0 all the time, so henceforth we will refer to the underlying gauge field as $\widehat{A}$, and a chosen classical solution as A, and we will write the expansion as $\widehat{A} = A + \mathsf{a}$. To linear order in a, the classical equations become

$$D_\mu(D_\mu \mathsf{a}_\nu - D_\nu \mathsf{a}_\mu) + [F_{\nu\lambda}, \mathsf{a}_\lambda] = 0 \tag{7}$$

where D_μ and $F_{\mu\nu}$ are the covariant derivative and field strength of the background solution A. We can write this equation as $L\mathsf{a} = 0$ where

$$(L\mathsf{a})_\nu = -D_\mu D^\mu \mathsf{a}_\nu + D_\nu D_\mu \mathsf{a}^\mu - 2[F_{\nu\lambda}, \mathsf{a}^\lambda]. \tag{8}$$

The corresponding action, to quadratic order in a, is

$$I' = -\frac{1}{2} \int_X \mathrm{d}^D x \, \mathrm{Tr} \left(D_\mu \mathsf{a}_\nu D^\mu \mathsf{a}^\nu - (D_\mu \mathsf{a}^\mu)^2 + 2F_{\mu\nu}[\mathsf{a}^\mu, \mathsf{a}^\nu] \right). \tag{9}$$

The operator L is not elliptic. Its leading symbol is the matrix-valued function $\sigma(p)_{\mu\nu} = p^2 \delta_{\mu\nu} - p_\mu p_\nu$ (tensored with the identity operator on the Lie algebra $\mathfrak{g}$ of G), and this matrix annihilates any vector that is a multiple of p_ν. This failure of ellipticity is an inevitable consequence of the underlying gauge invariance, which in terms of the linearization becomes

$$\mathsf{a}_\mu \to \mathsf{a}_\mu - D_\mu \varepsilon, \tag{10}$$

for any $\mathfrak{g}$-valued gauge parameter ε. (We will abbreviate Eq. (10) as $\mathsf{a} \to \mathsf{a} - \mathrm{d}_A \varepsilon$.) The gauge invariance implies that $\mathsf{a}_\mu = -D_\mu \varepsilon$ is a solution of $L\mathsf{a} = 0$ for any ε, so L has an infinite-dimensional kernel and cannot possibly be elliptic.

To restore ellipticity, we need a suitable gauge condition. A very natural one is $S = 0$, where

$$S = D_\mu \mathsf{a}^\mu. \tag{11}$$

When supplemented with the gauge condition $S = 0$, the equation $L\mathsf{a} = 0$ becomes $L'\mathsf{a} = 0$ where

$$(L'\mathsf{a})_\mu = -D_\nu D^\nu \mathsf{a}_\mu - 2[F_{\mu\lambda}, \mathsf{a}^\lambda]. \tag{12}$$

The symbol of L' is $p^2 \delta_{\mu\nu}$, and this is invertible for nonzero real p, so L' is elliptic. We note that

$$(L'\mathsf{a})_\mu = (L\mathsf{a})_\mu - D_\mu S. \tag{13}$$

Now let us discuss why $S = 0$ is a good gauge condition, first from the point of view of classical partial differential equations and then in terms of quantum perturbation theory.

From the first point of view, we usually want to describe the solutions of the gauge-invariant equation $L\mathsf{a} = 0$, modulo gauge transformations. The claim is that such equations are in natural correspondence with solutions of the gauge-fixed equation $L'\mathsf{a} = 0$. In one direction, if we are given an a that satisfies $L\mathsf{a} = 0$, we look for a gauge-equivalent $\mathsf{a}' = \mathsf{a} - \mathrm{d}_A\varepsilon$ with $S(\mathsf{a}') = 0$. The equation $S(\mathsf{a}') = 0$ is equivalent to

$$P\varepsilon = -S(\mathsf{a}), \tag{14}$$

where $P = -D_\mu D^\mu$ is the gauge-invariant Laplacian. Equation (14) will have a unique solution if the operator P is invertible, and more generally if the right hand side is orthogonal to any zero-modes of P. We observe that

$$-\int_X \mathrm{d}^D x \sqrt{g} \sum_\mu \mathrm{Tr}\,(D_\mu \varepsilon)^2 = -\int_X \mathrm{d}^D x \sqrt{g}\, \mathrm{Tr}\,(\varepsilon P \varepsilon). \tag{15}$$

The left hand side is strictly positive unless $D_\mu\varepsilon = 0$, in which case it vanishes. But if ε is an eigenfunction of P with zero or negative eigenvalue, then the right hand side is zero or negative. This shows that P is positive-definite — and therefore invertible — except for possible zero-modes that must be covariantly constant. But $S(\mathsf{a}) = D_\mu \mathsf{a}^\mu$ is orthogonal to any covariantly constant mode, since $\int_X \mathrm{d}^D x \sqrt{g}\, \mathrm{Tr}\,\varepsilon S(\mathsf{a}) = -\int_X \mathrm{d}^D x \sqrt{g}\, \mathrm{Tr}\, D_\mu \varepsilon \mathsf{a}^\mu$, which vanishes if $D_\mu\varepsilon = 0$. The right hand side of Eq. (14) is thus orthogonal to the kernel of P, so Eq. (14) always has a unique solution for ε.

Thus, a solution of $L\mathsf{a} = 0$ is gauge-equivalent to a unique solution of the gauge-fixed equation $L'\mathsf{a} = 0$. In the opposite direction, we would like to prove that any solution of $L'\mathsf{a} = 0$ actually obeys $L\mathsf{a} = 0$. To show this, we first observe that the gauge-invariant operator L satisfies a Bianchi identity, which can be written

$$D_\mu((L\mathsf{a})^\mu) = 0. \tag{16}$$

This is proved by linearizing the underlying Bianchi identity (6). Comparing L' and L, it follows that

$$D_\mu((L'\mathsf{a})^\mu) = -D_\mu D^\mu(D_\nu \mathsf{a}^\nu) = PS(\mathsf{a}), \tag{17}$$

and therefore a solution of $L'\mathsf{a} = 0$ satisfies $PS(\mathsf{a}) = 0$. As we have just seen, the equation $PS = 0$ implies that S is covariantly constant, $D_\mu S(\mathsf{a}) = 0$. But in view of Eq. (13), $L'\mathsf{a} = 0 = D_\mu S(\mathsf{a})$ implies that a satisfies the gauge-invariant equation $L\mathsf{a} = 0$. So it is equivalent to consider solutions of $L\mathsf{a} = 0$ up to gauge transformation or to consider solutions of $L'\mathsf{a} = 0$.

For a fuller understanding, let us consider BRST quantization. In BRST quantization, we introduce a ghost field c that represents a generator of gauge transformations, but with fermionic statistics (and ghost number 1). The BRST transformations of a and c are

$$\delta \mathsf{a}_\mu = -D_\mu c, \qquad \delta c = \frac{1}{2}[c, c]. \tag{18}$$

We also introduce an antighost multiplet consisting of an antighost field $\bar{c}$ and an auxiliary field B in the adjoint representation, with

$$\delta \bar{c} = B, \qquad \delta B = 0. \tag{19}$$

All this is consistent with $\delta^2 = 0$. The gauge-fixed action is obtained by adding $\delta \int_X d^D x \sqrt{g} \, V$, for some convenient choice of V, to the gauge-invariant action (4). Taking $V = \text{Tr} \left(\frac{1}{2} \bar{c} B - \bar{c} D_\mu a^\mu \right)$, we get

$$\delta \int_X d^D x \sqrt{g} \, V = \int_X d^D x \sqrt{g} \left(\frac{1}{2} B^2 - B D_\mu a^\mu - \bar{c} D_\mu D^\mu c \right). \tag{20}$$

Upon integrating out the auxiliary field B, we get the gauge-fixing action

$$I'' = \int_X d^D x \sqrt{g} \, \text{Tr} \left(-\frac{1}{2} (D_\mu a^\mu)^2 - \bar{c} D_\mu D^\mu c \right). \tag{21}$$

The gauge-fixed action $\widehat{I} = I' + I''$ is

$$\widehat{I} = -\int_X d^D x \sqrt{g} \, \text{Tr} \left(\frac{1}{2} D_\mu a_\nu D^\mu a^\nu + F^{\mu\nu} [a_\mu, a_\nu] + \bar{c} D_\mu D^\mu c \right). \tag{22}$$

The kinetic operator for a is the gauge-fixed operator L'.

The fact that a BRST-invariant action can be written with L' as the kinetic operator for gauge fields does *not* imply that L' is elliptic. Rather, a good BRST gauge-fixing — suitable for perturbation theory — is one in which V is chosen, as we have done in this example, to ensure that the resulting kinetic operators are elliptic. Since ellipticity is an open condition, this means roughly that V must be sufficiently generic.

In the above example, when we integrate out B, we get $B = S(a)$, and therefore the BRST transformations become

$$\delta \bar{c} = S(a). \tag{23}$$

Of course, we do not have to integrate out B; we could develop a formalism with both a and B present in the theory. For the purposes of the present paper, this would lead to a somewhat more involved discussion, leading to the same conclusions. That is because a obeys a second order classical differential equation, while the equations of motion involve only first derivatives of B; thus we would be in the situation described in footnote e of Sec. 2.1. To proceed in this way, we would have to change the definition of the "leading" term in a differential equation by saying that B has degree 1, just like a derivative $\partial/\partial x$. To avoid these complications, we will consider the theory with B integrated out — as is usually done in constructing Yang–Mills perturbation theory. A similar remark applies later when we come to gravity.

2.3. *Yang–Mills theory on a manifold with boundary*

Now let us suppose that X has a boundary M, and try to extend this analysis to make a BRST-invariant and elliptic gauge-fixing in Yang–Mills theory in the presence of the boundary. We assume that the gauge-fixed action is as above in the bulk, and we will discuss what we can do along the boundary.

We will try to implement a very natural boundary condition, in which one specifies the restriction of the gauge field A to the boundary. We locally model X

by $x^D \geq 0$; we denote the coordinates as x^i for $i < D$ and we write $x^\perp$ for x^D. Likewise we denote the gauge field components as A_i, $i < D$ and $A_\perp = A_D$. For a boundary condition, we specify the boundary values of A_i, $i < D$ but not of $A_\perp$. In other words, we specify the gauge connection that would be used for parallel transport within M. In terms of the field a that describes small fluctuations around a classical solution, this means that $\vec{a} = (a_1, a_2, \cdots, a_{D-1})$ will vanish on M, but the condition on $a_\perp = a_D$ will be different.

If we want to impose a condition $\vec{a}| = 0$, then in view of the gauge invariance a → a − $d_A\varepsilon$, we must constrain the generator ε of a gauge transformation to vanish along the boundary M. Since the ghost field c is always the generator of a gauge transformation (with statistics reversed), it will also have to vanish along the boundary. Thus c must obey Dirichlet boundary conditions:

$$c| = 0. \tag{24}$$

Once we impose Dirichlet boundary conditions for c, we have to do the same for $\bar{c}$. The reason is not that we want to be able to interpret $\bar{c}$ as the complex conjugate of c, but that once we impose Dirichlet boundary conditions for c, we will not be able to define a sensible Green's function for the c-$\bar{c}$ system if we do anything else for $\bar{c}$. Recall first of all that if we do impose Dirichlet boundary conditions on both c and $\bar{c}$, then there is a standard Green's function $G(x, y) = \langle c(x)\bar{c}(y)\rangle$ that obeys the differential equations $\Delta_x G(x, y) = \Delta_y G(x, y) = \delta^D(x, y)$ (where Δ_x and Δ_y are the gauge-invariant Laplacians acting on the x or y variables) along with the Dirichlet boundary conditions $G(x, y)|_{x \in M} = G(x, y)|_{y \in M} = 0$. But actually, just the equation $\Delta_x G(x, y) = \delta^D(x, y)$ plus the Dirichlet boundary condition in the x variable $G(x, y)|_{x \in M} = 0$ uniquely determines $G(x, y)$. The unique possibility is the standard Green's function that corresponds to imposing Dirichlet boundary conditions also on $\bar{c}$:

$$\bar{c}| = 0. \tag{25}$$

We should not expect to be able to set a_D to zero along M, because a_D is not invariant under gauge transformations whose generator ε vanishes on M. To make $a_D|$ invariant under a → a − $d_A\varepsilon$, we would want to require $D_\perp\varepsilon| = 0$. This would entail restricting the generator ε of a gauge transformation to satisfy $\varepsilon| = D_\perp\varepsilon| = 0$, so we would want c to satisfy $D_\perp c| = 0$ as well as $c| = 0$; but we cannot impose both Dirichlet and Neumann boundary conditions on a field that obeys a second-order wave equation.

We can easily deduce from Eq. (25) what boundary condition $a_\perp$ must satisfy. Given that $\bar{c}$ vanishes along M and given the BRST transformation law $\delta\bar{c} = S(a)$ (Eq. (23)), BRST invariance requires

$$S(a)| = 0. \tag{26}$$

If X is the half-space $x_\perp \geq 0$ in $\mathbb{R}^D$, then given our condition $\vec{a}| = 0$, $S(a)|$ reduces to $D_\perp a_\perp|$ and thus $a_\perp$ satisfies a gauge-invariant version of Neumann boundary conditions:

$$D_\perp a_\perp| = 0. \tag{27}$$

In the general case of a curved manifold, $S(\mathsf{a})| = 0$ corresponds to a boundary condition on $\mathsf{a}_\perp$ that is similar locally to Neumann boundary conditions on a scalar field, plus a lower-order term. We will loosely refer to this as Neumann boundary conditions for $\mathsf{a}_\perp$.

The fact that the leading symbol of L' is $p^2\delta_{\mu\nu}$ means that it behaves at short distances as a system of decoupled scalar Laplace equations for the components $\mathsf{a}_1, \mathsf{a}_2, \cdots, \mathsf{a}_D$. Dirichlet and Neumann boundary conditions on the scalar Laplace equation are elliptic. So the combination of Dirichlet boundary conditions for $\vec{\mathsf{a}}$ and Neumann for $\mathsf{a}_\perp$ comprises an elliptic boundary condition for the operator L', and thus we have found a BRST-invariant and elliptic gauge-fixing for Yang–Mills theory on a manifold with boundary.

Now let us discuss how one might motivate this boundary condition from the point of view of differential geometry, without mentioning the ghosts. In that framework, one may want to study solutions of the gauge-invariant equation $L\mathsf{a} = 0$, with $\vec{\mathsf{a}}| = 0$, modulo gauge transformations that are trivial along the boundary. One wishes to show that such solutions are in one-to-one correspondence with solutions of the gauge-fixed equation $L'\mathsf{a} = 0$, with boundary conditions $\vec{\mathsf{a}}| = S(\mathsf{a})| = 0$. In one direction, given a solution of $L\mathsf{a} = 0$, we look as in Sec. 2.2 for a gauge transformation $\mathsf{a} \to \mathsf{a} - \mathrm{d}_A\varepsilon$ that will set $L'\mathsf{a} = 0$. Given the relation between L' and L (Eq. (13)), this means that we want to set $D_\mu S(\mathsf{a}) = 0$. Since we also want to satisfy $S(\mathsf{a})| = 0$, we actually need $S(\mathsf{a})$ to vanish identically. So we have to find a gauge parameter ε such that $\varepsilon| = 0$ and $P\varepsilon = -S(\mathsf{a})$. Essentially the same argument as in Sec. 2.2 shows that with Dirichlet boundary conditions, P is invertible. So there is a unique solution of $P\varepsilon = -S(\mathsf{a})$.

In the opposite direction, if we are given a solution of $L'\mathsf{a} = 0$ that also satisfies the boundary condition $S(\mathsf{a})| = 0$, we want to show that actually $S(\mathsf{a})$ vanishes identically, so that the gauge-invariant equation $L\mathsf{a} = 0$ is satisfied. As before, the Bianchi identity (17), together with $L'\mathsf{a} = 0$, implies that $PS(\mathsf{a}) = 0$, and this, together with the boundary condition $S(\mathsf{a})| = 0$, implies that $S(\mathsf{a})$ vanishes identically.

3. Elliptic boundary conditions in gravity

In discussing elliptic gauge-fixing and elliptic boundary conditions in gravity, we will be brief on points on which there is a very close parallel with what we have already described for Yang–Mills theory.

3.1. *General Relativity on a closed manifold*

The action of classical General Relativity with a cosmological constant Λ is

$$I = -\frac{1}{\kappa^2} \int_M \mathrm{d}^D x \sqrt{g}\,(R - 2\Lambda)\,. \tag{28}$$

We can of course add matter fields, but we do not do so explicitly. The field equations read

$$R_{\mu\nu} - \frac{1}{2}g_{\mu\nu}R + \Lambda g_{\mu\nu} = 0, \tag{29}$$

and are governed by a Bianchi identity

$$D^\mu \left(R_{\mu\nu} - \frac{1}{2}g_{\mu\nu}R + \Lambda g_{\mu\nu} \right) = 0. \tag{30}$$

We expand around a background classical solution g_0 with $g_{\mu\nu} = g_{0\mu\nu} + h_{\mu\nu}$. However, again it is clumsy to always write $g_{0\mu\nu}$ for the background field, so we will instead write $\hat{g}_{\mu\nu}$ for the full metric, and $g_{\mu\nu}$ for the background, so that the expansion reads $\hat{g}_{\mu\nu} = g_{\mu\nu} + h_{\mu\nu}$. Covariant derivatives and curvatures will refer to the background metric, which is also used in raising and lowering indices. It is convenient to define $\mathsf{h} = h^\mu_\mu = g^{\mu\nu}h_{\mu\nu}$.

The quadratic part of the action for h is, of course, an important input in semiclassical quantization [6, 8]. The D-dimensional version of the formula, from for example Eq. (2.3) of Ref. [12], is[h]

$$I' = -\frac{1}{\kappa^2} \int_X \mathrm{d}^D x \sqrt{g} \left(\frac{1}{4}h^{\mu\nu}(D_\lambda D^\lambda + 2\Lambda)h_{\mu\nu} - \frac{1}{8}\mathsf{h}(D_\lambda D^\lambda + 2\Lambda)\mathsf{h} + \frac{1}{2}(D^\nu h_{\mu\nu} - \partial_\mu \mathsf{h})^2 \right.$$
$$\left. + \frac{1}{2}h^{\mu\nu}h^{\lambda\rho}R_{\mu\nu\lambda\rho} + \frac{1}{2}\left(h^{\mu\lambda}h^\nu_\lambda - \mathsf{h}h^{\mu\nu}\right)R_{\mu\nu} + \frac{1}{8}(\mathsf{h}^2 - 2h^{\mu\nu}h_{\mu\nu})R \right). \tag{31}$$

The gauge-invariant linear wave equation satisfied by h is $(Lh)_{\mu\nu} = 0$ where we set $2\kappa^2\delta I'/\delta h^{\mu\nu} = (Lh)_{\mu\nu}$. We will not write explicitly the rather unilluminating formula for L. Gauge invariance implies of course that the operator L is not elliptic, but it satisfies a Bianchi identity that descends directly from the underlying Bianchi identity (30):

$$D^\mu((Lh)_{\mu\nu}) = 0. \tag{32}$$

To restore ellipticity and carry out quantum perturbation theory, we need a gauge condition. The form of the action suggests a convenient and widely used choice of gauge (variously known as harmonic, de Donder, or Bianchi gauge), namely $T_\mu(h) = 0$ where

$$T_\mu(h) = D^\nu h_{\mu\nu} - \frac{1}{2}\partial_\mu \mathsf{h}. \tag{33}$$

To decide if this is a good gauge condition, we have to ask if it can be implemented by an infinitesimal coordinate transformation

$$\delta h_{\mu\nu} = D_\mu \varepsilon_\nu + D_\nu \varepsilon_\mu. \tag{34}$$

---

[h]In the absence of matter fields, the following could be simplified slightly using the equations of motion for the background field to replace $R_{\mu\nu}$ with a multiple of $g_{\mu\nu}$. The reason that we do not do that is that we wish to make statements that remain valid if matter fields are included. (Admittedly, we are following a hybrid logic, since we do not add explicitly the matter contributions to the action or the equations of motion.) In Eq. (31), the terms proportional to Λ are those that come from the Λ term in the original action (28).

We observe that

$$\delta T_\mu(h) = D_\nu D^\nu \varepsilon_\mu + R_{\mu\nu} \varepsilon^\nu. \tag{35}$$

So to set $T_\mu(h) = 0$, we need to solve

$$(P\varepsilon)_\mu = -T_\mu(h), \tag{36}$$

where the operator P is defined by

$$(P\varepsilon)_\mu = -D_\nu D^\nu \varepsilon_\mu - R_{\mu\nu} \varepsilon^\nu. \tag{37}$$

We note that this operator is elliptic and thus has a discrete spectrum. If P is invertible, there will be a unique solution ε of Eq. (36) and thus the gauge condition is good. The operator P is invertible in pure gravity with $\Lambda < 0$. This makes $R_{\mu\nu}$ negative-definite, and since the Laplace-like operator $-D^\mu D_\mu$ is positive semi-definite (by the same argument as in Eq. (15)), P is then strictly positive. Even if the cosmological constant is not negative or matter fields are present, one can reasonably expect that in expanding around a generic classical solution, P will have no zero-mode.[i] However, if P does have a zero-mode, then the gauge fixing procedure needs to be slightly modified to treat this mode correctly. Being elliptic, P will only have a finite number of zero-modes on a compact manifold. It is technically inconvenient to have to slightly modify the gauge-fixing condition, but as this only affects a finite number of modes, it does not really affect any questions of principle. Actually, on a manifold with nonempty boundary, which is our main interest in this paper, this complication does not arise, in the following sense. It is shown in Lemma 2.2 of Ref. [1] that, acting on vector fields that are required to vanish on the boundary, the operator P is always invertible, regardless of Λ. (A key step in the proof is the fact that a Killing vector field that vanishes along the boundary is identically zero.)

To implement the gauge condition $T_\mu = 0$ in the BRST framework, we first add the ghosts, which are a fermion field c^μ that represents the generator of a diffeomorphism (that is, c^μ transforms as a vector field), with BRST variations

$$\delta h_{\mu\nu} = D_\mu c_\nu + D_\nu c_\mu, \qquad \delta c^\mu = c^\nu \partial_\nu c^\mu. \tag{38}$$

One also needs an antighost multiplet consisting of the antighost field $\bar{c}^\mu$ and an auxiliary field f^μ, with

$$\delta \bar{c}^\mu = f^\mu, \qquad \delta f^\mu = 0. \tag{39}$$

A convenient gauge-fixing term is

$$\frac{1}{\kappa^2} \delta \int_X \mathrm{d}^D x \sqrt{g} \left(-\frac{1}{2} \bar{c}_\mu f^\mu + \bar{c}^\mu T_\mu(h) \right). \tag{40}$$

After integrating out the auxiliary field, this generates a correction to the gravitational action

$$I'' = \frac{1}{2\kappa^2} \int_X \mathrm{d}^D x \sqrt{g}\, T_\mu(h)^2 = \frac{1}{2\kappa^2} \int_X \mathrm{d}^D x \sqrt{g} \left(D^\nu h_{\mu\nu} - \frac{1}{2} \partial_\mu h \right)^2. \tag{41}$$

---

[i]With positive cosmological constant, for $X = S^4$, P has zero and negative eigenvalues that correspond to Killing vectors and conformal Killing vectors [8].

The gauge-fixed gravitational action is

$$I' + I'' = -\frac{1}{\kappa^2} \int_X d^D x \sqrt{g} \left(\frac{1}{4} h^{\mu\nu} (D_\lambda D^\lambda + 2\Lambda) h_{\mu\nu} - \frac{1}{8} h(D_\lambda D^\lambda + 2\Lambda) h \right.$$
$$\left. + \frac{1}{2} h^{\mu\nu} h^{\rho\sigma} R_{\mu\nu\rho\sigma} + \frac{1}{2} \left(h^{\mu\lambda} h^\nu_\lambda - h h^{\mu\nu} \right) R_{\mu\nu} + \frac{1}{8} (h^2 - 2h^{\mu\nu} h_{\mu\nu}) R \right). \quad (42)$$

The action for the ghosts is

$$I_{gh} = \frac{1}{\kappa^2} \int_X d^D x \sqrt{g}\, \bar{c} P c = \frac{1}{\kappa^2} \int_X \bar{c}^\mu \left(-g_{\mu\nu} D_\lambda D^\lambda - R_{\mu\nu} \right) c^\nu. \quad (43)$$

The gauge-fixed linear kinetic operator L' that governs metric fluctuations is defined by $2\kappa^2 \delta(I' + I'')/\delta h^{\mu\nu} = (L'h)_{\mu\nu}$. From Eq. (42), we see that the leading symbol $\sigma_x(p)$ of L' is invertible, but not positive-definite. (This lack of positivity was first pointed out and discussed in Ref. [7].) In fact, $\sigma_x(p)$ acts on the traceless part of $h_{\mu\nu}$ as a positive multiple of p^2 (tensored with the identity matrix on the $\mu\nu$ indices) and on the trace h as a negative multiple of p^2. Thus L' is elliptic even though not positive-definite.

Actually, in the absence of matter fields, the traceless and trace parts of $h_{\mu\nu}$ decouple (as one sees by using the classical equations of motion to replace the background $R_{\mu\nu}$ with a multiple of $g_{\mu\nu}$) and can be treated separately in constructing propagators and determinants. In this case, the eigenvalues of L' are almost all positive on traceless modes and almost all negative on trace modes. In the presence of matter fields, the traceless and trace modes do not decouple in general, but ellipticity ensures that L' still has a discrete spectrum. Large positive eigenvalues correspond to wavefunctions that are almost traceless, and large negative eigenvalues correspond to wavefunctions whose traceless part is very small.

Ellipticity guarantees that L' has at most a finite number of zero-modes. As usual, these modes must be treated specially in constructing perturbation theory. Ellipticity guarantees that L' always has a parametrix, that is, a propagator suitable for perturbation theory.

To construct perturbation theory, in addition to a propagator and a renormalization procedure, one requires a one-loop determinant, since the one-loop path integral formally includes a factor $\det P/\sqrt{\det L'}$. Here as the operators in question are elliptic, ζ-function regularization can be straightforwardly used to define the absolute values of the determinants, but the presence of negative eigenvalues — infinitely many of them in the case of L' — means that it is not straightforward to understand the phase of the one-loop path integral. This issue has been discussed in several papers [7,10], but its status is not entirely clear. We will not discuss these questions here except to note that existing computations rely on the decoupling of the traceless and trace parts of the metric that holds in pure gravity, so at a minimum some generalization is needed.

From the form (41) of the gauge-fixing part of the gravitational action, one can work out an explicit formula relating L' and L:

$$(L'h)_{\mu\nu} = (Lh)_{\mu\nu} + D_\mu T_\nu(h) + D_\nu T_\mu(h) - g_{\mu\nu} D_\lambda T^\lambda(h). \quad (44)$$

From this and the Bianchi identity (32), one finds that

$$D^\mu((L'h)_{\mu\nu}) = (PT(h))_\nu. \tag{45}$$

Thus the equations of motion of the gauge-fixed theory imply

$$(PT(h))_\mu = 0. \tag{46}$$

This statement remains valid when matter fields are included (assuming that the gauge-fixing takes the form of Eq. (41)), since the Einstein equations with matter fields included still satisfy a Bianchi identity, which reflects the underlying general covariance. (The proof of the general Bianchi identity requires the equations of motion for the matter fields as well as the metric.)

From the point of view of BRST quantization, the procedure that we have described is satisfactory if L' and P have no zero-modes, as one may expect in expanding around a generic classical solution. In general, the procedure needs to be slightly modified to treat properly a finite-dimensional space of zero-modes.

Let us now discuss how would one motivate this procedure from the point of view of differential geometry, without reference to quantization. From that point of view, one would like to compare the solutions of the gauge-invariant equation $Lh = 0$, modulo the gauge equivalence (34), to the solutions of the gauge-fixed equation $L'h = 0$. In one direction, we have already seen that if P is invertible, then every solution of $Lh = 0$ can be uniquely put in a gauge with $T_\mu(h) = 0$. Then Eq. (44) shows that $L'h = 0$. In the opposite direction, if $L'h = 0$, then from Eq. (45), we have $(PT(h))_\mu = 0$, which (if P is invertible) implies that $T_\mu(h) = 0$. Then using Eq. (44) again, we see that the gauge-invariant equation $Lh = 0$ is satisfied.

3.2. General Relativity on a manifold with boundary

We now consider General Relativity on a manifold X with boundary M. We start by analyzing the most direct analog of the boundary condition for Yang–Mills theory that was discussed in Sec. 2.3.

In this boundary condition, we keep fixed the boundary metric of M and allow fluctuations in the interior. Thus if X is locally defined by $x^D \geq 0$, while M is parametrized by[j] x^i, $i = 1, \cdots, D-1$, we specify the boundary values of g_{ij}. In terms of the metric perturbation $h_{\mu\nu}$, this means that part of the boundary condition will be

$$h_{ij}| = 0, \qquad i, j = 1, \cdots, D-1. \tag{47}$$

The boundary conditions on $h_{i\perp}$ and $h_{\perp\perp}$ are still to be specified.

To learn what the remaining boundary conditions will have to be, we first consider the gauge symmetries $\delta h_{\mu\nu} = D_\mu\varepsilon_\nu + D_\nu\varepsilon_\mu$. If ε, restricted to $M = \partial X$, has a nonzero component in the normal direction, then it does not really generate a

---

[j] As before, we will write $\vec{x}$ and $x_\perp$ for tangential coordinates $x^1, \ldots, x^{D-1}$ and the normal coordinate x^D.

symmetry of X, as it tries to move the boundary of X normal to itself. Thus the diffeomorphism group of X is generated by vector fields that are constrained by

$$\varepsilon^{\perp}| = 0. \tag{48}$$

In addition, if we wish to impose a boundary condition $h_{ij}| = 0$, we must restrict ourselves to vector fields with

$$\varepsilon^{i}| = 0. \tag{49}$$

Combining these two statements, we see that we should consider only diffeomorphisms generated by vector fields that satisfy

$$\varepsilon^{\mu}| = 0. \tag{50}$$

In BRST quantization, this means that the ghost field c^{μ} should satisfy Dirichlet boundary conditions

$$c^{\mu}| = 0. \tag{51}$$

Now let us assume that the gauge-fixing away from the boundary is carried out by the procedure of Sec. 3.1. For the same reason as in Sec. 2.3, the antighost field $\bar{c}^{\mu}$ must likewise satisfy Dirichlet boundary conditions:

$$\bar{c}^{\mu}| = 0. \tag{52}$$

On the other hand, after eliminating the auxiliary field, the BRST variation of $\bar{c}^{\mu}$ is

$$\delta\bar{c}^{\mu} = T^{\mu}(h), \tag{53}$$

which is the direct analog of Eq. (23) for Yang–Mills theory. Therefore, BRST invariance forces us to impose

$$T_{\mu}(h)| = 0, \tag{54}$$

similarly to Eq. (26) in gauge theory.

Equation (54) is a boundary condition for $h_{\perp\perp}$ and $h_{i\perp}$, somewhat analogous to Neumann boundary conditions. Together with Eq. (47), it gives the right number of conditions to make a boundary condition for metric fluctuations. For brevity we will call this the Dirichlet boundary condition. However, this boundary condition is not elliptic [1,3]. We will first show this by a short computation and then give a less computational explanation.

Since the considerations are local and only depend on the leading symbol of the linearized Einstein equations and the leading behavior of the boundary condition at short distances, we can take X to be a half-space $\mathbb{R}_{+}^{D}$ in a flat Euclidean space $\mathbb{R}^{D}$, say the half-space $x_{\perp} \geq 0$. A general plane wave solution with nonzero momentum $\vec{k}$ along the boundary that decays exponentially for large $x_{\perp}$ takes the form

$$h_{\mu\nu} = \alpha_{\mu\nu}e^{i\vec{k}\cdot\vec{x}-|\vec{k}|x_{\perp}}. \tag{55}$$

To show that the boundary condition is not elliptic, we have to show that it is possible for a solution of this kind with real nonzero $\vec{k}$ to satisfy the boundary conditions. (Since the boundary conditions are invariant under scaling of $\vec{k}$, if we can satisfy them for any nonzero $\vec{k}$ we can do so with arbitrarily large $|\vec{k}|$.) Dirichlet boundary conditions $h_{ij}| = 0$ imply that we should set $\alpha_{ij} = 0$ in Eq. (55). Let us write $\vec{\alpha}$ for the $(D-1)$-vector with components $\alpha_{i\perp}$, and β for $\alpha_{\perp\perp}$. A short computation reveals that the equations $T_{\perp\perp} = 0$ and $T_{i\perp} = 0$ become

$$i\vec{k}\cdot\vec{\alpha} - \frac{1}{2}|\vec{k}|\beta = 0 \tag{56}$$

and

$$-|\vec{k}|\vec{\alpha} - \frac{i}{2}\vec{k}\,\beta = 0. \tag{57}$$

We can satisfy both of these equations with

$$\vec{\alpha} = -\frac{i}{2|\vec{k}|}\vec{k}\beta, \tag{58}$$

and therefore Dirichlet boundary conditions for gravity are not elliptic.

At first this may look like an unlucky accident, and one may wonder if using a different bulk gauge condition would have avoided the problem. This is not the case, as is shown in several ways in Ref. [1]. One argument makes use of the second-order nature of the Hamiltonian constraint equation of General Relativity. A second argument is as follows. We will describe a compact X with boundary such that the linearized Einstein equations on X, without any gauge-fixing, and with Dirichlet boundary conditions $h_{ij}| = 0$ (but no boundary condition placed on $h_{i\perp}$ or $h_{\perp\perp}$), has infinitely many zero-modes, modulo gauge transformations. Any correctly gauge-fixed version of the linearized Einstein equations on X would have the same infinite-dimensional kernel, contradicting ellipticity.

We take X to be a product $T \times I$, where T is a torus with flat metric, parametrized by periodic variables $\vec{x} = (x_1, \cdots, x_{D-1})$, and I is the unit interval $0 \leq x_\perp \leq 1$. Now we pick a function $\varepsilon(\vec{x})$ and perturb X so that its boundaries are $\varepsilon(\vec{x}) \leq x_\perp \leq 1$. Since the extrinsic curvature of ∂X vanishes, the boundary geometry of X is unchanged to first order in ε, and therefore these perturbations satisfy $h_{ij}| = 0$. On the other hand, these perturbations cannot be eliminated by a diffeomorphism. So with Dirichlet boundary conditions, the kernel of the linearized Einstein equations on X, modulo its subspace induced by diffeomorphisms of X, is infinite-dimensional. Even though this is a very special example, it is enough to show that linearized Einstein equations with Dirichlet boundary conditions, and with any choice of gauge-fixing, cannot be elliptic.

It is instructive to see how to put the space of zero-modes that we found in this example in the form of a change in the metric $(g \to g + h)$ rather than a change in the range of the coordinates. Since the perturbation preserves the flatness of X, it must take the form

$$h_{\mu\nu} = \partial_\mu v_\nu + \partial_\nu v_\mu \tag{59}$$

for some vector field $v^\mu(\vec{x}, x_\perp)$. However, v^μ will not vanish on the boundaries of X. Rather, we take v^μ to vanish at $x_\perp = 1$, but at $x_\perp = 0$, we impose

$$\vec{v} = 0, \quad v_\perp = \varepsilon(\vec{x}). \tag{60}$$

(In other words, $v^\mu \partial/\partial x^\mu|_{x_\perp=0} = \varepsilon(\vec{x})\partial/\partial x_\perp$.) There is a unique v^μ that satisfies these boundary conditions and also satisfies

$$Pv = 0. \tag{61}$$

This condition ensures that $h_{\mu\nu}$, defined in Eq. (59), obeys $T_\mu(h) = 0$. With such a choice of v, the perturbations (59) are nontrivial zero-modes of the linearized Einstein equations on X in the gauge $T_\mu(h) = 0$. Of course, the connection between the two descriptions is that a diffeomorphism generated by v^μ, to first order, maps the interval $0 \le x_\perp \le 1$ to $\varepsilon(\vec{x}) \le x_\perp \le 1$. In particular, because $v_\perp \ne 0$ at $x_\perp = 0$, v does not generate a diffeomorphism of X.

We will discuss one last topic before moving on to the conformal boundary condition. Assuming as above that the bulk gauge-fixing is carried out by adding $g^{\mu\nu}T_\mu(h)T_\nu(h)$ to the action, the boundary condition $T_\mu(h)| = 0$ has another virtue that we have not yet explained: it is needed to make the gauge-fixed linearized Einstein operator L' hermitian, in the following sense. Let $\langle h, \tilde{h}\rangle$ be the obvious inner product on the space of metric deformations,

$$\langle h, \tilde{h}\rangle = \int_X \mathrm{d}^D x \sqrt{g}\, g^{\mu\nu} g^{\mu'\nu'} h_{\mu\mu'} \tilde{h}_{\nu\nu'}. \tag{62}$$

Then L' is hermitian in the sense that

$$\langle h, L'\tilde{h}\rangle = \langle L'h, \tilde{h}\rangle. \tag{63}$$

When one tries to prove this by integration by parts, one runs into surface terms. However, the surface terms cancel with the help of the boundary conditions that we have assumed.

It is not hard to prove this by hand, but a better explanation is as follows. First of all, the gauge-invariant linearized Einstein operator L satisfies the same identity

$$\langle h, L\tilde{h}\rangle = \langle Lh, \tilde{h}\rangle. \tag{64}$$

The most natural way to prove this is to use the fact that either the left or the right hand side can be interpreted as the quadratic part of the action, expanded around a chosen classical solution. However, for this to be true, one must add a boundary term to the action. The boundary term is chosen to ensure that, when one varies the action to derive the equations of motion, the boundary terms in the variation of the action vanish, once the boundary conditions are imposed. Of course, the boundary term that will make this work, if there is one, depends on what boundary condition one wants. With Dirichlet boundary conditions, the relevant boundary term is the Gibbons–Hawking–York (GHY) term [5, 13]; with the conformal boundary condition, a slightly different boundary term is appropriate [15], as we will explain in Sec. 3.3. The variation of the Einstein–Hilbert action

$$I = -\frac{1}{\kappa^2} \int_M \mathrm{d}^D x \sqrt{g}\,(R - 2\Lambda) \tag{65}$$

under $\delta g_{\mu\nu} = h_{\mu\nu}$ has the usual bulk term related to Einstein's equations and also a boundary term[k]

$$\delta_{\text{bdry}} I = -\frac{1}{\kappa^2} \int_{\partial M} d^{D-1} x \sqrt{g_\partial} \left(-2\delta\mathsf{K} - \mathsf{K}^{ij} h_{ij} \right). \tag{66}$$

Here K_{ij} is the extrinsic curvature of the boundary and $\mathsf{K} = g^{ij}\mathsf{K}_{ij}$ is its trace. $\delta\mathsf{K}$ is the variation of K under $\delta g_{\mu\nu} = h_{\mu\nu}$. We do not need the explicit formula for $\delta\mathsf{K}$, since this contribution to the variation of the Einstein–Hilbert action is canceled by adding the GHY term

$$I_{\text{GHY}} = -\frac{2}{\kappa^2} \int_{\partial M} d^{D-1} x \sqrt{g_\partial}\, \mathsf{K}. \tag{67}$$

Dirichlet boundary conditions $h_{ij}| = 0$ ensure the vanishing of the remaining term $\mathsf{K}^{ij} h_{ij}$ in $\delta_{\text{bdry}} I$ and the absence of any contribution from varying $\sqrt{g_\partial}$ in I_{GHY}. So with Dirichlet boundary conditions, there is no boundary term in the variation of the combined action $I + I_{\text{GHY}}$. This also means that there is no boundary term in proving Eq. (64).

In gauge fixing, we added to the gravitational action another term

$$I'' = \frac{1}{2\kappa^2} \int_X d^D x \sqrt{g}\, g^{\mu\nu} T_\mu(h) T_\nu(h). \tag{68}$$

When we vary this under $\delta g_{\mu\nu} = h_{\mu\nu}$, upon integrating by parts to derive the bulk equations of motion, we generate additional surface terms. But because I'' is bilinear in $T_\mu(h)$, these new surface terms are all proportional to $T_\mu(h)$ and so vanish if the boundary condition includes $T_\mu(h) = 0$. With this being so, the left and right hand sides of Eq. (63) are both equal to the gauge-fixed quadratic action, so in particular they are equal.

3.3. *A boundary condition that works*

Though the Dirichlet boundary condition is not elliptic, there is a simple elliptic boundary condition for Einstein's equations [1]. Instead of specifying the boundary metric, we specify only the conformal structure of the boundary. Differently put, we specify the boundary metric $\hat{g}_{ij}|$ only up to a Weyl transformation $\hat{g}_{ij} \to e^\phi \hat{g}_{ij}$. We write $\bar{g}$ for the conformal structure of the boundary, that is, for the equivalence class of the boundary metric, modulo a Weyl transformation.

In terms of the expansion $\hat{g}_{\mu\nu} = g_{\mu\nu} + h_{\mu\nu}$, specifying only the conformal structure of the boundary means that only the traceless part of the perturbation $h_{ij}|$ of the boundary metric is required to vanish, so that

$$h_{ij}| = g_{ij}\gamma \tag{69}$$

for some function γ.

---

[k]In the following, $g_\partial = g|$ is the induced metric of the boundary and so $d^{D-1}x\sqrt{g_\partial}$ is the natural Riemannian measure of the boundary.

We assume that the bulk gauge-fixing is that of Sec. 3.1 and therefore, as in Sec. 3.2, part of the boundary condition will be

$$T_\mu(h)| = 0. \tag{70}$$

We need one more boundary condition, to compensate for relaxing the constraint on the trace of $h_{ij}|$. For this, we impose a constraint on the trace of the extrinsic curvature. We will write K_{ij} for the extrinsic curvature in the metric g, and $\widehat{K}_{ij}$ for the extrinsic curvature in the metric $\widehat{g} = g + h$. We also write $\widehat{K} = \widehat{g}^{ij} K_{ij}(\widehat{g})$ for the trace of the extrinsic curvature in the full metric $\widehat{g} = g + h$, and similarly $K = g^{ij} K_{ij}(g)$ for the trace of the extrinsic curvature computed using the background metric g. Then we complete the boundary condition by requiring

$$\widehat{K} = K. \tag{71}$$

In other words, the condition is that the perturbation does not change the trace of the extrinsic curvature. We call the combination of Eqs. (69)–(71) the conformal boundary condition.

Explicitly, the linearization of Eq. (71), in coordinates in which the background metric satisfies $g_{\perp\perp} = 1$, $g_{\perp i} = 0$, is

$$D_\perp h^i{}_i - 2D^i h_{\perp i} + 2h^{ij} K_{ij} = 0. \tag{72}$$

The term $2h^{ij} K_{ij}$, being of lower order, does not affect the discussion of ellipticity.

To show that the conformal boundary condition is elliptic, it suffices again to take $X = \mathbb{R}^D_+$ and to analyze solutions of the equation $L'h = 0$ that propagate like a plane wave along the boundary:

$$h_{\mu\nu} = \alpha_{\mu\nu} e^{i\vec{k}\cdot\vec{x} - |\vec{k}|x_\perp}. \tag{73}$$

We have to show that for any nonzero real $\vec{k}$, a solution of this kind can satisfy the boundary condition only if $\alpha_{\mu\nu} = 0$.

As a first step, we see that Eq. (69) implies that $\alpha_{ij} = \delta_{ij}\gamma$ for some γ. As before, we write $\vec{\alpha}$ for the $(D-1)$-vector with components $\alpha_{i\perp}$, and β for $\alpha_{\perp\perp}$. In this geometry, Eq. (72) reduces to

$$2i\vec{k}\cdot\vec{\alpha} + |\vec{k}|(D-1)\gamma = 0. \tag{74}$$

The equations $T_0(h)| = 0$ and $T_i(h)| = 0$ become

$$i\vec{k}\cdot\vec{\alpha} - \frac{1}{2}|\vec{k}|\beta + |\vec{k}|\frac{D-1}{2}\gamma = 0 \tag{75}$$

and

$$-|\vec{k}|\vec{\alpha} - \frac{i}{2}\vec{k}\,\beta - \frac{i}{2}\vec{k}(D-3)\gamma = 0. \tag{76}$$

For nonzero real $\vec{k}$, these equations imply that $\vec{\alpha} = \beta = \gamma = 0$, so the conformal boundary condition is elliptic.

The gauge-fixed linearized Einstein operator L' with conformal boundary conditions is not just elliptic but self-adjoint. This can be proved by modifying the discussion at the end of Sec. 3.2. With conformal boundary conditions, a different normalization is needed for the GHY boundary term in the action [15]. The boundary variation $\delta_{\mathrm{bdry}} I$ of the Einstein–Hilbert action is given by Eq. (66) irrespective of the boundary conditions. However, with the conformal boundary condition, $\delta \mathsf{K} = 0$ (since the conformal boundary condition is defined by keeping K fixed) but we no longer have $h_{ij}| = 0$; instead $h_{ij}| = g_{ij}\gamma$ for some scalar function γ. This means that $\delta_{\mathrm{bdry}} I$ is now equal to $(1/\kappa^2) \int_{\partial M} \mathrm{d}^{D-1}x \sqrt{g_\partial}\, \mathsf{K}\gamma$. To cancel this, for a conformal boundary we need a boundary term I_{CB} that is a multiple of the usual GHY term:

$$I_{\mathrm{CB}} = \frac{1}{D-1} I_{\mathrm{GHY}} = -\frac{2}{D-1} \frac{1}{\kappa^2} \int_{\partial M} \mathrm{d}^{D-1}x \sqrt{g_\partial}\, \mathsf{K}. \tag{77}$$

The identity $\langle h, L\widetilde{h}\rangle = \langle Lh, \widetilde{h}\rangle$ now holds, just as before, because the left and right hand sides are both equal to the quadratic action derived from $I + I_{\mathrm{CB}}$. And likewise, after adding the usual bulk gauge-fixing term to the action and imposing the boundary condition $T_\mu(h)| = 0$, the gauge-fixed operator L' obeys the same identity.

Self-adjointness gives a natural identification between the kernel and cokernel of L', which in particular have the same dimension, generically zero. Self-adjointness also means that the absolute value of the one-loop determinant can be straightforwardly defined using zeta-function regularization. (As remarked in Sec. 3.1, the phase of the determinant is more subtle.)

3.4. *Expanding or contracting metrics*

In General Relativity, the metric on an initial value surface and the extrinsic curvature are canonically conjugate variables. We seem to have learned that at least in Euclidean signature, it is better to fix the conformal structure of the boundary and the trace of the extrinsic curvature, rather than constraining all of the boundary metric. This suggests that in quantization, one should consider a wavefunction $\widehat{\Psi}(\bar{g}, \mathsf{K})$ that depends on the conformal structure of a hypersurface and the trace of the extrinsic curvature, rather than a wavefunction $\Psi(g)$ that depends on the metric of the hypersurface. (See Ref. [13] for early ideas along these lines.) There is, however, a further important detail that may change the picture, at least for many applications. Even though Dirichlet boundary conditions are not elliptic, some of the important consequences of ellipticity do hold for Dirichlet boundary conditions, for a fairly wide class of metrics.[1]

Let us replace Eq. (69) with

$$h_{ij}| = K_{ij}\gamma, \tag{78}$$

with an unspecified function $\gamma(\vec{x})$, where again K_{ij} is the extrinsic curvature of the background metric. We leave Eqs. (70) and (71) unchanged. In the special case that K_{ij} is an everywhere nonzero multiple of the background metric g_{ij}, this

---

[1]See Sec. 3 of Ref. [2] for the boundary condition that we are about to describe and its properties. For antecedents of some of the ideas in a different context, see Ref. [4], pp. 187–93.

new boundary condition is just a different way of writing the conformal boundary condition that we already studied in Sec. 3.3, so in particular it is elliptic. Lacking a better name, we will call what we get with Eq. (78) the alternate boundary condition.

As was remarked at the end of Sec. 2.1, ellipticity is an "open" condition, preserved by any sufficiently small perturbation of a boundary condition. Since the alternate boundary condition is elliptic if K_{ij} is everywhere a nonzero multiple of g_{ij}, there must be an open set in the space of symmetric second-rank tensors on M such that the alternate boundary condition is elliptic if K_{ij} is everywhere in that open set.

To find this open set, we can proceed almost as before. We have to determine the large momentum behavior of a solution of $L'h = 0$ that looks like a plane wave along the boundary. For this, we can take the usual flat model with $X = \mathbb{R}^D_+$, $\partial X = \mathbb{R}^{D-1}$, and treat the tensor K_{ij} that appears in the boundary condition (78) as a fixed constant symmetric tensor. Of course, in order for K_{ij} really to be the extrinsic curvature of the boundary, X and its boundary cannot really be flat. But their curvature does not affect the high momentum behavior, which we can calculate using the flat model.

Proceeding in this way, it is straightforward to compute that with the alternate boundary condition, Eqs. (75) and (76) are replaced by

$$\mathrm{i}\vec{k} \cdot \vec{\alpha} - \frac{1}{2}|\vec{k}|\beta + \frac{1}{2}|\vec{k}|\gamma \mathsf{K} = 0, \tag{79}$$

$$-|\vec{k}|\alpha_i - \frac{\mathrm{i}}{2}k_i\beta + \mathrm{i}\gamma\left(k_j K_{ij} - \frac{1}{2}k_i\mathsf{K}\right) = 0. \tag{80}$$

Likewise Eq. (74) is replaced by

$$2\mathrm{i}\vec{k} \cdot \vec{\alpha} + |\vec{k}|\gamma \mathsf{K} = 0. \tag{81}$$

Ellipticity is the statement that (at every point on $M = \partial X$) Eqs. (79), (80), and (81) have no common solutions with real nonzero $\vec{k}$.

Comparing Eqs. (79) and (81), we see that a solution must have $\beta = 0$, and once we know this, we can eliminate $\vec{\alpha}$ to find that a nonzero solution has

$$\sum_{i,j} k_i k_j M_{ij} = 0, \tag{82}$$

where M_{ij} is the quadratic form

$$M_{ij} = g_{ij}\mathsf{K} - K_{ij}. \tag{83}$$

The condition for Eq. (82) to have no nonzero real solution is simply that the quadratic form M_{ij} should be positive-definite or negative-definite. For this it is sufficient, though not necessary, that the extrinsic curvature K_{ij} of the background metric should be positive- or negative-definite.

Now let us suppose that the quadratic form M is positive- or negative-definite, so that the alternate boundary condition is elliptic. What does this say about the linearized Einstein equations with Dirichlet boundary conditions?

In what follows, we will write L'' for the gauge-fixed linearized Einstein operator with alternate boundary conditions and L' for the same operator with Dirichlet boundary conditions. If L'' is elliptic, it has in particular a finite-dimensional cokernel. This means that, given a symmetric tensor f on X, imposing a finite number of linear constraints on f suffices to ensure that the gauge-fixed equation

$$L''h = f \tag{84}$$

has a solution, with h obeying the alternate boundary conditions of Eqs. (70), (71), and (78). The h that satisfies these boundary conditions does not in general satisfy Dirichlet boundary conditions, of course. According to Eq. (78), the Dirichlet boundary conditions are violated because $h_{ij}|$, instead of vanishing, is instead

$$h_{ij}| = \gamma K_{ij} \tag{85}$$

for some function γ on $M = \partial X$. However, we can compensate for this by shifting

$$h_{\mu\nu} \to h_{\mu\nu} + D_\mu v_\nu + D_\nu v_\mu \tag{86}$$

for a suitable vector field v^μ. We require first of all that

$$Pv = 0, \tag{87}$$

so that the shift (86) does not disturb Eq. (84) or the gauge condition $T_\mu(h) = 0$. Second, we require that v satisfies the boundary conditions

$$v^i| = 0, \quad v^\perp| = \gamma, \tag{88}$$

analogously to what we did previously in Eq. (60). Invertibility of P on a manifold with boundary, as proved in Lemma 2.2 of Ref. [1], means that such a v exists. (If this argument were not available, we would say that ellipticity of P implies that v exists after possibly placing a finite number of additional linear constraints on γ. This would be enough for what follows.) Equation (88), together with the fact that K_{ij} is the normal derivative to the metric of M, means that the shift (86) eliminates the right hand side of Eq. (85) and sets $h_{ij}| = 0$. Thus (as in Proposition 3.5 of Ref. [2]) ellipticity of L'' implies that after imposing a finite number of constraints on f, the equation $L'h = f$ can be satisfied with an h that obeys Dirichlet boundary conditions; in other words, it implies that L' has a finite-dimensional cokernel, as if it were elliptic.

Once we know that the cokernel of L' is finite-dimensional, it follows immediately from Eq. (63) that its kernel is also finite-dimensional. Suppose that $L'\widetilde{h} = 0$ for some $\widetilde{h}$ that satisfies Dirichlet boundary conditions. This implies that we cannot solve Eq. (84) unless $\langle f, \widetilde{h} \rangle = 0$, since if we can solve Eq. (84), then

$$\langle f, \widetilde{h} \rangle = \langle L'h, \widetilde{h} \rangle = \langle h, L'\widetilde{h} \rangle = 0. \tag{89}$$

Thus every element of the kernel of L' gives a constraint on f, so the dimension of the kernel of L' can be no greater than the dimension of the cokernel, and in particular the kernel has finite dimension if the cokernel does.

Actually, we can be more precise here. Let us think of the space of all metric perturbations as a Hilbert space $\mathcal{H}$ with inner product $\langle\ ,\ \rangle$. Once we know that the cokernel of L' is finite-dimensional, it follows[m] that the image of L' is a Hilbert subspace $\mathcal{H}' \subset \mathcal{H}$ and that the cokernel of L', which is $\mathcal{H}/\mathcal{H}'$, can be identified with the orthocomplement of $\mathcal{H}'$. Thus, we can identify the cokernel of L' with the space of all metric perturbations $\widetilde{h}$ obeying Dirichlet boundary conditions that are orthogonal to $L'h$ for any h that obeys Dirichlet boundary conditions. But this orthogonality together with Eq. (63) gives

$$0 = \langle L'h, \widetilde{h} \rangle = \langle h, L'\widetilde{h} \rangle. \tag{90}$$

Since this is supposed to be true for all h that obey Dirichlet boundary conditions, it follows that $L'\widetilde{h} = 0$. We can also read this backwards to show that if $L'\widetilde{h} = 0$, then $\widetilde{h}$ is orthogonal to the image of L'. Thus the kernel of L' is the orthocomplement of $\mathcal{H}'$ and so is isomorphic to the cokernel of L'.

All this is as if L' were elliptic when L'' is elliptic. That is certainly not true, since the failure of ellipticity of L' is universal. But it seems plausible (though apparently not known) that when L'' is elliptic, L' has the necessary properties for perturbation theory — notably the existence of a suitable parametrix or propagator. It seems doubtful that L' has reasonable properties in general, without ellipticity of the alternate boundary condition. But little seems to be known about this.

Acknowledgments

Research supported in part by NSF Grant PHY-1606531. I thank R. Mazzeo and especially M. T. Anderson for advice about boundary conditions in gravity and helpful suggestions, and M. J. Duff, G. Horowitz, J. M. Maldacena, and D. Stanford for a variety of discussions.

References

1. M. T. Anderson, "Boundary value problems for Einstein metrics," *Geom. Top.* **12**, 2009–45 (2008), arXiv:math/0612647.
2. M. T. Anderson, "Extension of symmetries on Einstein manifolds with boundary," *Selecta Math.* **16**, 343–375 (2010), arXiv:0704.3373.
3. I. G. Avramiki and G. Esposito, "Lack of strong ellipticity in Euclidean quantum gravity," *Class. Quant. Grav.* **15**, 1141–1152 (1998), arXiv:hep-th/9708163.
4. R. S. Hamilton, "The inverse function theorem of Nash and Moser," *Bull. Am. Math. Soc.* **7**, 65–222 (1982).

---

[m]By constrast, if V is a linear subspace of $\mathcal{H}$ that is not of finite codimension, then in general V is not a Hilbert space and $\mathcal{H}/V$ cannot be naturally identified with $V^\perp$. Rather, $V^\perp$ can be identified with $\mathcal{H}/\overline{V}$, where $\overline{V}$ is the Hilbert space closure of V.

5. G. Gibbons and S. W. Hawking, "Action integrals and partition functions in quantum gravity," *Phys. Rev.* **D15**, 2752–6 (1977).
6. G. Gibbons and M. Perry, "Quantizing gravitational instantons," *Nucl. Phys.* **B146**, 90 (1978).
7. G. W. Gibbons, S. W. Hawking, and M. Perry, "Path integrals and the indefiniteness of the gravitational action," *Nucl. Phys.* **B138**, 141–50 (1978).
8. S. M. Christensen and M. J. Duff, "Quantizing gravity with a cosmological constant," *Nucl. Phys.* **B170**[FS1], 480–506 (1980).
9. J. Hartle and S. W. Hawking, "The wavefunction of the universe," *Phys. Rev.* **D28**, 2960–75 (1983).
10. J. Polchinski, "The phase of the sum over spheres," *Phys. Lett.* **B219**, 251–7 (1989).
11. I. G. Avramidi, G. Esposito, and A. Yu. Kamenshchik, "Boundary operators in Euclidean quantum gravity," *Class. Quant. Grav.* **13**, 2361–2374 (1996), arXiv:hep-th/9603021.
12. F. Bastianelli and R. Bonezzi, "One-loop quantum gravity from a worldline viewpoint," arXiv:1304.7135.
13. J. W. York, "Role of conformal three-geometry in the dynamics of gravitation," *Phys. Rev. Lett.* **28**, 1082 (1972).
14. E. Witten, "Canonical quantization in anti de Sitter space," in *20 Years Later: The Many Faces of AdS/CFT*, Princeton Center for Theoretical Science, October 31–November 3, 2017.
15. M. T. Anderson, "On quasi-local Hamiltonians in General Relativity," *Phys. Rev.* **D82**, 084044 (2010), arXiv:1008.4309.

www.ingramcontent.com/pod-product-compliance
Lightning Source LLC
Chambersburg PA
CBHW081510190326
41458CB00015B/5331